ADVANCES IN MACHINE LEARNING

ADVANCES IN MACHINE LEARNING

Edited by:

Ivan Stanimirović

ARCLER
PRESS

www.arclerpress.com

Advances in Machine Learning

Ivan Stanimirović

Arcler Press
2010 Winston Park Drive,
2nd Floor
Oakville, ON L6H 5R7
Canada
www.arclerpress.com
Tel: 001-289-291-7705
001-905-616-2116
Fax: 001-289-291-7601
Email: orders@arclereducation.com

ISBN: 978-1-77361-067-2 (Hardcover)

Library of Congress Cataloging-in-Publication Data

A CIP record for this book has been applied for and is available from Library and Archives Canada.

DECLARATION

Some content or chapters in this book are open access copyright free published research work, which is published under Creative Commons License and are indicated with the citation. We are thankful to the publishers and authors of the content and chapters as without them this book wouldn't have been possible.

ABOUT THE EDITOR

Ivan Stanimirovic gained his PhD from University of Niš, Serbia in 2013. His work spans from multi-objective optimization methods to applications of generalized matrix inverses in areas such as image processing and computer graphics and visualisations. He is currently working as an Assistant professor at Faculty of Sciences and Mathematics at University of Niš on computing generalized matrix inverses and its applications.

TABLE OF CONTENTS

LIST OF CONTRIBUTORS

Ivan Stanimirović

Ken Takiyama
Tokyo University of Agriculture and Technology, Department of Engineering, 2-24-16, Nakacho, Koganei, Tokyo 184-8588, Japan

Yutaka Sakai
Brain Science Institute, Tamagawa University, 6-1-1, Tamagawa-gakuen, Machida, Tokyo 194-8610, Japan

Naoya Oosugi
Laboratory for Adaptive Intelligence, BSI, RIKEN, Saitama, Japan
Graduate School of Arts and Sciences, University of Tokyo, Tokyo, Japan

Keiichi Kitajo
Laboratory for Advanced Brain Signal Processing, RIKEN Brain Science Institute, Japan

Naomi Hasegawa
Laboratory for Adaptive Intelligence, BSI, RIKEN, Saitama, Japan

Yasuo Nagasaka
Laboratory for Adaptive Intelligence, BSI, RIKEN, Saitama, Japan

Kazuo Okanoya
Graduate School of Arts and Sciences, University of Tokyo, Tokyo, Japan

Naotaka Fujii
Laboratory for Adaptive Intelligence, BSI, RIKEN, Saitama, Japan

Diego G. Loyola R
German Aerospace Center (DLR), Oberpfaffenhofen, 82234 Wessling, Germany

Mattia Pedergnana
German Aerospace Center (DLR), Oberpfaffenhofen, 82234 Wessling, Germany

Sebastián Gimeno García
German Aerospace Center (DLR), Oberpfaffenhofen, 82234 Wessling, Germany

Victoria J. Hodge
Advanced Computer Architecture Group, Department of Computer Science, University of York, York, YO10 5GH, UK

Simon O'Keefe
Advanced Computer Architecture Group, Department of Computer Science, University of York, York, YO10 5GH, UK

Jim Austin
Advanced Computer Architecture Group, Department of Computer Science, University of York, York, YO10 5GH, UK

Alicia Costalago Meruelo
Institute of Sound and Vibration, University of Southampton, Southampton, UK

David M. Simpson
Institute of Sound and Vibration, University of Southampton, Southampton, UK

Sandor M. Veres
Department of Autonomous Control and Systems Engineering, University of Sheffield, Sheffield, UK

Philip L. Newland
Centre for Biological Sciences, University of Southampton, Southampton, UK

Soeren Strauss
Centre for Computational Neuroscience and Cognitive Robotics, School of Psychology, University of Birmingham, Birmingham B15 2TT, United Kingdom

Philip J.W. Woodgate
Centre for Computational Neuroscience and Cognitive Robotics, School of Psychology, University of Birmingham, Birmingham B15 2TT, United Kingdom

Saber A. Sami
Centre for Computational Neuroscience and Cognitive Robotics, School of Psychology, University of Birmingham, Birmingham B15 2TT, United Kingdom

Dietmar Heinke
Centre for Computational Neuroscience and Cognitive Robotics, School of Psychology, University of Birmingham, Birmingham B15 2TT, United Kingdom

Tamas Madl
School of Computer Science, University of Manchester, Manchester M13 9PL, UK
Austrian Research Institute for Artificial Intelligence, Vienna A-1010, Austria

Ke Chen
School of Computer Science, University of Manchester, Manchester M13 9PL, UK

Daniela Montaldi
School of Psychological Sciences, University of Manchester, Manchester M13 9PL, UK

Robert Trappl
Austrian Research Institute for Artificial Intelligence, Vienna A-1010, Austria

M. Zavagli
Wellcome Trust Centre for Neuroimaging, University College, London WC1N 3BG, UK
Department of Electronics, Computer Science and Systems (DEIS), Via Venezia 52, 47023 Cesena, Italy

R.T. Canolty
Helen Wills Neuroscience Institute, University of California, Berkeley, CA, USA

T.M. Schofield
Wellcome Trust Centre for Neuroimaging, University College, London WC1N 3BG, UK

A.P. Leff
Institute of Neurology and Institute of Cognitive Neuroscience, UCL, 17 Queen Square, London WC1N 3AR, UK

M. Ursino
Department of Electronics, Computer Science and Systems (DEIS), Via Venezia 52, 47023 Cesena, Italy

R.T. Knight
Helen Wills Neuroscience Institute, University of California, Berkeley, CA, USA

W.D. Penny
Wellcome Trust Centre for Neuroimaging, University College, London WC1N 3BG, UK

PREFACE

Recently, a new field of computer science was derived, including methods and techniques of problem solving that cannot be easily described by traditional algorithms. This field, called "cognitive computing" or "real-world computing", has a varied set of working methodologies, such as: fuzzy logic, approximate reasoning, genetic algorithms, chaos theory, and the Artificial Neural Networks (ANN). The objective of the present work is to introduce the problematic of the latter: definitions, principles and typology, as well as concrete applications in the field of information retrieval.

During the past decade in the field of information retrieval has been experimented with artificial Intelligence (AI) techniques based on rules and knowledge. These techniques seem to have many limitations and difficulties of application, so that already in the present decade work has begun with the more recent. AI techniques, based on inductive learning: symbolic learning, genetic algorithms and neural networks (Chen, 1995).

The earliest work in neural computing dates back to the early 1940s, which neuro-physicist Warren McCulloch and mathematician Walter Pitts proposed, based on their system studies. A formal neuron model implemented by electrical circuits (McMulloch, 1943), whose enthusiasm aroused the neuronal model drove research in this line during the 1950s and 1960s. In 1957 Frank Rosenblatt developed the Perceptron, a network model that possesses the generalization capability, so it has been used to this day in various applications, generally in recognition of patterns. In 1959 Bernard Widrow and Marcial Hoff of Stanford University developed the model ADALINE (ADAptative LINear Elements), first ANN applied to a real problem (noise filters in lines phone calls).

In 1969 Marvin Minsky and Seymour Papert, of MIT, published a work in which they attack the neural model and consider that any research along these lines was sterile (Minsky, 1969). Due to this criticism the works on ANN stop to a new impetus during the 80's. Despite this pause, several researchers continued to work in that direction during the 1970s. Such is the case of the American James Anderson which develops the BSB (Brain-State-in-a-Box) model, or Finnish Teuvo Kohonen who does the same with one based on self-organizing maps.

As of 1982 the interest for the neuronal computation began to take force again. The progress made in hardware and software, methodological advances around learning algorithms for ANN, and the new techniques of artificial intelligence, favored this rebirth. The same year, the first conference between neuronal computing researchers from the US and Japan. In 1985 the American Institute of physics establishes annual meeting Neural Networks for Computing. In 1987 the IEEE held the first conference on ANN. That same year the International Society of Neural Networks was created (INNS).

In this book, an automatic learning system that identifies the expressions of denial and speculation in biomedical texts is presented. The objective of the work is to compare the efficiency of this approach centered in automatic learning with which it is based on regular expressions. The results show the superiority of the approach based on automatic learning regarding the use of regular expressions. Some application of Artificial Neural Networks (ANN) are also provided, reflecting the recent trend of applying ANN to different fields of knowledge: classification robotics, artificial vision, signal processing, writing and speech recognition. Finally, a self-coding method in the identification of non-linear systems is presented.

Editor

CHAPTER 1

MACHINE LEARNING VERSUS REGULAR EXPRESSIONS IN NEGATION AND SPECULATION DETECTION IN BIOMEDICINE

Ivan Stanimirović

INTRODUCTION

Medicine is increasingly incorporating greater extent, the support of clinical evidence on the decisions of the voluntary practice habitual. The availability of data of scientific articles allows to obtain these evidences. However, their huge dimensions also make it difficult to access relevant articles. On the other hand, some hospitals have electronic records of the clinical histories of their patients and others many are proceeding to digitize them. This makes it easier for doctors to perform clinical studies that allow progress in the evidence based medicine. But nevertheless, as in the case of access to literature medical practitioners need to have efficient tools for accessing information.

This concern gives a good account of the notable number of initiatives, in the form of congresses, which are organized in the mastery of biomedicine around the challenges of very different characteristics. Some of the most relevant studies that have been carried out continue to organize are: BioCreative, i2b2, bioNLP Shared Task or Genomic TREC

Track. Advanced tools for accessing the information or mining of texts to which we have referred cannot be based on a simple approach like the "bag of words". An analysis of the text in greater depth. This analysis should include the detection of denial and speculation. This is especially necessary in the biomedical domain where, on the type of documents, the number of speculative or negated phrases varies between 13 and 20% (Vincze et al., 2008).

Denial converts an affirmative sentence in negative, as for example: "it is not appreciated pneumothorax or fractures." Speculation is used to express that some fact is not knows with certainty: "can correspond with an incipient pneumonic process." The identification of denial or speculation can be divided into two phases. In the first, the expressions that indicate denial / speculation. In the second, would determine the scope of the say, which words are affected by expression of denial or speculation. In this work we focus on the first phase.

The rest of the article is organized by the following form. First, they are presented the most relevant work related to the detection of denial / speculation, both the based on regular expressions such as automatic learning. In Section 3, presents the collection of BioScope documents, used for the training and evaluation of the system that is proposed. In section 4, describe the main characteristics of the system. The following is the framework for evaluation, and then discusses the results. It concludes with the conclusions and future work. 2 Related work the detection of denial and speculation has received much attention in the in recent years, especially in the field of biomedicine. Proof of this are the numerous recent research related to which are based on two approaches different: regular expressions (RE) and automatic learning (AA).

Among ER-based that emphasized by (Chapman et al. al., 2001). Its algorithm, NegEx, implements a set of patterns denoting denial. Filters phrases that falsely appear to be denial and limits the scope of these. For that, relevant phrases are identified making use of UMLS and the rules of negation are applied. Although the algorithm is qualified by itself author as simple, has proven to be potent in the detection of the negation in abstracts doctors. The results show an accuracy of 84.5%, coverage of 77.8% and specificity of 94.5%. However, when NegEx applied on a set of documents belonging to a different domain for which was conceived, performance gets worse (Mitchell et al., 2004).

On the other hand, (Mutalik et al., 2001), develop an algorithm called negfinder. This algorithm consists of an analyzer, as well as an analyzer syntax based on an LALR (1) grammar. The input of the algorithm is a medical text which is preprocessed to recognize concepts UMLS. The system displays 95.7% and a specificity of 91.8%.

Recently, (Elkin et al., 2005) describe a system based on rules that assigns the concepts a level of certainty as part of the generation of a syntactic tree in two phases. First, you preprocess each sentence, dividing it into text and operators. The text is analyzed and the concepts are represented on the basis of the greatest coincidence within SNOMED-CT for the fragment of analyzed sentence. In the second phase, the system decides whether each concept is a statement positive, negative or uncertain. The evaluation of the system shows an accuracy of 91.2%, coverage of 97.2% and specificity of 98.8%. (Huang and Lowe, 2007) propose a system hybrid to detect denials in reports clinical radiology. The algorithm uses ER accompanied by a grammar analyzer. With this approach, the authors accuracy of 98.6%, coverage of 92.6% and 99.8% specificity.

Many of the recent work in the field of negation identification are based on the AA approach. The work of (Averbuch et al., 2004) is example of denied concept detection in medical narrative using AA techniques. This algorithm is based on the gain of information to learn patterns of denial. It has been demonstrated that this algorithm is superior to the traditional algorithms of classification. In particular, the coverage is 95.45%, the specificity is 97.47% and obtains an F1 measure of 99.57%.

(Goldin and Chapman, 2003) describe extension of NegEx using AA techniques, with the aim of distinguishing the present phrases in clinical reports, where the word “not” deny medical observation. As classification algorithms are used Naïve Bayes and decision trees. With this approach, performance is slightly improved of the baseline algorithm, NegEx.

For its part, (Morante, Liekens and Daelemans, 2008) study the problem of detection of denial in medical narrative focusing on the identification of the scope. He system proposed by the authors, consists of two classifiers. A primer classifier identifies the expressions negation present in each sentence. One second classifier, determines which words in each phrase, are affected by the negation. He sample system, for the collection of summaries of scientific articles, a measure F1 of 80.99% and correctly identifies the 50.05% of the ranges.

(Morante and Daelemans, 2009a), present an improvement of the system described above. In this case, they use four classifiers to identify the scope of the expressions of denial. First, three classifiers predict whether each word in a sentence is the first of a range sequence, the last one or does not belong to the scope of the expression. The fourth classifier uses the prediction of the three previous classifiers to predict the scope. The set of documents used for experimentation is more extensive since the complete collection of documents is used BioScope (Vince et al., 2008). For the documents, the F1 measure in the case of the negation is 84.2% while it correctly identified the 70.75% of the reaches. For scientific articles, the system obtains an F1 measure of 70.94% and 41% of scopes is correctly identified. For article summaries scientists, the F1 measure is 82.6% and the percentage of reaches correctly classified is 66.07%.

This research is expanded by these authors including the detection of speculation and its scope in (Morante and Daelemans, 2009b). This is shown, as the same approach can be applied both to the denial as to speculation. The F1 measure in the detection of speculation for clinical documents is 38.16% and 26.21% of the ranges are correctly identified. For scientific articles, the F1 measure is of 59.66% and the correct identification of the 35.92% of the ranges. Finally, for the collection of abstracts of scientific articles, the system obtains an F1 measure of 78.54% and 65.55% of the reaches are correctly classified.

THE BIOSCOPE DOCUMENT COLLECTION USED FOR MACHINE LEARNING

The collection of documents that we have used to train and evaluate our system is provided in the collection of BioScope documents. It should be noted that the new and recent compilation of documents. Novel because it is the first corpus with annotations about negative or indicative keywords uncertainty as well as the scope of these. Recent because this work was presented in June 2008 at BioNLP (Columbus, Ohio).

The collection consists of three types of documents:

Clinical documents. Represents the most of the collection since consists of 1954 documents. In concrete, radiology reports, each of which contains the parts of clinical history and radiologist impressions. This same collection of reports, used for the task of labeling with ICD-9-CM codes each of the documents (Pestian et al., 2007), organized by the Center for computational Medicine in Cincinnati (Ohio) in 2007. Scientific articles. 5 items from "FlyBase" and 4 articles from "BMC Bioinformatics ". This documents are useful for assessing the classification of denial and speculation, since the different parts of an article show different properties in the use of phrases denied or speculative.

Summaries of scientific articles. Specifically, 1273 abstracts scientists obtained from the corpus Genia (Collier et al., 1999). East type of documents are the main objective of many applications of text mining due to its accessibility.

The collection of documents compose it in more than 20,000 sentences along with their annotations. This size is considered by the authors large enough to serve as standard assessment compilation for denial / speculation in the biomedical field.

In articles and abstracts of scientific articles, the number of phrases speculative is greater than the number of phrases denied. This is because the ambiguity in this type of documents is greater. Set of clinical documents, for its part, is the dense, with respect to the number of words, than the number of negation expressions and speculation refers.

In the BioScope collection, each sentence is annotated with information on speculation and denial and the scope associated with them. During the annotation process, the authors follow a min-max strategy. When mark the expressions follows a strategy minimalist, marking as an expression the minimum unit that expresses speculation or denial. At the time of marking the scope, this extends to the largest syntactic unit possible.

DESCRIPTION OF THE SYSTEM USED

The system architecture can be divided into two parts, a first part dedicated to the training and construction of the classification and a second test phase. Both phases have been implemented using methods supervised, trained and evaluated separately for each of the subcollections that make up the collection of BioScope documents.

Figure 2 shows the architecture of the system in the training phase. As can be observed, the set of documents used for training is preprocessed in order to obtain a valid representation for the classification algorithm. In this representation, each instance is a word of the text together with a set of attributes associated with it. Finally, and through the use of the classification algorithm corresponding, the model of final classification.

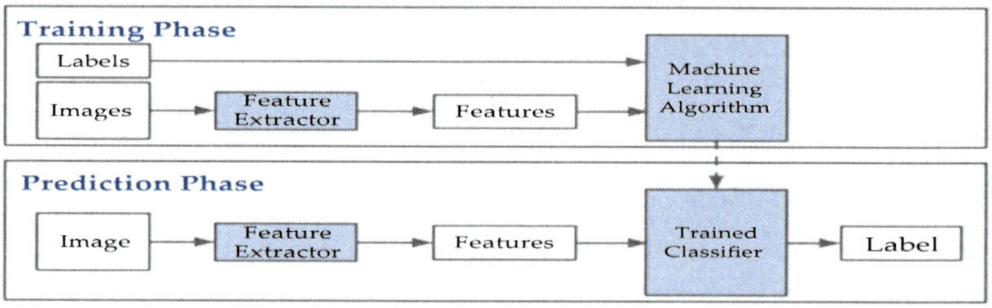

Machine Learning Phases

In the lower part of Figure 2, the architecture of the system in the test phase, which evaluates the performance of the system by using the classification model generated in the training. In this phase, the classifier decides whether the words in each sentence are found at the beginning, inside or outside an expression of negation / speculation. This allows the system to detect expressions complex by more than one word. In both phases, we use the algorithm implemented in Weka (Witten and Frank (2005). We also experiment with the algorithm Naïve Bayes but showed worse performance in general.

The algorithm C4.5 (Quinlan, 1993) generates a decision tree from the data through partitions made recursively in the same way as ID3 (Quinlan, 1986), using the concept of entropy. The algorithm has parametrized using a factor of confidence of 50%, which influences the size and predictive capacity of the tree constructed.

For its part, Weka, is a well-known software written in Java, robust and freely available under the GNU General Public License that application, analysis and evaluation of any set of user data, the most relevant techniques of data analysis, mainly from the AA.

This is a non-class problem balanced, in which the algorithms of classification have a tendency toward class majority. The techniques of

Selecting instances, applied to the data set training, allow you to solve this problem by increasing classes minorities and a reduction of the class majority, using a random in our case, for each of the collections, we have experimented with the resample and Spread Subsample techniques implemented in Weka as well as with combinations. The resample technique produces a random sample of the data set using sampling with replacement. To highlight that the supervised version of this technique. The most important parameter to define is the distribution of classes, which can take values between 0 and 1, where 1 would correspond to a sample with distribution class uniformity. The Spread Subsample technique produces a sample of the data set.

The Resample technique, has been used in its supervised version. In this case, the most important parameter is the in the sample between the majority class and the minority class. For the collection of clinical documents, the use of these techniques has proven not to be cash. This is due both to the characteristics of the documents such as the one in question collection in expressions of denial / speculation, so that the imbalance is lower than in the rest of collections.

The technique used in the case of summaries of scientific articles has been resample as it is with which you get the best results. After testing with different values of the class distribution parameter, the most efficient has been proven to be 0.15. Other problem that presents this collection, in addition of the problem of class imbalance is its big size. Therefore, when applying the selection of instances was reduced the set of training data at 10%.

Finally, for the collection of articles scientists, the best results are obtained using the Spread Subsample technique, with relationship between the majority class and the minority classes of 500 to 1. In this case addition, it is added to the collection used for training, denials and speculation present in the abstracts collection of scientific articles since both collections resent similar characteristics. This technique has been very effective especially in the case of speculation, where an increase in the F1 measure of 0.2174 attributes used the set of BioScope documents used in our experimentation is organized in phrases. That is why we decided to work at the phrase, which implies considering the words of each sentence independently. Each word in the text has been represented through a set of attributes that were selected, from an initial set, applying the InformationGain and ChiSquared techniques implemented at Weka. This final set of attributes is formed by the following 13 characteristics:

Token: Lemma, grammatical category, boolean attribute that indicates whether a word is at the beginning of a phrase or not, Boolean attribute that indicates whether a word is at the beginning of a document or not as a label you can take the values BN, IN, BE, IE, O according to whether a word is at the beginning of an expression of negation, within an expression of denial, at the beginning of expression of speculation within an expression of speculation or out of the expression.

Token Context: Motto, category grammar, a Boolean attribute that indicates whether a word is at the beginning of a sentence or not, attribute boolean that indicates whether word is at the beginning of a document or do not. All these features for the first word on the left and the first word to the right of the word that is being analyzed. Attributes selection techniques show that the characteristic that contributes most information is the motto of the word followed by lemma of words before and after it. For the sub-collection of abstracts of scientific articles are added to the previous ones two more attributes, since this improves the system performance. These attributes are the bigrams, formed by the word in analysis and the word immediately after it and a boolean feature that will be true if the word being analyzed contains the prefix "in" or "un" or false otherwise.

Finally, the grammatical category of Words have been obtained using the grammar tagger developed by the Stanford University. It is characterized by being a Java implementation of the tagger grammar based on Maxima entropy described by (Kristina Toutanova et al 2000). Reviews to measure effectiveness, that is, to have an estimate of the "goodness" of the system, have used as a measure the accuracy and coverage. In order to obtain a single value that allow to compare the different experiments, if uses the measure F of (Van Rijsbergen, 1979). In our case, the measure F is used which balances accuracy and coverage. It is known as F1 and is calculated as, being P precision and R coverage.

RESULTS AND DISCUSSION

The objective of the system is to identify expressions of denial and speculation present in the document collection BioScope. The results of experimentation have been obtained by training and evaluating each Sub-collection separately. In particular, randomly divided each sub-collection into 3 parts, of which 2/3 have been used to train the system and the remaining 1/3 has used to carry out the assessment.

The results obtained by our system, based on AA, have been compared to different ER-based systems in order to contrast the efficiency of both techniques the same set of data. In the first place, two baseline algorithms. A first line base has been built directly tagging as expressions of denial or speculation the two expressions of negation / speculation more frequent occurrences in each set of training. In the second baseline we use the 10 most frequent expressions, except in the case of denial for the collection of clinical documents, where we used the 8 most frequent expressions. This is because all other expressions have frequency of occurrence equal to 1.

In addition, the results are compared obtained by NegEx. This is the algorithm of reference to the identification of the denial refers. In fact, it has been recently included in the tool Metamap Transfer Information (Aronson, 2001). Both evaluations are not directly comparable as NegEx performs an evaluation at the UMLS concept level. Do not however, it gives us an idea of how efficient it is our system with respect to a system based on ER as popular as NegEx. The comparison with NegEx for the case of the detection of speculation has not been carried out since this system has not been designed to identify this type of expressions. At the exit of the classifier a post processing algorithm. The procedure it carries out is the following: when the expression is formed by single word, the algorithm changes the class of the words classified as within expression, by the beginning of the expression.

When the expression consists of more than a word and the classification of these does not agrees on type, the made for this expression up to now and the class is replaced by the one with the highest frequency of appearance. In in the event that it has not yet been classified, modifies its type by speculation since in all sub-collections, the number of expressions of speculation is greater than the number of negation expressions. I only know show the results of applying the post-processing algorithm in those where this improves the results obtained without postprocessing. At case of denial, the first baseline and gets quite good results. This is due to the fact that the two most of negation represent 88.1% of the expressions of negation present in the collection of training data. This does not happen with speculation, where, as the last column shows, the performance ratio is lower. In this case, the two most frequent speculative expressions represent only 31.9% of the total.

The second baseline shows how, with larger list of expressions, it is possible improve the values obtained by the first base line. NegEx gets the worst results in the case of negation. This can be due to the fact that the system was not designed to work with radiology. For denial, our system and baseline algorithms get results similar, this result being comparable to those who would get a human performing the same task. This is due to the characteristics of the collection, where almost 90% of the expressions, they include the total expressions of this type present in the text. This does not happen with speculation, where the detection is made more difficult because the expressions of speculation are not concentrated in a small number of expressions, as was the case with denial. Therefore, in this case, it seems more an approach based on AA, such as shows the important difference between our system and the baseline algorithms.

The results for the collection of scientific articles worsen with respect to the collection of clinical documents. This is because the ambiguity in this type of documents is greater and therefore the detection becomes more complicated. For denial, the difference between the baseline algorithms is significant, so it is not enough to take into counts only the two most frequent expressions. NegEx obtains mediocre results and does not succeed in this case either none of the baseline algorithms. The difference between our system and the second baseline, as shown in the fourth column. However, this difference makes greater in the case of speculation, where the results obtained by our system are much better than those obtained by the second baseline, as shown in the shown in the last column. However, both results are improved. Once again, with a collection of different documents nature, it is demonstrated how a based on AA allows better performance than an ER-based approach.

The characteristics of this collection of abstracts of scientific articles are similar to that of scientific articles, although the class imbalance in the latter is greater. This will be especially reflected in the case of speculation, where the imbalance is makes more patent. For denial, our system and the second baseline obtain the better results. Difference between the two. NegEx, for its part, returns to get the worst results. At case of speculation, our system obtains the best results, being these good, more so if we consider that only uses 10% of the collection for the training. This difference is important with regard to baseline algorithms, such as can be seen in the last column of the table.

Therefore, the comparisons made with the same collection of documents, put manifest the fact that an approach based on in AA is more appropriate for problems of this type than algorithms based on ER. Probably, NegEx is one of the algorithms most used for the identification of the denial of documents in the biomedical field. Although it behaves in a homogeneous way for all types of documents, in the denial, the results are improved, there is an average difference of between 20 and 30% when compared with the results obtained by our system.

CHAPTER 2

THE APPLICATION OF ARTIFICIAL NEURAL NETWORKS (ANN)

Ivan Stanimirović

During the 1990s, ANNs began to be applied to different fields of knowledge: classification robotics, artificial vision, signal processing, writing and speech recognition, and so on. In this in this context, documentary applications have not been able to remain outside this process.

DEFINITION AND CHARACTERISTICS

ANN can be defined as an information processing system composed of a large number of processing elements (neurons), profusely connected to each other through communication channels (Reguero, 1995). These connections establish a hierarchical structure and allow the interaction with objects of the real world trying to emulate the biological nervous system. Unlike computing based on predictable algorithms, neural computation allows solve complex problems whose mathematical formalization is extremely difficult. This is achieved thanks to the principles of ANN functioning, of which we cite below the five most important (Row, 1995):

- Adaptive learning: this is perhaps the most important feature of ANN, since they can behave based on a training with a series of illustrative examples. In this way, no it is necessary to elaborate an a priori model, or to establish probabilistic functions. An ANN is adaptive because it can be constantly modified in order to adapt to new conditions of work.
- Self-organization: while learning is a process where the internal information of the ANN, self-organization consists of modifying the entire complete network in order to carry out a specific objective. Self-organization means generalization, so a network can respond to data or situations that you have not experienced before, but which you can infer based on your training. This feature is very useful especially when the input information is unclear or is incomplete.
- Fault Tolerance: In traditional computing the loss of a small piece of information can commonly lead to system disablement. ANNs have a high capacity of tolerance to failures. Fault tolerance is understood here in two senses: first, networks can recognize information patterns with noise, distortion or incompleteness (fault tolerance with respect to of the data). Secondly, they can continue working (with some degradation) even if it is destroyed part of the network (fault tolerance with respect to the structure). While traditional computing stores information in unique, localized spaces and addressable, neural networks do so in a distributed and high degree of redundancy.

Real-time operation: of all existing methods, ANN are the most suitable for the recognition of patterns in real time, because they work in parallel updating all their instances simultaneously. It is important to note that this feature is only appreciated when implemented networks with hardware specially designed for parallel processing.

Easy insertion in existing technology: it is relatively easy to obtain specialized chips for networks that improve your ability in certain tasks. This facilitates modular integration in systems existing ones.

ELEMENTS OF AN ANN

ANNs are information processing models based on the functioning of the human brain. They are formed by a large number of units or processing elements called neurons, which we can divide into three different groups: 1) those that receive information from outside, also called as input neurons; 2) those that transmit information to the outside, denominated neurons and finally those who have no contact with the outside and 48 with other neurons in the network, these are called hidden neurons. In any type of ANN neurons are strongly interconnected with each other, organized in layers and forming different topologies. There are different topologies to solve different types of problems.

In Figure 1 we can see a schematic diagram of a neuron (Regueiro, 1995). It has a variable quantity of inputs that come from the outside (in case our neuron is in the input layer of the network), or other neurons (if it is in some hidden layer or output). In turn it has a single output (X) that will transmit the information to the outside (in case our neuron is in the output layer of the network), or towards other neurons (if it is in the layer of entry or in some concealed). The output signal will be calculated according to the inputs, for which each one of them is affected by a certain weight. These weights vary freely as a function of time and in each of the neurons that are part of the network, and this is extremely important because the knowledge that contains the network will be in said weights. All input signals are combined by called combining function (fj), which may vary according to the type of neuron but will always generate a vector whose components will determine the final output value by the activation function (g). As well there are different types of activation functions, the most common are step function, linear or mixed function, sigmoidal, and Gaussian function. Each function has its own characteristics, and will be used according to the type of information that we are going to manage (e.g. binary or analog, etc.)

As we have already stated, as in the human nervous system, knowledge is found in the weights of connections between neurons, and not in the neurons themselves. These weights have no constant value, but which vary depending on a certain algorithm called learning rule. A rule of learning makes the value of the weights of a network vary until they adopt a constant value, when this happens it is said that the network has already "learned". There are many different learning rules, but we could split them in two large groups: 1) those of supervised learning, and 2) those of unsupervised learning. In the first, there is a sample of information that serves as an example for the network to learn, while in the latter does not exist this information and the weights of the network fluctuate freely without any type of reference, until they stabilize.

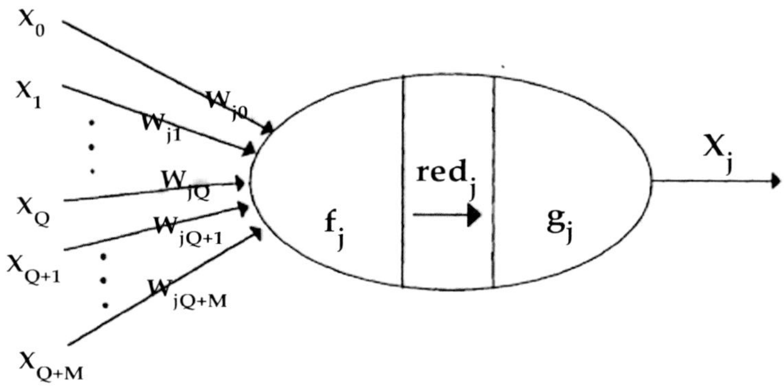

Figure 1. A schematic diagram of a neuron

Within each of these two groups, there are several algorithms that we will not name here and which are associated with different network models. Finally we can say that there is another criterion of classification of learning rules, which is related to the form of operation of the instruction: 1) off-line, and 2) on-Une. In the first case, the network must be disconnected from its operation until the learning ends, to then set its weights and start operating. In the second case, learning takes place during the normal operation of the network, so that here the learning phase of the operation is not different. The main difference between the two will be in off-line networks will be more stable in their operation than on-line.

STRUCTURE OF AN ANN

We have already seen how this compound is a processing element or neuron (microstructure). Connecting several neurons in a certain way we get a network (mesostructure). There are infinite variations of topologies, but we can classify them according to four criteria: 1) number of levels or layers, 2) number of neurons per level, and 3) forms of connection.

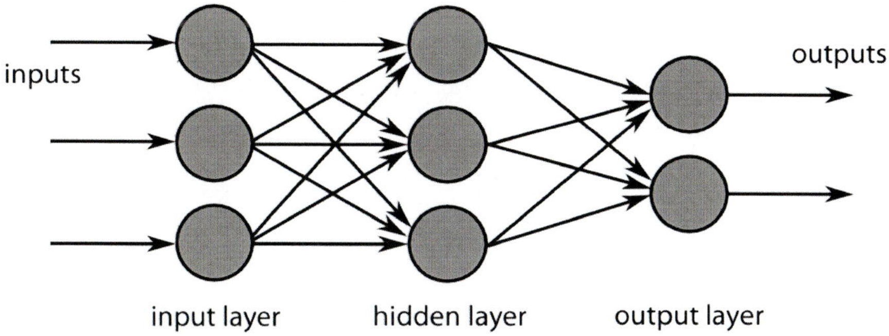

Figure2. Neural network with multiple layers

In level of entry levels hidden level of departure as for the number of layers, we can say that a network can be formed by a minimum of one layer and a maximum of n layers. When there is only one layer, this officiates both inbound and outbound. In the case of a two-layer network, the first will be input and the second output. In the case of networks with more than three layers, as the one of figure 2: the first one will be of entrance, the last of exit and the rest will be the layers hidden as we mentioned above. The number of neurons per level can vary from 1 to n, and depends of the topology of each particular model.

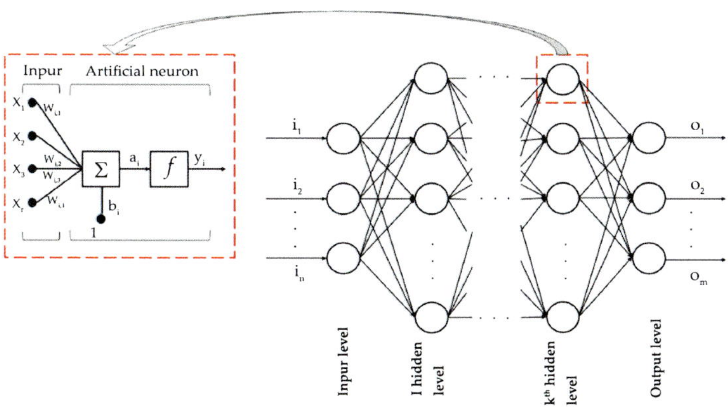

Figure 3. Neural network with feedback of the outputs to the previous inputs or to the neurons of the same level

The way in which the output of a certain layer of neurons is connected to the input of another, determines the connectivity of the network. Here are two basic forms:

- When no output from a layer is input of the same level or of previous levels, called forward propagation (figure 2),
- When there is some type of feedback of the outputs to the previous inputs or of the same level, denominated propagation to the rear (figure 3). Each of the features listed does not say too much on its own, so it should the network model that adopts it, however, such an analysis goes beyond the objectives of this article.

ANN MODELS

As we have repeated throughout the work, there are several different models of neural networks. Each of them has been designed for more or less specific purposes, however, there are several who have won great popularity.

The Kohonen model

Before entering fully into the applications of ANN in the field of Documentation we will stop briefly to analyze one of the models of the previous table most used in information retrieval, is the one developed by Teuvo Kohonen. At the beginning of the 1980s, Kohonen showed that information by itself, assuming its own structure and a functional description of the behavior of the network, was sufficient to force the formation of topological maps (Kohonen, 1982). These maps have the feature of organizing the input information by automatically sorting it.

This model has two variants:

1) The so-called "Vector Quantification of Learning" (Learning Vector Quantization) or LVQ, and
2) The so-called "Topology-Preserving Map" or TPM. The latter is also commonly referred to as a "Self-Organizing Map" (SOM). The difference between the two models is that while the LVQ works with outputs in a single dimension, the output in the SOM is two-dimensional. The latter is the most widespread.

The principle of operation of this model is simple, it is a matter of establishing a correspondence between the input information and a two-dimensional output space, or topological map. In this way, the input data with common features will activate areas near the map. It is very common to represent this network model with its output neurons arranged in a two-dimensional way, as we can see in Figure 4. When a data is entered into the network it reacts in such a way that only a neuron of the layer of output is activated. This neuron is called the winner-take-all unit and determines a point on the two-dimensional map. What the network is really doing is sorting the input information, since the winning neuron represents the class to which the input belongs, in addition to that before similar inputs the same neuron will always be activated. Therefore, the network is highly indicated to establish relations, unknown previously, between a certain set of data.

SOM model with its output neurons forming the map the learning method of the SOM model is called competitive and is of the unsupervised and off-line type, so it will have a previous stage of training and a later stage of operation. The Kohonen model is one of the most useful ANNs that have been developed, although it has two limitations: 1) the process of learning is often long and arduous, and 2) to learn new data it is necessary to repeat the learning process completely. However, the versatility of this type of network is very broad, allowing you to classify all type of information, from literary (Honkela, 1995) to economic (Kaski, 1995). Later they will be seen different applications of this model.

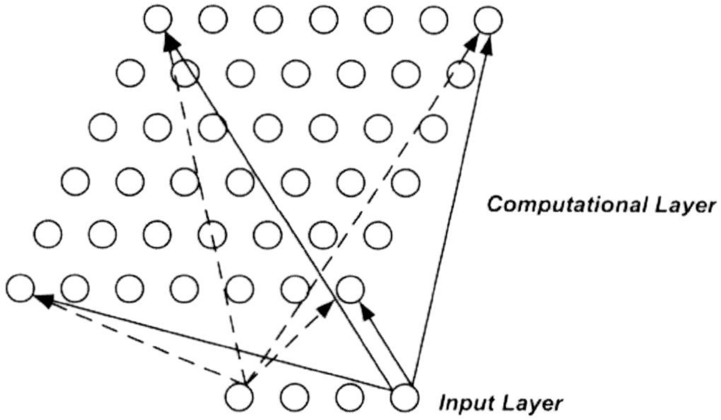

Figure4. Network model with its output neurons arranged in a two-dimensional way

ANN within the Specialized Documentation Literature

The topic of ANNs is relatively new in the field of Documentation and yet it has experienced a great boom in recent years. A study conducted in the Library and Information Science database of abstracts (LISA), in its edition of winter 1996, allows us to observe the abrupt growth of the specialized literature in this subject. References were retrieved by searching for the term "neural" in free text, which yielded about 200 entries. From there, the entries corresponding to investigations in course and only articles, reviews and conference papers were left: 170 records. They were then ordered from time to time from 1981 to 1996, and the graph of Figure 5 was constructed. The exponential growth of the literature on the subject, especially in the periods 1992-93 and 1994-95. The fall of 1996 should not be taken into account since the information on this year is incomplete, which will be available in later editions.

Another important fact is the titles of journals involved in this study. In table 2 we can to appreciate the names of the publications ordered by their frequency of appearance. It is clear that this distribution resembles the one of the dispersion of Bradford, since it counts on a nucleus of few titles that monopolize a large percentage of articles while on the other hand we have a large number of journals with only one or two articles.

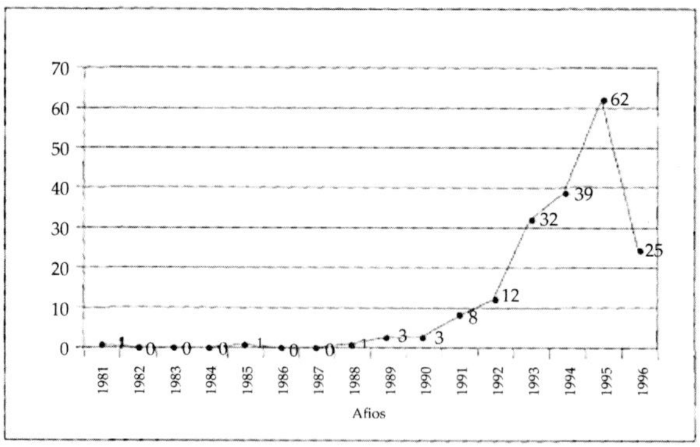

Graph 1 - Articles on ANN in the base LISA

The curious thing about the study may be that the most frequent publication is not specialized in neural networks nor in artificial intelligence, but comes from the chemical field. The Journal of Chemical Information and Computer Sciences covers about 25% of the references on the subject, so we can say that the chemical information retrieval appears as the most active segment in the study of neural networks. Behind it appear different titles specialized in expert systems and artificial intelligence, thematically more closely related to ANN, covering a range of 1.5 to 7.5% (Experts Systems 7.64%, Knowledge-Based Systems 5.88%, IEEE Expert 3.53%, Artificial Intelligence Review 2.94%, Artificial intelligence 1.76%). However, if we added the references of all these journals, we would not yet reach the numbers of the first. In the others we find titles of Informatics in general (Telematics and Informatics 4.11%, Byte 2.35%), specializing in Information Retrieval and management 3.53%, Information Systems Management 1.76%, and Library Hi Tech 1.76), one specialized in Biology and Medicine (Computer in Biology and Medicine 2.35%), and another in Bibliometrics (Scientometrics 2.94%). Behind them comes a platoon composed of 48 magazines that present at most a couple of articles about the theme.

The last issue that we must highlight is the fact that most of the titles specialized in documentation and retrieval of information, and which we could call the hard core of the specialty, do not appear or do so with one or two references. This would indicate that ANNs are not yet a topic fully addressed by the literature the discipline and that if we want to make a study about its application to the information retrieval we should pay close attention to the works that are in the border between diverse disciplines (Chemistry, Biology, Artificial Intelligence, etc.) and documentation.

For a more in-depth study of the literature on neural networks, see the bibliometric Van Rann (Raan, 1993). It presents a map of science developed using the method of co-occurrence of terms in the thematic field.

APPLICATIONS OF ANN

ANN applications can be classified by various criteria. We will use an adaptation of the scheme of Scholtes (1995), which contemplates an ordering by type of application:

- Management of data
- Classification of information (information clusteríng)
- Interface design
- Information filtering

- Incomplete searching
- Information discovery (data mining)

The enumeration of applications is not intended to be exhaustive at any time, and the level of detail of each one varies depending on the information available.

Applications in Data Management and Data Mining

There are few neural applications aimed at library management. Perez (1991) cites some, but all they are only packages aimed at accounting applications that have been used for forecasting budget in libraries. For Scholtes (1995), the experiences in this field focus on the management of loans and periodicals. However, it indicates that work in this field is increasingly rare. This is because it has been concluded that the information to be worked on is too precise and structured, whereas the optimal performance of ANNs is obtained with unstructured data and incomplete.

The automatic classification of information is the field where the most ANNs have been applied, particularly in the generation of two-dimensional maps of concepts.

As we have seen, one of the variants of the Kohonen model is that of self-organized maps by characteristics. Xia Lin, a researcher at the University of Kentucky and a specialist in the graphic representation of the information (Lin, 1991), uses these maps in order to generate a "displayable" output (map display) of a collection of documents (Lin, 1995 and 1997). Lin presents three different examples:

1) A series of documents extracted from the INSPEC database through DIALOG,
2) A collection of personal documents, and
3) A group of papers from the SIGIR conference (1990-93). For each one of them, it generates a determined map through the following procedure:
    - A list containing all the terms appearing in titles and summaries of all documents of the collection.
    - Irrelevant terms are removed by a list of empty words (stop list).

- A steaming algorithm is applied to the list to bring the terms to their root and reduce their number. Then duplicates are removed.
- The terms with high and low frequency of appearance are eliminated, preserving the middle zone of the spectrum.
- A vector is created for each document of n dimensions, such that n is the number of terms that have remained on the list.

Each vector is completed with the weight associated with each term-document. The weight is calculated as proportional to the frequency of the term in the document and inverted to the frequency of the term throughout the collection. In some examples the weights are dispensed with and only the presence or not of the term is indicated, through '1' and '0'.

These vectors are used as input to train a Kohonen network with n input elements (such as which is equal to the number of terms of each vector), and a variable number of outputs (e.g. 10x14 = 140 neurons), which will be the size of the output map.

This output map will also be represented by a vector of weights to which it is initially assigned small and random values.

During the training process a random vector-document will be taken. The output neuron whose vector looks more like the vector-input document, it will be declared a winning neuron. Then the neurons (and those of their neighbors), will be adjusted to approximate vector-document.

This process is repeated for a number of cycles. When the training process has finalized each term is compared to the weights of the output neurons, in this way it is associated the closest term (best-match term) to each of the outputs. Connecting the output neurons with similar associated terms will generate the different areas of the map.

With this output matrix a map is constructed, and that corresponds complementary visualization tools can be added to the output map to obtain different resolution levels or pop-up menus for each node. The purpose of these add-ons is to facilitate the examination and visualization of the information represented in the map.

Lin establishes the limitations of the model, among which are: inability to work with high volumes of information and high cost of information processing. However, the application of this model seems to be one of the most promising in automatic classification using neural networks. There are currently some applications that present a visualization map for Internet nodes. On the subject, including a set of Web pages classified by this method.

WEBSOM

On the same line as Xia Un has worked a Finnish group belonging to the Research Center in Neural Networks of the Technological University of Helsinki, directed by the own Teuvo Kohonen. The group developed by WEBSOM (Kaski, 1996; Honkela 1996a and b; Largus, 1996), a system that allows automatic sorting large masses of information in full text, in order to facilitate their exploration and navigation.

The potential of this system has been demonstrated in a case study where group articles are organized. The treatment of this type of information is somewhat complicated due to that, unlike the maps of Xia Lin where formal information is organized (articles, presentations, patents, etc.), in Usenet groups are colloquial messages, without a specific format, usually with misspellings and no style correction.

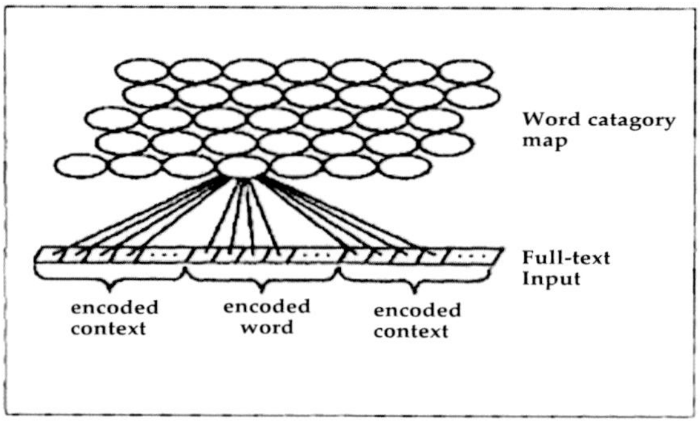

Figure 6. A grid of two dimensions with relations, such that words that tend to appear on the same node (or near one), form the same category

Word category map represents a full-text input encoded context which contain about 1,200,000 words. To do this, it uses a method that consists of the following steps:

- Messages are taken and non-textual information is deleted. Special codes and numeric expressions are treated by heuristic rules.
- To reduce computational processing, we discard those words that have a low occurrence (Less than 50 times) and are treated as empty entries.
- In order to emphasize the theme of each message and reduce the erratic variations caused by different discussion styles, common words that make thematic discrimination difficult are eliminated. In the example of a total of 2,500 disappears 800.
- These words generate a first word category map. These are ordered in a grid or map of two dimensions according to their relations, in such a way that words that tend to appear on the same node (or near one), form the same category (Figure 6).

This map is used as a histogram that filters information from documents. The authors define to the histogram as the "fingerprint" of documents. With filtered and blurred information {Blurred), a second document map is generated, as can be seen in Figure 7.

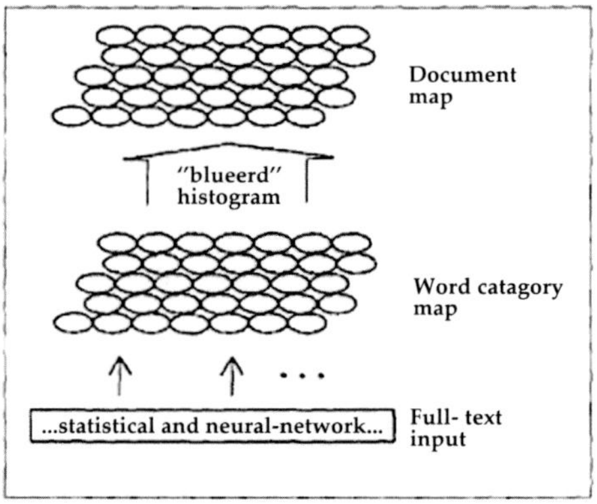

Figure7. A second document map generated with filtered and blurred histogram

In this way, the final product is a map like that of Figure 7, where each document occupies a place in the space according to its thematic contents. Each map area reflects specific content and topics vary slightly along the same. The different shades indicate the density of documents, the darker the more documents we will find.

The user interface to consult the map has not been neglected. This has been implemented in HTML and allows the exploration in four levels

1) Global map,
2) Augmented zone,
3) Map of nodes, and
4) View of the message. Level 1 presents a sensitive map with the whole of the documentary space. If we select a region of the same, it will appear enlarged in a similar map (level). At this level, it is also possible to "move" to neighboring areas without having to return to the general map and thanks to a tool that indicates with arrows the course to take. When we select a region this is second map, we access a list of the messages or nodes of that region. It is also possible at this level like the arrows at level 2.

Finally, it is important to note that to generate these maps a great power of computation is needed.

In this direction, a public domain package for the simulation of LVQ networks is also available (Kohonen, 1996b). This is a prototype of a system of categorization of Internet homepages. It has been developed by the Artificial Intelligence Group of the University of Arizona, funded by several US government entities. The system tries to provide, automatically and scalable, a thematic approach to categorization and search for information on the network. This project has analyzed the content of more than 110,000 homepages related to the field of entertainment. With this information, and using the Kohonen SOM model, a map of two levels with different thematic regions. In regions with more than 100 URLs, it is possible access to a new, more detailed map. In the case of regions with less than 100 URLs, they are only accessed by a list of them. The map is constantly updated thanks to some splits or agents that they trace the network, in a similar way as the search services (Altavista, Lycos, etc.).

Maps of Scientific Publications Networks

A special application of the maps analyzed in the previous points is the organization of publications scientific research. The study of the relations between the different magazines of a certain thematic field, is realized through analysis of the Journal of Citation Report (JCR). This repertoire, published by the Institute of Scientific Information (ISI), collects quotations between different titles. This information is analyzed with different methodologies: cluster analysis, multidimensional scaling, principal component analysis, etc. Bell tower proposes the study of these relations through the use of the Kohonen model (Campano, 1995).

Similar to the Lin procedure, a vector is constructed for each journal title. The resulting vector has n elements, such that n is equal to the number of titles to be analyzed, and each element will contain the number of citations of each publication. The size of the output matrix is calculated according to the number of titles to be analyzed. Then the network is trained with the input information and a map called "map of relations », as can be seen in Figure 9-a. In opposite, another map is obtained called the "domains map", and which represents only the neurons that are found strongest response to a given journal (Figure 9-b).

Only one title appears once on the map. Both maps are complementary and allow, through analysis of the size and proximity of areas, to study the relationships between two or more scientific journals.

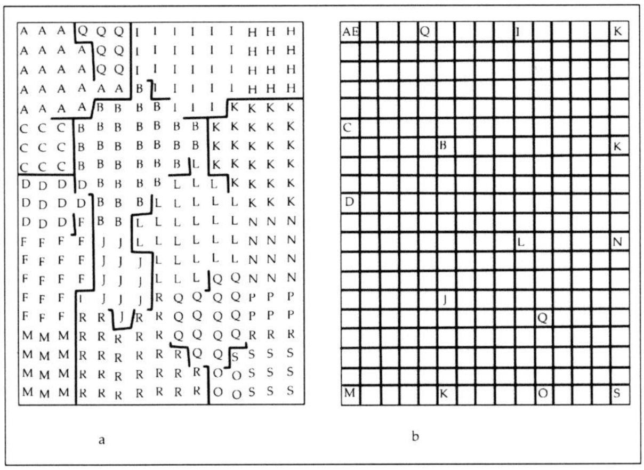

Figure 9. Map of relations and domains map

Macleod's Algorithm

Self-organizing maps (SOM) are not the only method of neuronal clustering. Macleod has designed a specific algorithm for document clustering, using unsupervised learning (Macleod, 1991). In this work, the characteristics of the algorithm are analyzed and experimented through two small tests. Based on the results of these tests, it is demonstrated that the MacLeod algorithm is superior to hierarchical (sequential) clustering algorithms. Finally, it should be noted that the clustering is output independent of the order of entry of the documents.

ANLI: Adaptive Network Library Interface

ANLI is a program that acts as an interface between the user of an online catalog and the catalog itself (Kantor, 1993). The term Network of the name refers to a network of relationships between documents, articulated through recommendations. These recommendations are provided by users and represent a certain value added to catalog information. We can browse the network using a hypertextual tool contained in ANLI, which allows different users to share information through "cooperative work computer-supported collaborative work (CSCW). The system is based, among other techniques, in neural networks, but unfortunately Kantor does not go into detail about the nature and characteristics of the type used. It limits itself to saying that ANLI is a great "adaptive network" and that the neural model has been very useful in the overall development of the system.

Application in the Filtering of Information

Usenet newsgroups daily generate a volume of information that is impossible to assimilate. The secret in the use of these groups lies in the advance selection of groups and profiles of interest. Even so, the reading of this mass of messages is usually very arduous. To avoid this, Jennings and Higuchi designed a system that filters Usenet information and is based on ANN (Jennings, 1992).

The operating principle of the system is quite simple, as can be seen in figure 10. At the beginning, each article is analyzed to extract a series of relevant terms. These terms have associated certain weight according to the place they occupy in the message (header, body, subject, etc.), and with them the network of relationships is built. The terms of each new message are submitted to the network, if they keep any correspond to it, will tend to activate or "trigger" certain nodes, and these in turn will be able to do so with others. In case the document generates a certain activation, it is considered that it has the minimum energy to activate the network and is therefore accepted for reading. Otherwise, energy is not enough and therefore is discarded without being read.

The network is not maintained as a fixed structure, as it is modified by each input document. From similarly, with the passage of time non-activated nodes tend to lose weight, they may even disappear. This is to try to make the network dynamic and adapt to variations in the profile of the user with the passing weather.

Although this application is practical and interesting, Scholtes (1995) thinks that this system is not completely automatic, which is not adaptive and does not exploit the typical characteristics of ANN: generalization and association.

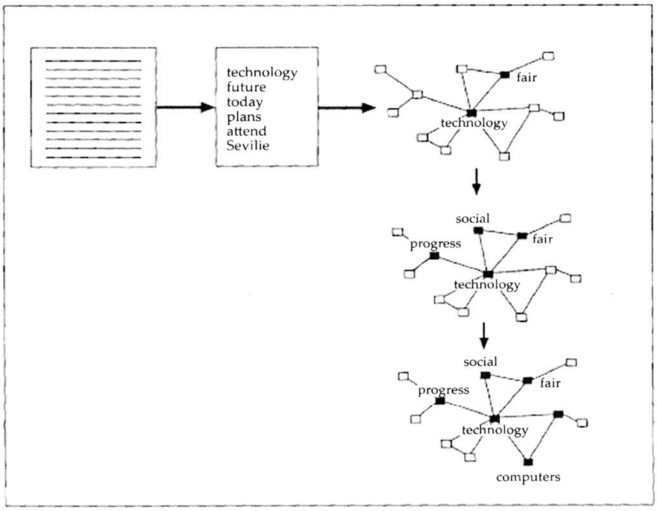

Figure 10. The operating principle of the system, where at the beginning, each article is analyzed to extract a series of relevant terms

Incomplete Search Application within Neurodoc Project

The search and retrieval of information through natural language is a very important field of research. The degree of tolerance to typographical errors, the use of synonyms, etc., facilitates the human-machine relationship and make this type of systems open to a wide range of users. However, we cannot still talk about an anthropometric dialogue with the computer, because certain problems still exist not resolved. The main one is perhaps the inability of the system to supply the human interlocutor who, knowing of the nature and scope of the information contained in the database, allows guiding the user in the search of only that which can be found at the base. In order to provide an answer to this problem the Neurodoc Project has been developed, whose objective is to develop a recovery system that expansion of feedback questions (Lelu, 1991). The system is based on neural networks, although clearly establishes the type of network topology used. A prototype of the system was programmed for Macintosh environment (Lelu, 1992).

Intelligent Negotation Neural Network (INN)

Research on information retrieval indicates that the search for topics is the most common and less satisfactory search modalities. In the same line as the previous system and with the in order to approach a solution to the problem of thematic recovery, the INN (Meghabghab, 1994). This system behaves as an information specialist capable of learning to negotiate the question of the user, in order to transform it into a correctly constructed search formulation and with greater possibilities of success.

The design of the INN is based on a multilayer ANN architecture using the backscattering algorithm (Back-propagation). A network called ACN (Attentional Connectionist Network) is used for the separation of concepts, agents and relationships, a knowledge base and analogical reasoning.

The network must be trained with at least 200 searches selected in advance. Once trained, the program is trained to intelligently process user questions. Meghabghab does not indicate any specific study on the performance of the system, but states that it significantly improves the percentage of hits for thematic search.

INFORMATION DISCOVERY (DATA MINING)

It is called data mining to study the recovery of information in adverse conditions (much noise, incomplete searches), and with various types of data (numbers, free text, structured records, etc.). As we have already stated, ANNs are extremely useful for the processing and search of heterogeneous information, incomplete and with high levels of noise; So that these will find in the data mining an area of application which is very important (Scholtes, 1995).

This is a complete data mining package developed by Integral Solutions Ltd. A resemblance to so-called executive information systems (EIS). This product presents a series of advanced information processing tools, such as: rule induction, neural networks, regression models, information clustering using Kohonen networks, graphical analysis of information, etc.

Emphasis is placed on the fact that having a large amount of data does not guarantee obtaining the relevant information. This is also called data richness, information poverty. This is why the added value of information is the main objective of the program, and attempts to achieve it through the following functions:

- Data access: the system can directly access the contents of various commercial databases, such as Oracle, Ingres, Sybase and Informix. You can import data from these databases and combine them with others from other applications, eg spreadsheets.
- Data manipulation: it is possible to select records using a search criterion, and with them create series of data that can be manipulated, exported and processed.
- Visualization of data: information can be seen in different ways, such as histograms, clouds of points, distribution tables, networks of relations, etc. All presentations are interactive, being able to expand and reduce by zones.
- Learning: through the neural networks and the induction of rules it is possible that the system "learns" make certain decisions without human supervision.
- Visual programming: this is a complex program, however, it can handle with relative ease due to which any task is reduced to dragging icons and establishing visual relationships.

The system is not only addressed to the business world, but also to the scientist, and to any field where the decision making has a significant weight.

As we have seen in the present work, there are several applications that exploit the characteristics of ANNs and apply them to the field of information retrieval. However, they are still very large limitations. The main problem is the amount of information processing required. It is important to bear in mind here that ANN is a model of information processing, and that the systems that emulate them are just that: emulators of an ideal model.

This clarification is necessary in the case of ANNs, since most of the applications mentioned here simulate the operation of a massively parallel network using a sequential computer with Von Neumann's architecture. These simulations do not exploit the main feature of networks, parallel processing, making it impossible to develop applications that handle masses of real information. They are only limited to what Scholtes (1995) calls "toy problems".

The solution to this question is not to have more powerful sequential machines, but to have computers specifically designed for massive parallel computing, the application of the aforementioned method which has such a resource is the Kohonen WEBSOM.

Another issue to consider is the limitation of ANNs as to the nature of the information they can drive. The rise of these techniques may suggest that they are useful for solving information problems, but it is not like this.

The supremacy of ANNs is appreciated when it comes to processing information in natural, noisy language and incomplete. In the context of information retrieval, it is common to see ANN applications to problems of structured information, which could be solved by other, more traditional methods that are easy to implement (Scholtes, 1995).

In spite of these limitations, ANN-based techniques applied to the retrieval of information often perform better than some other similar methods. The results from ANN are usually confronted with ones determined using the spectral polarization difference (SPD) method, as well as the HUT model-based iterative inversion. Generally, the results obtained using ANN method are better than, or equal to results given using other techniques when trained with data that is previously simulated. Performances are also very good.

CHAPTER 3

ADVANCES IN PRE-TRAINING TECHNIQUES IN DEEP LEARNING

Ivan Stanimirović

INTRODUCTION

The first precedent in the successful use of deep learning is due to Geo King Hinton [29, 30] who introduced the Deep Belief Networks using in each layer Boltzmann Restricted Machine (RBM) for the initial assignment of the free weights.

The general principle of the operation of a deep architecture is to guide the training of the intermediate layers of representation using unsupervised learning spec For each layer; For this, different techniques are used such as Autoencoders [32], RBMs [31] and DBNs [29] (Deep Belief Networks). Same neural network, have been used to initialize the weights of deep architectures of neural networks with connections are now monitored.

In the present work the application of pre-training techniques is explored, not supervised in deep architectures in the task of comparing the science of these techniques with the results obtained by means of fully supervised systems.

A deeper architecture is seen as a composite of several stages of processing in the case of a neural network the main problem of using an architecture deep is the following: If a good representation of the entry has been found in each layer of the network, it can be used to initialize and train the layers subsequent by means of a supervised optimization based on downward gradient. However, the so-called "good representation" is difficult to achieve through of habitual methods of allocation of free weights [9,21].

It has been found that the representation of knowledge in the human brain is carried to be carried out in a distributed manner [12, 28], which means that the learning is distributed in a large number of neurons instead of being focused on a particular region; In addition, a scattered representation is used, that is to say, although the knowledge is distributed in many neurons, also spread among them in such a way that only between one and four percent of human neurons are active instant given.

In this chapter there is a general notion of the motivation and importance that have deep architectures in the solution of problems of classification and regression. Some theoretical and simulation results are mentioned from the literature, which suggest advantages in the performance of deep topologies front to topologies composed of a few layers of representation.

The main algorithms of initial adjustment of weights and of supervised adjustment for regression models are presented, and some neural identification topologies are introduced. Then, the Boltzmann learning procedure is introduced, which serves as a theoretical framework for the understanding of pre-training algorithms subsequently submitted. In turn, stochastic techniques are described for the choice of the model of hyper-parameters.

The method of auto-coding is later explained, introducing autoencoders as building blocks of a deep architecture. Learning through restricted Boltzmann machines is described, detailing the considerations necessary for their training. Are presented the results obtained solving the problems of classification of patterns MNIST and identification of non-linear systems, the accuracy obtained in both applications exposing the differences and similarities obtained.

DEEP LEARNING

A problem solved by means of the application of diverse algorithms belonging to Artificial Intelligence consists in the representation of a function by means of learning architecture. We say that the representation of a function is compact when you have few computational elements or degrees of freedom that must be adjusted using a learning procedure [4, 28, 3].

In this context a function can be represented by computational architectures very diverse. Functions that can be represented compactly by one architecture may require an exponentially large number of elements to be represented by an architecture of depth k1. A character fundamental is that the greater the number of elements used to represent a function, the greater the number of patterns needed for the process of learning with the aim of obtaining a good generalization. We denote a set of computational elements such as the set of cells which must be made to solve a particular problem. For example, the set of all the cells that can be made using only the basic operations of addition, subtraction, product and division.

A function can be expressed by the composition of computational elements of a given set. This composition is generally by means of a graph representing each computational element by a node of the graph. There has been talk of the depth of a model architecture but has not been given a formal statement of the same, the concept of depth of an architecture of representation refers to the depth of the corresponding graph or drawing, i.e., the longest path from an input node to an output node. When the set of computational elements is the set of calculi that can lead to an artificial neuron, the depth corresponds to the number of layers of the network neuronal.

Example 3.1. Consider the function without if the set of elements computational allowed as and use the values we can obtain a representation of the function f (x) as shown in Figure 3.1:

The depth of a model is studied not only in the computational field but also constitutes an element of assessment in the area of neurosciences. Some research suggests that the cerebral cortex can be seen as an architecture deep layer that involves 5 to 10 layers during processing of the visual information [23].

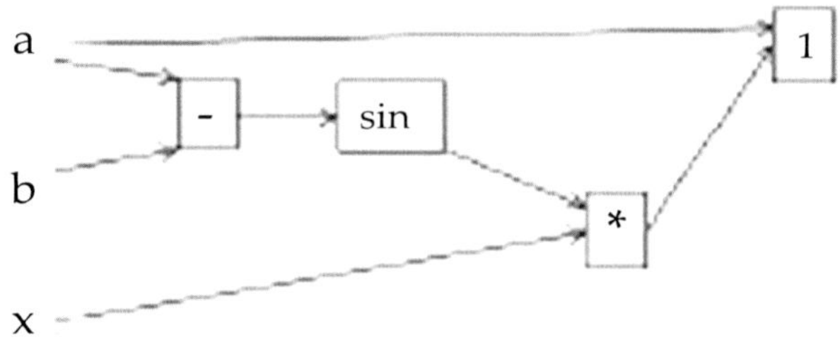

Figure 3.1: Depiction of network architecture with depth 4

An artificial neuron can be seen as the combination of an analogous transformation (Application and sum of unpopular weights) followed by a non-linear transformation (function of activation). By placing artificial neurons as computational elements of a deep architecture we obtain artificial neural networks. The better way a network of such networks consists of the choice of only a related hidden layer with an exit layer with which we get a depth of 2.

Consider a multitasking model, suppose a system with multiple outputs where each corresponds to a goal that must be achieved, if many of these characteristics are shared and have correlation with each other is achieved robustness of the system.

Learning Objectives Using Intelligent Artificial Network

For any AI-based training or learning system is sought to fulfill the following characteristics [18, 19]:

1. Ability to learn complex functions highly variable with relatively few examples of training.
2. Ability to learn in an unsupervised way (without human interaction).
3. Ability to learn from a vast number of examples.
4. Ability to take advantage of synergy of phenomena, that is, to have a multi-objective apprenticeship.
5. Ability to achieve a good generalization in the case of non-Supervised learning.

Computational Complexity

It has been seen previously that when a function can be represented by a deep architecture compactly, this usually requires a large number of elements to be represented by shallow architecture.

The size of an architecture is de ne as the number of computational elements which constitute it.

Consider the following case: Suppose a monotonically weighted threshold circuit with depth k-1 by calculating a function f_k $F_{K,N}$ has at least one size 2^{cN} for some constant c> 0 and N> $N_{0.}$

The function class $F_{K,N}$ contains functions with N^{2K-2} nested inputs for K. The circuit can be seen as Boolean sheets are the input variables (not negated) and the value of the function descends to the root [14]. The least *i*-level or layer from the base consists of gates AND when *i* is even and OR gates if *i* is odd. The number of inputs supported by a computational element in the layer of the cup and in the base of the tree is N and in the remaining layers is N^2 [15].

The above example is a concrete case of the exponential increase in the number of computational elements in a representation architecture if a depth is used less than appropriate for a function. This does not verify that all require a representation using a deep architecture but suggests that there is adequate depth for each function. If a function has many variations and these variations are not related between s through a subjacent regularity, then neither a learning algorithm would not work much better than estimators in the local domain. The decision trees are one of the most used tools in the design of systems learning. One of their most outstanding features is that they need at least as many examples of training as variations of the objective function.

If cooperatively working sets of trees are used, this is equivalent to add a third layer in the architecture that allows to discriminate between a large number of regions within the space of parameters. These regions form implicitly distributed representation using the output of all trees in the forest. For example, consider a set of three binary trees, each tree is capable of dividing the space of parameters in two regions, so that the three trees separately divide the space into 6 zones; Suppose the trees are found collaborating in a forest of decision, with this architecture the trees divide the space in 7 regions instead of 6, which suggests that the greater the amount of trees, the number of regions in which space is divided increases (See Figure 3.2).

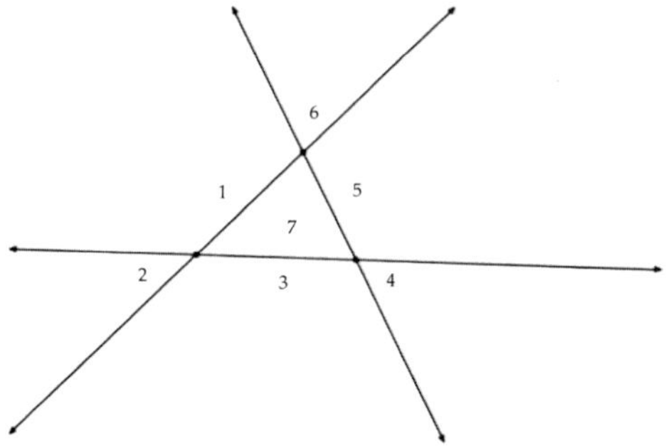

Figure 2.2.1

Learning Distributed Representations

As detailed above, a distributed representation can be much more compact than one local because the distribution can be achieved by increasing the depth of the architecture that is being used and therefore is reduced exponentially the number of computational elements used.

In a distributed representation the input pattern is represented by a set of characteristics that are not mutually exclusive but can be statistically independent. In a local representation the process of clustering is characterized because the clusters do not have common regions and therefore their intersection is null.

Unlike local clustering, in distributed clustering, the number of classes potential is comparatively immense and its behavior is similar to overlapping described in the "decision trees" [15]. In the context of artificial neural networks, Boltzmann multilayer networks and machines are aimed at learning of a distributed representation of the input vector by means of the layers of hidden neurons.

It is said that a function is highly variant when an approximation to pieces of the function requires the use of many pieces. A function of this style can be represented by a deep architecture compactly while generally it will take a lot of computational elements to be represented by an architecture of depth 2.

Some cognitive psychologists have studied the idea of neuronal computation by proposing in the brain a hierarchy of levels of representation corresponding to different levels of abstraction with a representation distributed at each level.

Local Generalization vs. Non-Local Generalization

Local estimator in the input space gets a good generalization for an input vector x by exploiting examples of training near the neighborhood of x. Local estimators "split" the input space into regions and require many parameters to model the objective function in each region. When many regions are necessary because of the great variability of the objective function, the number of parameters will grow, generating the need for a high number of examples for produce a good approximation.

The problem of generalization is closely related to dimensionality of the function. However, the most important factor in the process of generalization is the number of variations that the objective function we want to have by our estimator.

There are architectures based on the pairing of local templates forming two levels of abstraction. At the first level, a template would obtain an output whose value indicates the degree of pairing with each example of training while the second layer uses these values to estimate a convenient output.

An example of this two-layered behavior is the Kernel Machine [4] denoted by equation 2.4.1.

$$f(x) = b + \sum_i \alpha_i K(x, x_i) \tag{2.4.1}$$

where b and α_i are in forms of the second level of decision, while the kernel function K (x, x_i) relates the input x to the training example xi. In this case also exist local and distributed kernel functions. A kernel is local when the relationship K $(x, x_i) > \rho$, with ρ being positive, it's true only for x' in a certain region connected around of x_i. An example of a local kernel is the Gaussian function denoted by the equation 2.4.2.

$$K(x, x_i) = e^{-(\|x - x_i\|/\alpha)^2} \tag{2.4.2}$$

where σ Controls the size of the region around x_i.

The Kernel Machines exploit the so-called softness priority, i.e., if in the network is presented as an input the pattern x_i is intelligent to expect the predictor to grant an exit close to given that this is exit for which he was trained. This technique is good in a large number of applications but is insufficient when dealing with to model highly variable functions due to the lack of a sufficient number of training patterns. For this we use another technique called Deep Belief Networks.

The characteristic space of an architecture of representation as internal representation of the training data in which their characteristics are used intrinsic and even more useful than the original data in recognition and estimation.

Using a multilayer neural network that calculates the characteristic space of the function, the architecture can be adjusted to optimize the error obtained by The Kernel Machine using the descending gradient method.

Deep Learning and Multiple Outputs

In a multilayer neural network, layer k calculates an output vector using the output Hk1 obtained from the previous layer beginning this process with the input x = h0. In a multilayer neural network, layer k calculates an output vector using the output Hk1 obtained from the previous layer beginning this process with the input x = h0. A classic example of the activation function of a neural network is the hyperbolic tangent in which case the output of the analyzed layer will be given by Equation 2.5.1:

$$h^k = \tanh\left(b^k + W^k h^{k-1}\right) \tag{2.5.1}$$

Where b^k is the vector of subsets and W^k is the weight matrix, the activation function can be changed depending on the problem being attacked, another very important function used is the sigmoid function given by equation 2.5.2.

$$\mathrm{sigm}(u) = \frac{1}{1+e^{-u}} = \frac{1}{2}\left(\tanh(u)+1\right) \tag{2.5.2}$$

If we denote the total output of the network per h^l, we can calculate, using additionally the information on the supervised objective, and the cost function L (h^l; y), which is typically convex for $b^l + W^l h^{l-1}$. A variation of the neural network is one in which the output layer has an activation function different from that used in the other layers, an example of this is equation 2.5.3

$$h_i^l = \frac{\exp\left(b_i^l + W_i^l h_i^{l-1}\right)}{\sum_j \exp\left(b_j^l + W_j^l h_j^{i-1}\right)} \tag{2.5.3}$$

where $\sum_i h_i^l = 1$ in h_i^l is positive.

The value h_i^l can be used as an estimator of $P(Y = y / x)$. In this case, a negative log-likelihood

$$L(h_i^l y) = -\log P(Y = y / x) = -\log h_y^l$$

is usually used and whose expected value over (x,y) must be minimized.

The Ability to Train a Neural Network with Deep Architecture

Prior to [35, 29], deep architectures had not been the subject of discussion within the scientific field Due to the poor results that had been in their learning process. This behavior was generally obtained using a random initialization of the net free weights. The only cases of success that had been reported up to that time had been the so-called convolutional networks [36].

Many observations suggest that training based on the gradient method descendant for deep neural networks (DNNs) supervised is stagnant in what has been called the apparent local average. In general, despite the better representation obtained using a deep architecture are obtained worse results than with solutions obtained through architectures with a single layer of hidden neurons.

The problem described above was attacked from different points of view. It was discovered that better results are obtained if each layer of the network is trained using an unsupervised pre-training algorithm on each of the layers beginning with the layer of input to the network (it is proposed to use with this n autoencoders or Boltzmann machines). After having all the neurons pre-trained, the network can be adjusted using a monitored criterion such as method of the descending gradient.

Under the foregoing considerations the idea arises that by injecting a training signal was not monitored at each layer, you can guide the layer parameters of the network to better regions within the space of parameters [35, 28, 21].

Within the architecture of the neural networks we can consider that the output of each layer of the network constitutes a new representation of the input vector x, of this way consider h^k (x) as the representation level k of x in the model.

Immersed in this context consider a local criterion of nest in each layer of the network which pushes the representations $h^k(x)$ and h^k (x~) either to the same region or far one of the other depending on whether the training patterns suggest whether they should be neighbors or not.

Unsupervised pre-training leads to a restriction on the region of space of parameters where a solution is allowed, which gives it a behavior regulatory framework.

Quality Regulator-Optimizer

When the last hidden layer is restricted to a small number of computational elements, the DNNs with initialization of random weights behave in a little way is aware of both identification and classification tasks. Nevertheless, if it allows the layer to grow indefinitely, there are cases where the error accumulated during the training phase is attractively small. This is because there is a lot of elements of the layer is sufficient to represent any pattern training.

Another technique that has been used during the training of neural networks with architectures is to train in a general way the neural network and only apply optimization algorithms to the training processes of the 2 layers of the network, this is much simpler than the optimization process of all networks and also provides better results than when implemented a global training algorithm. If your units exist in the last layer hidden, the training error may be very small even though the layers are not properly trained. It should be mentioned that although the error of training under this does not imply that the network has a good generalization what the output error with an input pattern different from those used during the learning phase in general would produce a considerably high error.

To the phenomenon of obtaining a low training error and a high generalization error is known as overfitting which can be understood as an over-learning of a particular region of the space of parameters specializing only in one region thereof.

Although the training error can be reduced by exploiting only the skill of the upper layers to fit the training examples, a good generalization on is notoriously difficult to achieve with what should be resorted to methods that fit all layers of the network appropriately minimizing some cost of reconstruction. Such an adjustment as stated above may be carried to some pre-training of the upper layers that aims to restrict the lower layers to capture patterns and regularities of the distribution of the input patterns.

A particular case of this ideology is the following: consider the random pair (X, Y). In this example such regularization is similar to the normative effect of maximize the plausibility of P(X, Y) against P(Y I X). If P(X) and P(Y I X) are functions X-related (i.e., chosen independently such that learning one provides no information on the other), then unsupervised learning of P(X) would not help in the learning of P(Y I X). But if they were related a neuronal network with deep architecture and the same parameters are involved in its estimation, then each vector of Input (X, Y) will provide information of P(Y I X) through P(X).

If you consider the problem of adjusting the lower layers while maintaining a small number of computational units in the last layer of hidden neurons and it is also required that the magnitude of the weights in the two last layers does not exceed certain threshold, then we are talking about an optimization problem within the architecture of the neural network.

If the hypothesis that has been presented is correct, we would expect that the pre-training supervised to have benefits on the initialization of random weights even would have at its disposal an infinite set of training patterns. If the effect of pre-training was only regulatory, it would be expected that the error during tests of generalization converged in approximately the same way or at the same rank of values with both pre-training.

When training a deep architecture without the effect of a pre-workout in the lower layers usually causes the training dynamics to remain trapped at some apparent local minimum and in this way, the addition of more examples training will not provide enough information to get out of that attractive region.

The above information suggests that the downward gradient method used in set with backpropagation techniques may not be effective in moving the parameters of a particular hidden layer made regions corresponding to desired solutions [21].

A different topology of ANN is the recurrent neural network, this type of networks are characterized by having feedback loops between the upper layers and the layers in this way, this type of network can be "unfolded over time" the output of each neuron at different times as a different variable turning the unfolded network into a deep architecture.

Unsupervised Training for Pro-Covers

Given the problem of the existence of local training, an unsupervised nesting learning criterion is proposed at the each layer that can be used to exit or avoid those local mornings than the method of the descending gradient cannot be exceeded. This is to be expected if the to discover a representation that captures statistical vector input.

Some techniques such as PCA (Principal Component Analysis) or ICA (Independent component Analysis) fail in the process of obtaining these characteristics in the called an overdetermined case where the number of outputs is greater than the number since the two techniques mentioned above require 1 to 1. In this case there are specially developed variants of ICA that solve this problem as related algorithms such as autoencoders and the Restricted Machines and Boltzmann (RBM) [30, 35].

We can conclude that unsupervised training algorithms can extract outgoing information from the input vector distribution. This information may be captured in a distributed representation, that is, a set of features in which are coded the outstanding factors of the variations in the vector of entry.

A deep architecture with unsupervised learning can be understood from the following way: the characteristics obtained by the first layer are seen as level of abstraction while those of the upper layers represent a level of abstraction.

Deep Degenerative Architectures

The training of a network using an unsupervised algorithm is of interest for obtaining a distributed representation of the input and generating samples of said distribution.

The models that generate the distribution can be visualized as graphs whose nodes represent random variables while the connector loops provide information about mutual dependence between variables. An example of architecture degenerative disease is the Sigmoid Belief Network (SBN), in this network the units (a random binary variable) belonging to each layer are independent of the units corresponding to any layer higher as shown in Figure 3.3:

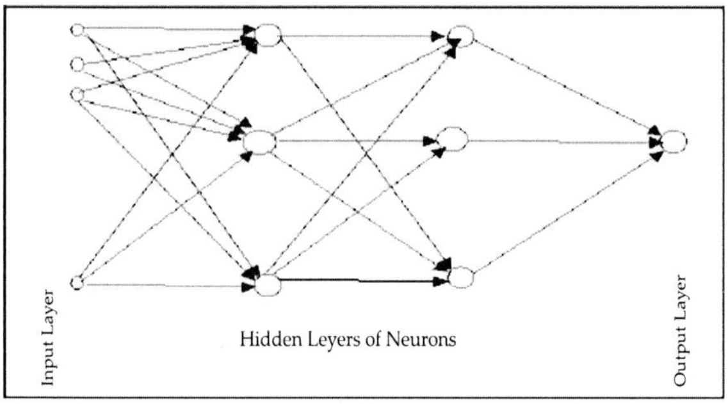

Figure 3.3.

A typical parameterization of this type of network is the equation of neuronal activation (Equation 2.6.1):

$$P\left(h_i^k = 1 / h^{k+1}\right) = sigm\left(b_i^k + \sum_j W_{i,j}^{k+1} h_j^{k+1}\right) \qquad (2.6.1)$$

where h_i^k Is the binary activation of hidden node *i* belonging to layer k, while h^k is the output vector of representation (h_1^k h_2^k ,.....) We denote as the vector of input x to the layer h^0. If we make P represent the distribution of training, it is the distribution of examples for the learning process. To use this distribution for the prediction of estimates. The lower layer generates a vector x in the input space and our goal is to obtain a good model of representation that gives a high probability of membership to the input data. Considering several layers of the network, the generator model is decomposed of the form:

$$P\left(xh^1,\ldots,h^l\right)=P\left(h^l\right)\left(\prod_{k-1}^{l-1}P\left(h^k/h^{k+1}\right)\right)P\left(x/h^1\right) \tag{2.6.2}$$

On the other hand, the great advance in the procedures of training of a network deep wine given by the presentation of the so-called Networks of Deep Belief (DBNs) that are similar in architecture to SBNs (see Figure 2.6.2) with a difference in the parametrization of the two upper layers obtained

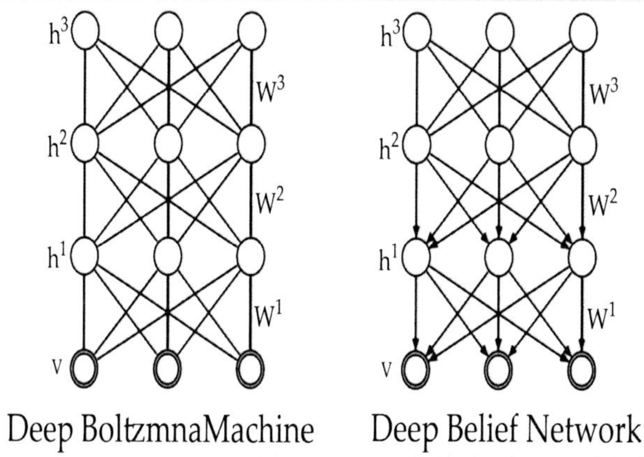

Figure 2.6.2: Architecture of a Deep Belief Network, the two upper layers constitute a Boltzmann Restricted Machine having dependencies in both directions

The probability is as in equation 2.6.3:

$$P\left(xh^1,\ldots,h^l\right)=P\left(h^{l-1},h^l\right)\left(\prod_{k-1}^{l-2}P\left(h^k/h^{k+1}\right)\right)P\left(x/h^1\right) \tag{2.6.3}$$

The joint action of the upper 2 layers $P(h^{l-1},h^l)$ is The Boltzmann Machine Restricted (RBM) whose behavior is denoted by equation 2.6.4:

$$P\left(h^{l-1},h^l\right)\propto e^{b^j k^{l-1}+c^j h^l+h^{1^\ell}}Wh^{l-1} \tag{2.6.4}$$

This change develops the notion of self-directed training applying to one layer of the network at a time. This quality is necessary for the pre-training of the hidden layers of the network.

Convolutional Neural Networks

The convolutional neural networks were inspired by the structure of the system human visual. In the Neocognitron of Fukushima [23], learning was based on that when the output of similar neurons is applied or incident in specific places of the layer of neurons, a form of translational variance is obtained. Currently, pattern recognition systems based on convolutional neural networks are the ones that have the best performance (Example: handwriting recognition raised).

The convolutional networks are organized in 2 types of layers: the convolutional layers and the subsampled layers. Each layer has a topographic structure associated with a two-dimensional position with respect to a receptive field. In every locality of each layer, there is a different number of neurons with each neuron related with a set of weights without input associated with neurons belonging to an area of the anterior layer. The same set of weights but with a new zone of rectangular shape is associated with neurons distributed in different places.

A small fan-in (each neuron accepts only a small number of input signals) In these neurons helps the descending gradient method to act and propagate to Multi-layered braking without its action fading so as not to act, the hierarchical connectivity structure conveys the parameters of the neural network complete in a favorable region in which the optimization based on the gradient descending works well.

An important point is that this type of networks is used with random weights in the first layers without training have a better performance than an architecture deep traditional fully trained using supervised learning. However, its performance improves considerably compared to a convolutional network fully trained. To emphasize, we must remember that these networks were the only case of successful training of a deep architecture even before the arrival of the Deep Belief Networks [4, 29].

Autoencoders

Autoencoders have been used as building blocks for training deep architectures, where each level is associated with a self-empowering coach initially the layer independently of the others [32, 31].

A self-encoder is trained to encode the input x into some corresponding representation C(x) such that the input can be reconstructed from this representation. The formulation used generalizes the criterion of the mean square error proposing the minimization of the logistic likelihood of reconstruction, then given the representation c (x) we have:

$$RE = -\log P(x / c(x))$$

If the distribution of xjc (x) is of Gaussian type we recover the criterion of error medium square The inputs xi are binary or considered as probabilities of binomial type, as well, the cost function is given by equation 2.6.5:

$$-\log P(x / c(x)) = -\sum_t x_t \log f_t(c(x)) + (1 - x_t)\log(1 - f_t(c(x))) \quad (2.6.5)$$

where f (.) is called a decoder and f (c(x)) is the reconstruction produced by the network, in this case a vector of numbers in the range (0, 1) obtained, for example, using a sigmoid type function self-encoder with an n-dimensional input and a coding dimension of less n can potentially learn exclusively the identity function given that there is no condition to avoid this type of codification; However, this case is to be avoided. Some techniques for training autoencoders are:

1. Training with stochastic descending gradient (Type 2 regularization of the parameters)
2. Addition of noise in autocalling
3. Establish a condition of sparsity in the code
4. Maximizing the likelihood of the generative model (denoising autoencoder)

CLASSIFICATION MODELS

Deep learning has its main application in the solution of the class problem of patterns, some of the techniques that basis for the constitution of a deep architecture that functions as a classifier. These algorithms constitute the fundamental pillar to attack the problem of regression and identification of the dynamics of systems.

Logistic Regression

Logistic regression is a probabilistic type linear classification model. The structure of the algorithm is parameterized by an array of weights W and a vector of Bias b. Classification is done by projecting each set of data on a set of hyperplanes, whose distance from each of them is the probability of membership to each class that was represented and corresponds to a specific hyperplane.

This description can be represented mathematically as:

$$P(Y = i / xW, b) = soft\max x_t (Wx + b) = \frac{e^{W_{tx+b_1}}}{\sum_j e^{W_j x + bj}} \quad (3.1.1)$$

The output of the classifier or its prediction is made simply by taking the argument maximum of the vector whose i-th element is

$$y_{pred} = \arg\max, P(Y = i / x, W, b) \quad (3.1.2)$$

In the task of the identification of a system, the output of the network would simply be the linear combination of the W-weights and the bias vector b.

A fundamental problem for the optimization of any model is the choice of a function of loss, the learning of the optimal parameters of the class model, it is done with the purpose of minimizing this cost function. In the case of multi-class logistic regression, it is very common to use negative log-likelihood as minimizing function. This is equivalent to maximizing the likelihood of the set of data D under the model parameterized by the set θ. This way you can the likelihood L and the loss function l:

$$L(\Theta = \{W, b\}, D) = \sum_{t=0}^{|D|} \log\left(P\left(Y = y^{(t)} / x^{(t)}, W, b\right)\right) \quad (3.1.4)$$

$$l(\Theta = \{W, b\}, D) = -L(\Theta = \{W, b\}, D) \quad (3.1.5)$$

There are many publications related to the minimization technique ([33, 42]) and although different methods have been proposed to achieve this goal, the algorithm of gradient, is by far the simplest method to minimize linear regressions. Here we will describe the method of stochastic downward gradient with mini-batches. For dynamic identification of a system, it is not possible to define a function of discrete verisimilitude as presented by Equation 3.1.4. Instead it is used the function of the mean square error is given by Equation 3.1.6.

$$l(\Theta = \{W, b\}, D) = \frac{1}{N} \sum_{t=0}^{|D|} \left(Y^{(t)} - y^{(t)}\right)^2 = \quad (3.1.6)$$

Multilayer Perceptron

A Multilayer Perceptron (MLP) can be interpreted as an extension of the algorithm logistic regression where the input is first transformed using a non-linear transformation? The purpose of this transformation is to project the data of entry to a space where they are linearly separable. This intermediate layer is known as the hidden layer and is characteristic of an MLP possessing two or more of them. A hidden layer is unique to make an MLP a universal approximation [26]. However, as noted earlier there are several advantages to using many layers hidden, that is, follow the philosophy of deep learning.

Formally, an MLP constitutes a function $f:R^D \rightarrow R^L$, where D is the size of the input vector x and L is the size of the output vector f(x). If p hidden layers then the output is determined by equation 3.2.1:

$$f(x) = \psi_P\left(b^{(P)} + W^{(P)}\left(\psi_{P-1}\left(b^{(p-1)} + W^{(p-1)}\left(\psi_{p-2}(\cdots)\right)\right)\right)\right)$$

bias vectors $b^{(0)}$, ..., $b^{(p)}$ and matrices of unequal weights $W^{(0)}$,....$W^{(p)}$. The functions $\psi_{p-1,....}\ \psi_0$ are called activation functions, the most common being the function tan $h(a) = (e^a - e^{-a}) / (e^a + e^{-a})$ and the sigmoid function $sigm(a) = 1/ (1 + e^a)$. From this in this way the input vector x enters the system by forming the output $\psi_0 (b^{(0)} + W^{(0)} x)$ this value is considered the output of the first hidden layer of the network, this value is propagates to the next layer by substituting the vector x for the corresponding vector ψ_i .

As in the case of logistic regression, the expected value of the ypred system is performed of the function p as sigmoid for classification or the identity function with systems identification.

Stochastic Descending Gradient Method

It is a simple algorithm in which small steps are repeatedly taken in direction decreasing on the error surface is caused by a loss function given by some parameters. For this purpose it is considered that the training data are evaluated within the function of loss. The pseudocode of this algorithm can be written as:

```
while True:
lost = f (parameters)
```

gradient = …

parameters = parameters - learning_rate* gradient

if <unemployment_condition>

return parameters

The stochastic descending gradient algorithm works exactly by the same principles than the ordinary descending gradient method, but works more quickly since it estimates the gradient of the function of loss from only some samples of data for a given time instant rather than evaluating the complete training set. In its simplest form, the gradient is estimated using only one example per event of estimation.

The variant of the most commonly used and recommended stochastic gradient method for its utilization in deep learning techniques, is the downward gradient method by mini-batches. In this case more than one example of training is used for each estimation of the gradient, where a crucial factor, would be the number of examples which will contain each parameter updater set. This technique reduces variance in the estimation of the gradient, and usually makes better use of the organization hierarchical memory of modern computers [6].

for each set of data (x_{batch}; y_{batch}) of the set of batch data:

lost = f (parameters; x_{batch}; y_{batch})

Gradient = :::

Parameters = parameters –learning_rate* gradient

If <unemployment_condition>

Return parameters

With this algorithm there is an exchange or balance between the reductive effect of the variance and the choice of minibatch of size B. The variance reduction factor improves notably when B increases from 1 to 2, but the improvement achieved, decreases as the size of B. increases. With large B, the time is wasted by reducing the variance in set of data generated by estimator of the gradient, but it is believed that time would be more profitable if spent in steps additional parametric updates [9].

Batch B is optimal if it is dependent on the model, dataset, and hardware used for the implementation of the algorithm and can have values from 1 or 2 to several hundreds, the election of the same is done arbitrarily. If you are training a model using a number of times, the size of the minibatch becomes really important because it controls the number of updates that are made to the parameters. Train the same model for 10 epochs using a batch size of 1 has completely different results compared to the obtained during the same 10 epochs but with a batch size of for example [20].

Supervised Method with Restricted Unemployment

Early unemployment aims to avoid over-adjustment by monitoring performance of the model on a validation set. A validation set is a set of examples that are never used during the training stage, but which are not are part of the test set. Examples of validation are considered as representative of the test examples that will be used in the future. If he performance of the model ceases to improve in the validation set, or even if it presents a performance lower than a threshold determined during the optimization process, then a heuristic is implemented that stops the optimization algorithm.

The choice of heuristic that makes the decision of when to stop depends a lot on the application, but a very common strategy is the one based on the geometric increase of a pair called patience. The pseudocode of this algorithm is as follows:

while (time <total_time) and (exit = false):

epoch = time + 1

for each minibatch in the training set:

lost = f (parameters; x_{batch}; y_{batch})

gradient = ::: calculagradiente

parameters = parameters- learning_rate* gradient

iteration =(time – 1) * number_ batches + current_batch_index

cost_validation = calculates_cost :::

If cost_validation <best_cost

If cost_validation <best_cost* 0.95

patience = patience* 1,1

best_cost = cost_validation

best_parameters = current_parameters

If patience <= iteration

exit = true

return parameters

The percentage of improvement allowed and the percentage that is increased in patience are values proposed during the implementation and should be changed depending on the needs of the application.

Cross Validation

Cross-validation is a technique used to evaluate the results of an analysis cross-validation is a technique used to evaluate the results of an analysis and ensure that they are independent of the participation made between the data validation and testing [10,26]. It consists of repeating and calculating the arithmetic mean obtained from the cost functions between different proposed partitions.

The retention method or hold out method consists in dividing the data into two sets sample, an analysis of a subset called training set is performed and the analysis is validated in the other subset as well called test set. We take an independent sample as validation data, usually the model would not fit with equal accuracy to the test data than to the training data. This is known as over-adjustment and occurs when the size of training is small or when the number of model parameters is too high big.

Cross-Validation of K Iterations

The sample data are divided into k subsets. For each iteration one of the subset is used as test data and the rest as training data. The validation process repeats k iterations, with each of the possible subsets of test data. Finally we calculate the arithmetic mean of the results of each iteration to obtain a single result, see Figure 3.5.1.

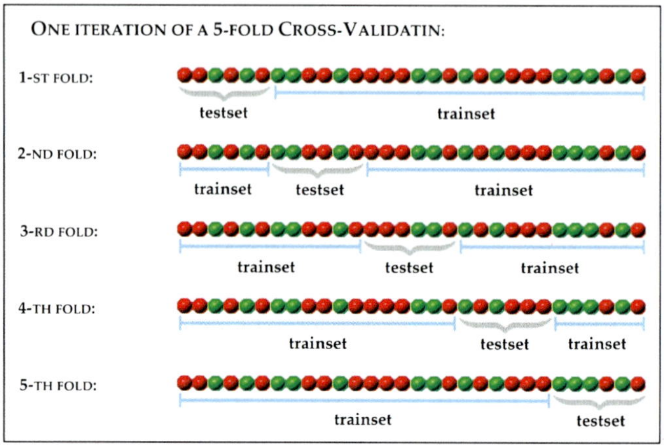

Figure 3.5.1: Cross-validation of k iterations with k = 4

Validation

The objective of cross-validation is to estimate the level of fit of a model to a certain set of test data independent of those used to train the model. These measures can be used to estimate any measure appropriate fit for the data and model. For example, in a model based on binary classification, each sample is labeled as correct or incorrect, so that in this case, you can use the "error rate of classification" to summarize the adjustment of the model. Same, other measures could be used as the predictive value positive.

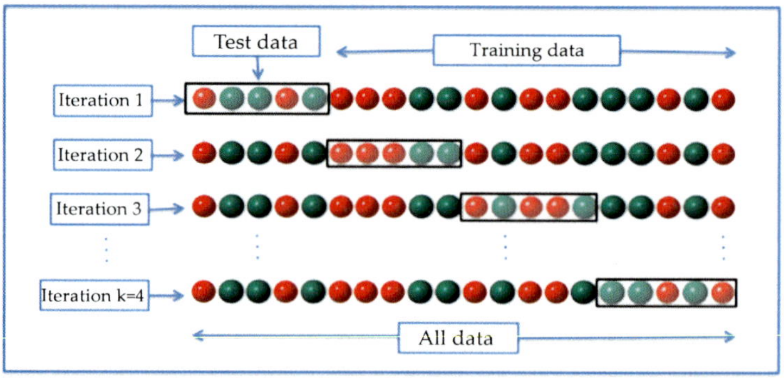

Figure 3.5.2

Unattended System Identification

The problem of system identification or regression problem has its main problem exponent in supervised learning [42, 33], in the unsupervised case, the idea is to divide the sample space into several subspaces, where data with are gathered. The procedure is expanded to the data set complete, including the output data.

The goal of nonlinear optimization is to look for the point, where the evaluation of a predefined objective function results in a maximum or minimum value considering some constraints which are given by appropriate equations. According to the value of cost function J, it is possible to establish the best candidate for the solution. Learning methods explore the solution space and finally choose the candidate of the best solution.

In the case of obtaining a solution based on measurements or observations, the objective function is always a function of observations. For example, consider that we look for a function f that minimizes the objective function J:

$$J = E\left\{\left(f(u) - y\right)^2\right\}$$

In this case, J is defined by pairs of data (u, y), which are generated by D. In this case it is possible to address the problem of minimization using the equation:

$$\hat{J} = \frac{1}{n}\sum_{k-1}^{n}\left(f(u) - y\right)^2 \qquad (3.6.1)$$

The function J was based on a number of examples N taken from a set distributed according to D. The methods to solve this problem can be divided into 3 groups:

1. Supervised learning methods:

They are based on a data set , and is used to establish a transformation, i.e. a function f of a set of functions which transforms the input data into functions of departure. The resulting function is then evaluated according to a predefined criterion, which consists of the difference between the transformed data and the output data of the original set.

One of the possible techniques used to obtain a solution is the optimization using the descending gradient method.

2. Unmonitored learning methods.

These methods are based on the supervised learning in which the output data are not differentiated a priori, for all data are treated as inputs. Unsupervised learning represents A method where the model is adjusted to the data in a optimal way, the learning outcome models the distribution of data in the space presented in the problem.

Unsupervised learning methods are used during preprocessing of the data before being used by supervised feature algorithms. These algorithms are also widely used in data compression. All methods are implicitly or explicitly based on the probability distribution in space by the input variables.

3. Reinforced learning methods:

These methods exploit quality information of the resulting model, although the reference output information based on some entrance is not available. The algorithms are mostly used in the development of long-term strategies, where a reference or measure of output is not found in an explicit way.

Identity Topology

In addition to the problem of classification of patterns, the next task that has been historically attacked by means of model and learning algorithms where the theory of systems and artificial intelligence is the identification of dynamic systems to test the neural identification task using deep architectures subject to pre-training processes, is considered a class of systems represented by equations in differences of form:

$$y(k) = f\left(y(k-1),\ldots,y(k-n),u(k-1)\right)$$

where y(k) represents the evolution of the system output that depends on n outputs and y(k-1) is the value of some signal that is interacting actively with the system. In the present work a serial-parallel type model is used to carry out the process of identification. On makes use of the output of the real system for the calculation and updating the free weights.

CHAPTER 4

CHOICE OF HYPER-PARAMETERS IN THE IDENTIFICATION OF NON-LINEAR SYSTEMS

Ivan Stanimirović

Boltzmann›s learning is a stochastic learning algorithm derived from ideas based on statistical mechanics. A neural network based on this principle he calls it Boltzmann›s Machine (BM).

In the Boltzmann machine neurons constitute a recurrent structure and operate in a binary manner, for example, they may be in a «on» state denoted By +1 or a «off» state described by a 1. This machine is characterized by a function of energy which is determined by the states occupied by each neuron, for example, a characteristic energy function is the one denoted by equation 4.1.1:

$$E = -\frac{1}{2}\sum_{j}\sum_{k} w_{kj} x_k x_j \; con\, j \neq k \tag{4.1.1}$$

where x_j is the state of the neuron j and wkj is the weight without apoptic that relates to the neuron j with neuron k. In The Boltzmann machine the case is not considered of self-feedback, the machine operates by picking a random neuron at some point of the learning process, then the state of the state neuron k changes Xk to the state -xk with some temperature T, this change in the state of the neuron occurs with a probability given by equation 4.1.2:

$$E = -\frac{1}{2}\sum_{j}\sum_{k} w_{kj} x_k x_j conj \neq k \qquad (4.12)$$

where ΔE_k is the change of energy in the neural network. Temperature T does not match at a real physical temperature but to a measure of the disorder or level of entropy. If the change of state of the neurons is chosen as the maximum probability, the system will reach the thermal equilibrium. Neurons are classified into two functional groups: visible and hidden. Neurons provide an interface between the network and the environment in which they operate, while the hidden neurons act freely without interacting directly with the environment. The Boltzmann machine has two modes of operation that must be considered:

1. Anchoring condition: Visible neurons are anchored in specific states. Cos determined by environmental stimuli.
2. Condition of freedom: Both visible and hidden neurons act freely without conditions imposed by the environment.

Let p_{kj}^{+} be the correlation between states j and k with the con guration of the network in its condition anchor and is p Kj the correlation but in condition of freedom. Both correlations are averaged over all possible states of the machine when it is found in thermal balance. Then according to Boltzmann's rule of learning, the change in the free weights during the learning phase is (equation 4.1.3):

$$\Delta w_{kj} = \eta\left(P_{kj}^{+} - P_{kj}^{-}\right) conj \neq k \qquad (4.13)$$

where μ Is a parameter known as the learning rate. Note that both correlations p_{kj}^{+} and p_{kj}^{-} have values in the range of -1 to +1.

The distinctive feature of the Boltzmann Machine is the use of connections without sympathetic apathy between neurons.

During the training phase the visible neurons are locked into specific states by the environment. Hidden neurons always operate freely and are used to find underlying features contained within of the input vectors, this is observed when the hidden neurons present correlations of high order. This method of learning is a supervised and its main objective is the modeling of a probability distribution specified by the patterns applied at the inputs of the visible neurons.

When the neural network input does not have complete information input pattern, ie only an incomplete subset of the visible neurons is being stimulated, the network manages to determine the values of the missing inputs provided that it has been appropriately trained within the distribution of probability. Let x be the state vector of the Boltzmann machine with the state of the neuron i. The synaptic connection of neuron i to neuron j is w_{ij} fulfilling thes symmetry character is fulfilled that $w_{ij}=w_{ji}$ for all 6 = j and w_{ii} = 0 for all i. the use of bias is permitted in this process using a weight wj0 with input equal to 1. The expression for the power of the network is given by: the Gibbs equation gives an estimate of the probability that the network is in a given state x, this probability is calculated by:

$$E(x) = -\frac{1}{2}\sum_{i}\sum_{j} w_{ji}x_i x_j coni \neq j$$

The Gibbs equation gives an estimate of the probability that the network is in a given state x, this probability is calculated by:

$$P(X = x) = \frac{e^{-E(x)/T}}{Z}$$

where Z is called the participation function and is determined by:

$$Z = \sum_{i} e^{-E_i/T}$$

The conditional probability that a neuron j is found in a determined state xj given a priori a particular state of all other neurons are (equation 4.1.4):

$$P\left(X_j = x_j \mid \{X_t = x_t\}_{t-1}^{k}, i \neq j\right) = \varphi\left(\frac{x}{T}\sum_{t \neq j} w_{ji}x_t\right) \quad (4.1.4)$$

1. Each input vector persists enough to allow the network to reach its thermal equilibrium.
2. There is no structure in the order in which the patterns coming from the medium environment are assigned to the input of the visible neurons of the network.

A set of weights is a perfect model of the environmental structure if gives as output the same probability distribution in states of neurons visible when they operate in the phase of freedom than when they operate during the phase of anchorage. This characteristic is usually achieved by using a number its large number of hidden neurons [31].

GIBBS SAMPLING

Suppose we have an arbitrary configuration if performs the following sampling algorithm:

is taken from the distribution of x_1, given

is taken from the distribution of x_2, given

is taken from the distribution of x_l, given

is taken from the distribution of x_l, given

This process constitutes an adaptive iterative scheme. After n iterations we obtain

There is an equivalence between the concepts of statistical mechanics and those used in the study of neural networks of stochastic behavior:

The Gibbs sampling process gives good results if the environment in which it is applied obeys the following Theorems:

Convergence rate theorem. The joint probability distribution of the Random variables converges to the true distribution of Joint probability of with a geomorphic approach rate in n.

Theorem of ergodicity. For any measurable function g of the random variables whose mathematical hope exists, we have:

with probability of 1

If the Gibbs process is carried out for its long time, the network will reach equilibrium temperature with temperature T.

SIMULATED COOLING PROCESS

1. Initial value: The starting value T0 is chosen to be high enough for initial value: The starting value T0 is chosen to be high enough to ensure transitions are accepted and thus avoid possible local minimas.

2. Decrement: It is done in a general way using an exponential decrease whose changes are relatively small.
3. Value: The process stops when the desired number of elements of allowed input is not reached for three consecutive temperatures.

In the Boltzmann Machine the tempering process is performed for a sequence temperature range T_0; T_1; ::::; T_{final}. The initial temperature T0 is high to allow that the thermal equilibrium is quickly achieved. In this way we get gradually the temperature T_{final}, when this temperature is reached the neurons would have reached their desired probability distributions. Boltzmann Learning Rule widely used performance index is the likelihood function. The goal of Boltzmann's learning is to maximize this parameter. Consider the following statements:

1. x state vector
2. x vector state of visible neurons
3. x state vector of hidden neurons

$$\lim_{n\to\infty} P\left(x_k^{(n)} \leq x / x_k(0)\right) = F_{xk}(x)\forall k = 1,2,...,K$$

Let F be the set of training examples of the probability distribution of interest. All values are assumed to be bivalve and more than one time. There are two phases of operation of the Boltzmann machine.

1. Positive Phase: The network operates in its anchoring condition on visible neurons.
2. Negative phase: The network can operate freely without the input derived from the environment.

$$\lim_{n\to\infty}\frac{1}{n}\sum_{t-1}^{n} g\left(x_1(i),(i),r_2(i),...,x_k(i)\right) \to E / g\left(x_1,x_2,...,x_k\right)$$

Given a vector of weights w. The probability that visible neurons have a state x is P (X '= x'). If all possible values of x2 F are statistically assumed independent, probability distributions, probability distributions is the factorial type distribution X 2F P (X = X). To formulate the function corresponding to the log-likelihood L (w) this distribution is taken and the weight as an unknown vector (equation 4.1.5).

$$L(w) = \log \prod_{x_\alpha \in F} P(X_\alpha = x_\alpha) = \sum_{x_\alpha \in F} \log P(X_\alpha = x_\alpha) \qquad (4.1.5)$$

It should be remembered that

$$P(X = x) = Z^{-1} e^{-E(x)T}$$

By definition x is the joint combination of x_α belonging to the visible neurons and x_β belonging to the hidden neurons. The probability of finding neurons visible in an x_α state given any state of hidden neurons x_β is given by:

$$P(X_\alpha = x_\alpha) = \frac{1}{Z} \sum_{z\beta} e^{-E(z)T}$$ where Z is defined by $$Z = \sum_x e^{-E(x)T}$$

Then we have

$$l(w) = \sum_{z_\alpha \in F} \log \frac{1}{Z} \sum_{x\beta} e^{-E(x)T} = \sum_{z_\alpha \in F} \left(\log \sum_{x\beta} e^{-E(x)/T - \log} \sum_z e^{-E(x)/T} \right) \qquad (4.1.5')$$

Differentiating L(w) with respect to w_{ij} we obtain the following result:

$$\frac{\partial L(w)}{\partial w_{ji}} = \frac{1}{T} \left[\sum_{x\beta} P(X_\beta = x_\beta \mid X_\alpha = x_\alpha) x_j x_t - \sum_x P(X = x) x_j x_t \right]$$

In order to simplify, two concepts are defined according to the two stages of which have been described previously (equations 4.1.6 and 4.1.7):

$$P_{ji}^{+} = \langle x_j x_t \rangle^{+} = \sum_{x_\alpha \in F} \sum_{x_\beta} P(X_\beta = x_\beta \mid X_\alpha = x_\alpha) x_j x_t \qquad (4.1.6)$$

$$P_{ji}^{-} = \langle x_j x_t \rangle^{-} = \sum_{x_\alpha \in F} \sum_{z} P(X = x) x_j x_t \qquad (4.1.7)$$

where p_{ji}^{+} is the correlation between states i and j with the network operating in its positive phase or anchor, while p_{ji}^{-} is the correlation of the states of neurons i and j in the negative phase or freedom. We can simplify the derivative of log-likelihood as:

$$\frac{\partial L(w)}{\partial w_{ji}} = \frac{1}{T} \left(P_{ji}^{+} - P_{ji}^{-} \right)$$

Recall that the goal is to maximize the function of L(w). We can use the method of the downward gradient to achieve the desired goal using:

$$\Delta w_{ij} = \eta\left(P_{ji}^{+} - P_{ji}^{-}\right)$$

where μ Is a parameter called rate or speed of learning. This law of learning causes that ΔL (w) is always positive giving this pair a growth sustained. This series of equations is what is known as the learning rule of Boltzmann.

The Boltzmann Machine's unequipped weights are adjusted using only available observations on 2 different conditions: the first one called phase positive or anchoring and a second phase of free or negative behavior. In this context the following distinction can be made:

p_{ji}^{+} represents a Hebbian rule of learning

p_{ji}^{-} represents a term of unlearning

It can be seen as a generalization of the repeated forgetting and relearning rule for the case of symmetric networks without hidden neurons.

The reason for conceiving a negative stage during the learning process is that direction of the maximum descent in the energy space is not the same as the direction of the maximum rise in the probability space which helps to avoid and in the latter case escape from the presence of local minimums during the training phase reaching the result is optimal in many case studies. Since the Boltzmann Machine requires the hidden neurons to know the difference between the two training phases, there is an external network of stimuli that signals neurons when they are being stimulated by environmental stimuli.

The main difficulties of the Boltzmann machine are the computational time notoriously increased compared to the time associated with deterministic topologies and also in the sensitivity that presents in front of statistical errors [26, 31].

ADJUSTMENT TECHNIQUES OF HYPERPARAMETERS OF A DEEP NEURAL NETWORK

In [37] it is said that the backpropagation algorithm is actually an optimization method on by means of descending gradient and therefore there is no guarantee that the absolute minimum is to be reached. Despite this theoretically accepted fact, the method of backward propagation works properly in many applications converging to the true global minimum, or at least makes it possible to achieve there is some criterion for stopping the algorithm.

In several texts [37] it is recommended not to nir neural networks with more parameters than the number of training examples available. The quality of the solution obtained depends on several factors among which stand out:

1. The complexity of the required approximation.
2. The size of the network in relation to the size required for an optimal solution.
3. The noise level of the data presented

A classic problem is one that minimizes the size of the neural network by maintaining a good performance. This solution can be accessed using two basic techniques.

Growth of the network: Begin with a network of small architecture and add a new neuron or a new layer only when we are unable to meet design specifications.

Starting with a network: it's large enough that it satisfies criteria and proceed to prune it by removing weights without a certain criterion of performance.

Random choice of Hyperparameters

In [11] the optimization of hyperparameters is observed as the optimization problem of a cost function on a congruation space structured by a graph. In this work, in turn, study spaces of congruation of type tree, understood such as those in which some leaf-type variables (e.g., the number of units in DBN) are only well-defined when specifically specified the node-type variables (e.g. the number of network layers).

An optimization algorithm should not only optimize variables that are discrete, ordinal or continuous, but must also choose which variables optimize. In [11], it is mentioned that the goal of a typical learning algorithm A is find a function f that minimizes an expected loss L(x; f) on examples (Independent and identically distributed) x of a natural distribution G_x. Training algorithm sometimes depends on variables known as hyper-parameters λ , and the actual learning algorithm is the one obtained after the choice this vector λ, Which can be denoted as A_λ , and $f = A_\lambda (X^{(train)})$For some vector of training $X^{(train)}$.

What is necessary in practice is a way to choose λ that minimizes the error of generalization

The optimization of hyper-parameters is reduced to:

$$\lambda^{(\bullet)} = \arg\min_{\lambda\in\wedge} E_{x\sim Gz}\left[L\left(x; A_\lambda\left(X^{(train)}\right)\right)\right]$$

In general, there are no known algorithms to perform such optimization. Respect to hope over Gx, the technique called cross-validation is used to estimate a possible optimization. Cross-validation is the technique of replacing the mathematical expectancy with an average of a validation set $X^{(valid)}$ whose elements are taken. Cross-validation is always impartial and when $X^{(valid)}$ is independent of any data used by A_λ. Then we can reformulate the problem of optimization of hyperparameters as:

$$\lambda^{(\bullet)} \approx \arg\min_{\lambda\in\wedge} mean_{x\in x(train)}\left[L\left(x; A_\lambda\left(X^{(train)}\right)\right)\right]$$

$$= \arg\min_{\lambda\in\wedge} \psi(\lambda) \qquad (4.2.1)$$

Equation 4.2.1 expresses the problem of optimization in terms of the function of response of hyperparameters ψ. Generally speaking, very little is known about response area ψ or search space Λ , The dominant strategy for finding good set Is to choose some number (S) of test points $\{\lambda^{(1)}......\lambda^{(s)}\}$ evaluate and returns ψ(λ) which behaved best as the problem then lies on how to choose the test set which is usually resolved using mesh search or manual search algorithms.

Since Λ is a set indexed by K configuration variables (learning rate, number of layers, number of neurons in each layer, etc.), then the search for mesh requires that a set of values be chosen for each variables ($L^{(1)}$ ::: $L^{(k)}$) in the mesh search the test set is formed by assembling every possible combination of values, as The number of tests in the mesh is

$$S = \prod_{k-1}^{K} | L^{(k)} |$$

East product suffers from the so-called dimensionality curse [30, 8] because the number of values increases exponentially with the number of hyperparameters. On the other hand, the manual search of hyperparameters is used to identify regions in which are promising and used to develop the intuition necessary to choose the sets L (k). The main disadvantage of this type of selection is the produce results which is very important for scientific progress. In the field of machine learning. Despite its limitations there are several reasons why manual and mesh searches prevail:

1. Manual optimization gives researchers some degree of visibility within from ψ .
2. There is no barrier or limit for manual optimization.
3. Search by mesh is simple to implement and parallelization is trivial.
4. The mesh is compact and small for dimensionally low spaces.

Given the great complexity that a complete meshing represents, it has been proposed in [11] random search for hyperparameters having independent selections of uniform space of the same congruation space similar to the mesh process with the goal of choosing $\{\lambda^{(1)}......\lambda^{(s)}\}$. In general the random selection has all the advantages of the search by mesh and exchange a small reduction in science in space of small dimensions by a great improvement in the science in spaces of high dimension.

In [11] it is also mentioned that in general the functions ψ have a dimensionality effective low, that is, they are more sensitive to changes in some dimensions than in others.

4.3.2. Opposite methods to randomness for the optimization of neural networks

When determining the performance of learning algorithms, it is useful to take into uncertainty generated by the set of values hyperparameters. In [11] we distinguish the difference between the performance estimates $\psi^{(\text{Valid})} = \psi$ and $\psi^{(\text{test})}$ based on test and validation sets.

$$\psi^{(valid)} = mean_{x \in X(valid)} \left[L\left(x; A_{\lambda}\left(X^{(train)}\right)\right) \right] \qquad (4.2.2)$$

The variance of these values must be determined in function of the chosen hyperparameters, using Bernoulli's variance:

$$V^{(valid)}(\lambda) = \frac{\psi^{(valid)}(\lambda)\left(1-\psi^{(valid)}(\ )\right)}{X\,|^{(valid)}|-1} \quad (4.2.4)$$

To resolve the difficulty of choosing a winning set of values, a weighted average of all the results of the test set, whose weighting corresponds to the probability that $\lambda^{(s)}$ is actually the best. In this point in view of the uncertainty due to $X^{(valid)}$) being a small sample of Gx, the score of the test set with the best model $\lambda^{(s)}$ is a random variable z. This z score is modeled by a Gaussian model whose S components have values $\mu_s=\psi^{(valid}(\lambda^{(s)}))'$, variances $\sigma_s^2=V^{(valid)}(\lambda^{(s)})$ and w_s defined by

$$w_{ß} = P\left(Z^{(ß)} < Z^{(ß')}, \forall g' \neq g\right) \quad (4.2.6)$$

$$Z^{(t)} \sim N\left(\psi^{(valid)}\left(\lambda^{(t)}\right), V^{(valid)}\left(\lambda^{(t)}\right)\right) \quad (4.2.7)$$

and weights of nests defined by

$$\mu_z = \sum_{ß-1}^{S} w_{ß}\mu_{ß}$$

$$\sigma_z^2 = \sum_{ß-1}^{S} w_{ß}\left(\mu_{ß} + \sigma_{ß}^2\right) - \mu_z^2 \quad (4.2.8)$$

The procedure for estimating weights is repeatedly selecting validation scores hypotheses $Z^{(s)}$ of normal distribution and it is taken into account that so frequent each test set generates a winning score, then, does not need to be estimated very precisely. This technique gives a higher estimate than the traditional technique of reporting the training error of the best model in validation.

The choice of hyperparameters that is made in [11] for a Multi-layer Perceptron Network is the following considering a learning by the method of the descending gradient:

1. Number of hidden layers: It is geometrically sampled from the range of 1 to 10.
2. Number of neurons for each layer: Geometrically samples from 5 to 1024 creating a new pair of this level for each hidden layer.

3. Learning rate: Geometrically sampled from 0.001 to 1.
4. Moment rate: Geometrically sampled from 0.001 to 1.
5. Type of preprocessing: None, standardized or using the technique of Main Component Analysis (PCA).

In the required case that is the Deep Learning, the particular use of DBNs (Deep Belief Networks) is analyzed, possessing the following properties:

- Number of network layers between 1 and 10 with equal probability.
- For each layer is chosen Number of hidden units (uniform logarithmically between 128 and 4000)
- Number of iterations for contrastive divergence during pre-training (Uniform logistic distribution between 1 and 10000)
- Initial learning rate for contrastive divergence (log distribution uniform thermal between 0.0001 and 1)
- Learning rate for the non-network approximation using the mean slope gradient (uniform logarithmic distribution between 0.001 and 10)
- Type of regularization of the weight matrices for each layer during adjustment not that it can be 0 with probability of 0.5 or a log-normal distribution uniform from 10^{-7} to 10^{-4}

The search by mesh is ineffective due to the large number of possible values for hyperparameters needing an enormous power of computation for their calculation.

UNSUPERVISED OPTIMIZATION WITH LAYER-EGO TECHNIQUES

The technique presented in [40] responds to the question that arises if the unsupervised learning process on a Deep Belief Network can be extended to the selection of the hyperparameters of the network. In [40] it is used as criterion of evaluation of the model called Reconstruction Error, which is formally obtained by establishing the visible units and some entries x, calculate the value of the hidden units after the consideration P, propagates backward this result makes the units visible and calculates the Euclidean distance between x and the conguration of the visible units that has been obtained.

Let the vector W_θ^* the weights of a Restricted Boltzmann Machine (RBM) trained with the data set D using the hyper-parameters θ. Then the Error of Reconstruction is nesting like:

$$l(D,\theta) =_{\rightarrow} L\left(D, W_{\omega}^{\bullet}\right) = \sum_{x \in D} \left\| x - g_{w_\theta^{\bullet}} \circ f_{w_\theta^{\bullet}}(x) \right\|^2 \tag{4.2.9}$$

The reconstruction error corresponding to the layers of the RBM can be directly of the nest of equation 4.2.9 as:

$$L\left(D, W_1, \ldots, W_t\right) = \sum_{x \in D} \left\| x - G_{l} \circ F_l(x) \right\|^2 \tag{4.2.10}$$

The procedure for choosing the set of hyperparameters using the information provided by the reconstruction error lies in selecting a suitable number of neurons for each of the layers of the network. The reconstruction error should decrease as the number of neurons increases, so it must use a regularization parameter to stop this process. Another possibility is waiting until the reconstruction error does not decrease significantly to a minimum threshold of change.

Figure 4.2.1 Example of rebuilt digits using an RBM with 50,000 examples per every time Bottom: Originals, Top: Reconstruction with 300 hidden units,

Figure 4.2.1 shows the change that is achieved by increasing the number of hidden neurons in the second layer of a RBM, this represents a reconstruction error much more under which it will stop the learning algorithm.

CHAPTER 5

SELF-CODING METHOD IN THE IDENTIFICATION OF NON-LINEAR SYSTEMS

Ivan Stanimirović

AUTOENCODERS

An autocoder takes as input a vector $x \in [0,1]^d$, and first perform a mapping (Using a coder) to a hidden representation $y \in [0; 1]^{d'}$ by means of a deterministic assignment characterized by free weights and bias:

$$Y = s(Wx + b), \quad (5.1.1)$$

where s is a non-linearity, such as a sigmoid. The hidden representation or latent and, also called code, it is then mapped back (using a decoder) to a reconstruction z is the same shape and size as x through a similar transformation:

$$Z = s(W'x + b') \quad (5.1.2)$$

where ' does not indicate a transpose, and z must be interpreted as a prediction of x given the code obtained in y. The weight matrix W' of the regressive mapping can be optionally restricted to simply being the transposed matrix of the matrix of weights of the codification stage: W' = W^T, this assignment will be known as association of weights.

The parameters of this type of model (within W, b, b' and if associated weights are not used, also in W') are optimized such that the error of average reconstruction is minimized. The reconstruction error can be measured from several ways, depending on some assumptions of probability distribution appropriate in the conditioned entrance to the code, that is, using the traditional error L(x, z) =Iix-zII^2 , or if the input is interpreted either as vectors binary probabilities using the cross-entropy of the reconstruction as:

$$L_H(x,z) = -\sum_{k-1}^{d} | x_k \log z_k + (1+x_k)\log(1-z_k) | \qquad (5.1.3)$$

This procedure is aimed to a distributed representation that capture the coordinates on the main factors of variation in the data (similarly as the projection of the main components captures the factors of variation). Generally and is seen as a compression with loss of X and therefore cannot be interpreted as a good compression (with losses small) for all x, as training results in a compression that works particularly well for training examples, but not for arbitrary entries. These are the main limitations in the use of autoencoders and as these generalize, these structures result in a small reconstruction error examples of probabilities belonging to the same probabilistic distribution of examples of training, but have as output a high error of reconstruction using inputs of other distributions.

Initial Considerations in Self-Encoders

If there is a linear hidden layer (the code) and a mean square error criterion is used to train the network, then the hidden units k learn to project the input in the span of the first k major components of the data. If the hidden layer is nonlinear, the self-behaving behaves differently from the analysis of principal components [6], with the ability to capture multi-modal aspects of data distribution of entry.

An event that may occur during autoencoders training is that if there are no other restrictions besides minimizing the reconstruction error then a self-encoder with n inputs and a coding dimension of at least n can potentially learn only the identity function, so many autoencoders they will be useful. Some experiments suggest that in practice, when training using stochastic descending gradient, nonlinear autoencoders with more units hidden that inputs (called over-complete) provide representations [32].

It has been suggested that to achieve a good reconstruction of entries with values continuous, single-layer autocoder hidden with non-linear hidden units initially require small weights in the coding layer (to ensure that learning begins at the approximately linear stage of nonlinearities) and weights of considerable magnitude in the decay layer.

As regularization implicit or explicit makes it difficult to reach solutions automatically with weights of high magnitude, the optimization algorithm works well for examples similar to those training set, which is the main feature of autoencoders. This means that the representation is exploiting present statistical regularities in the training set, instead of learning how to replicate the identity function.

Noisy Autoencoders

The main idea of using noisy autoencoders is simple. To force the layer to discover robust features and prevent the network from simply learning the identity function, we train self-encryption in such a way as to reconstruct corrupt version of it.

A loud self-censor is a stochastic version of traditional self-control. Intuitively, a noisy autocoder does two things: try to codify the input (preserves input information), and attempts to undo the effect of a process of corruption as applied to the input of the model.

In the literature, the process of stochastic reconstruction consists of randomly some of the entry as zero (including half of the total of them). As, the self-power noisy is trying to predict the corrupt values of noncorrupt values.This is important because if the autocoder is able to predict any subset of variables in the rest, this is a sufficient condition to fully capture the distribution of a set of training input vectors (as than the Gibbs sampling).

NOISY STACKED AUTOENCODERS

Noisy autoencoders can be stacked to form an architectural network by simply feeding the latent representation (or output noisy autocoder located in the lower layer as input to the current layer. Unsupervised pre-training of such an architecture is done one layer at a time. Each layer is trained as a noisy autocoder minimizing reconstruction error of its input (which is the output layer of the previous layer). Once the first K layers are trained, you can perform the training of the k + 1 layer due to that until that moment it becomes possible the calculation of the code or latent representation of the previous layer.

Once all layers are pre-drilled, the network passes through a second step known as setting no. Generally, for this last stage, it is considered an unsupervised setting where what is desired is to minimize the prediction error during a supervised task. For this purpose, as an object of classification or identification.

Supervised training is done in the way a perceptual is trained multilayer, that is, using standard downward gradient methods.

It can be analyzed to a noisy autoencoder stacked as if it had two differentiated parts, the first one is a list of autoencoders, while the second one is an MLP (Multilayer Perceptron). During the pre-training phase the first section is the one used, that is, the model is worked as a list of autoencoders using the ego layer paradigm so that the training of each self-power is performed separately. In the second phase of the training the MLP is used. The two structures (autoencoders and MLP) were connected because autoencoders and layers of the MLP share parameters and, autoencoders have as input latent representations of the intermediate layers of the MLP.

AUTOENCODING ALGORITHMS

Noisy Autoencoders

What is desired is to implement a self-encryption using some high-level language with ability to manage classes and objects, this is because you want to design a class which can then be used as a constructive element of a deep architecture. The first step is to nir the attributes that will have the autoencoder class that include W, B, and 0, since a configuration of weights is being used. A usual way of initializing the attributes of that class will be:

W = random (size = [size_visible_units, size_hidden_units], maximum)

bias = numberOfZeros (size = [visible_units])

hidden_bias = numberOfZeros (size = [hidden_units])

W_hidden = transposed (W)

Here, the process of random selection of weights is carried out on a uniform distribution with minimum and maximum values calculated by:

Minimum = -4* sqrt (6 / (hidden_units + visible_units))

Maximum = +4* sqrt (6 / (hidden_units + visible_units))

The cost function and the parameters corresponding to a step of the learning process by gradient downward stochastic can be computed as

cost = total (input *log (reconstr_values) + (1- entry)* log (1-reconstr_values)

gradient_x = calculates_gradient (cost, x)

The parameter x in the cost function of such it is possible to estimate an expression for its updating, this is done to each pair of the model, in this case the two bias vectors and the two matrices. It is also appreciated that the cost function corresponds to the entropy function.

In this way, an iterative function can be defined, such that it is applied, parameters so that the cost of reconstruction is minimized.

As described, the method most used during this iterative process is the downward stochastic gradient that requires movement throughout the whole of the training set batches, that is:

While (time <total_time) and (exit = false):

Epoch = time + 1

for each minibatch in training set:

lost = f (parameters; x_{batch})

Calc_gradient

parameters = parameters - learning rate* gradient

if lost <= target

exit = true

return parameters

The need to introduce some corruption process to prevent the model from simply learning the identity function which removes its abstract rendering ability in the hidden layer.

To convert a class of autoencoder type, as described above, in a noisy auto-encoder type, the only thing needed is to add a stage of stochastic corruption operating at the entrance of the model. The entrance can be corrupted in several ways, the present work uses the corruption mechanism in the masking of the entrance causing some elements of this to be randomly zero.

The following values should be considered: n is the value of the mask in case a particular element of the input is conserved and p the probability that an element of the input becomes zero, this probability is parameterized by the variable that usually has values between 0.1 and 0.2. The new entry is calculated by means of a product point to point of the entrance with the mask.

This new entry is the one to be used to calculate both intermediate and the model but the input with which the reconstruction error is calculated is the original entry, this is done with the objective of constructing a model that predict missing values of the input under the probability distribution of the set of training.

Stacked Autoencoders (SAE)

Boltzmann's autoencoders and restricted machines constitute the structures building techniques of so-called deep architectures, as mentioned, an ego-type training strategy is used during the adjustment of the initial non-apocalyptic weights of the model and, furthermore, the adjustment is effected considering the model as an MLP optimized by means of gradient downward techniques stochastic If you want the model to have n layers, n layers of sigmoid type (the MLP is formed by these layers) and n type autoencoder type noisy (with the characteristics of the class built in Section 5.3.1). We link dynamically the sigmoid layers such that they form an MLP, and each self-encryption is created such that it shares the matrix of weights and bias of its coding step with the corresponding sigmoid layer i.

Once these assignments have been made, a layer is added to the model that calculates the total network response, the structure of this last layer depends fundamentally of the task that the model is performing, it would be a linear layer in the case of identification task or a logistic regression layer if the objective is a classification of patterns.

With the above considerations, the parameters that are assigned to the architecture deep would be the weights and biases of the sigmoid layers and the layer; The parameters of autoencoders are not considered as model parameters because they are shared with the sigmoid layers and their updating is done together.

There are two stages during the training of a set of noisy autoencoders stacked, the first is a pre-training of the ego layer and the second is a supervised learning for the pre-training stage, iterations are performed on all the layers of the model.

For each layer a function is implemented that executes a gradient algorithm descending tool that optimizes the weights to reduce the reconstruction error of each layer. This function is applied using the training data set during the number of training periods.

from i = 1 to i = number_of_layers:

while(time <pre-training_periods):

epoch = time + 1

for each minibatch in the training set:

preentrena_layer[i](entry = current_batch)

return parameters

The adjustment cycle is not the same as the one to be used for training MLP, the only consideration is that the cost function should be appropriately chosen which you want to minimize according to the task or problem you are performing. In the case of using a logistic regression layer, the cost function would consist of the calculation of negative log-likelihood while for regression nodes, it is used the classic half-square error function in order.

$$E_c = \frac{1}{N}\sum\left(y_{ezt,t} - y_t\right)^2$$

To optimize the adjustment convergence speed, the conditions are not used of early unemployment where the updated parameters by means of the stochastic descending gradient are the parameters of the model (sigmoid layers and output layer).

APPLICATION IN MNIST NUMBER CLASSIFICATION

The MNIST database is an ordered agglomeration of handwritten words (see figure 5.4.1) and is commonly used for training various image processing. The database is also used during processes training and testing in the field of automated learning. The images in black and white were taken by reorganizing the datasets from the base of original NIST data that lacks practicality when presenting a test set with samples not representative of the training set.

The database consisted of 40,000 training images, 10,000 images for validation and 10,000 images for testing. Each of the images constitutes matrix of 28x28 pixels with a gray scale value normalized between 0 and 1. There are a lot of articles that explain attempts to reach the error rate of predicted lower by what has been database has become a tool is to measure the performance of any classification topology of patterns.

0 0 0 0 0 0 0 0 0 0 0 0 0 0 0
1 1 1 1 1 1 1 1 1 1 1 1 1 1 1
2 2 2 2 2 2 2 2 2 2 2 2 2 2 2
3 3 3 3 3 3 3 3 3 3 3 3 3 3 3
4 4 4 4 4 4 4 4 4 4 4 4 4 4 4
5 5 5 5 5 5 5 5 5 5 5 5 5 5 5
6 6 6 6 6 6 6 6 6 6 6 6 6 6 6
7 7 7 7 7 7 7 7 7 7 7 7 7 7 7
8 8 8 8 8 8 8 8 8 8 8 8 8 8 8
9 9 9 9 9 9 9 9 9 9 9 9 9 9 9

Figure 5.4.1

Then the performance of a deep topology based on autoencoders comparing with the results obtained by means of little architectures deep and without pre-training.

Logical Regression

The first classification technique is to be tested and that it is of importance in the development of deep learning techniques is the logistic regression, in the case of the problem the MNIST model consists of 28x28 inputs (784) that provide 10 outputs where each output corresponds to the probability that the input image corresponds to a certain amount.

The supervised learning rate was 0.1 with a maximum number of epochs of 1000. The training algorithm was applied using early stopping conditions to avoid the phenomenon of over adjustment. Each minibatch consisted of 500 examples and the early follow-up had a significant improvement of 0.5%.

With these restrictions it is achieved that the algorithm stops at the time 96 with an error of validation of 7.22% and a test error of 7.69%. The execution time of the algorithm was 230 seconds.

There is an oscillatory behavior in the error value training, this is because each sample of the error was taken once the parameters of the model were updated due to the in minibatch in particular, and since the MNIST is a database not uniformly distributed one in particular may represent an increase in error next training iteration, although as it is also observed, the trend is the decrease of the error.

On the other hand, the validation error and the test error show a clear improvement as the number of times increases.

Validation and test errors show similar behavior but with a clear tendency that the error of validation was always better than the test error, this is because the MNIST has as a set of validation examples belonging to the same sample set of the training examples come from while the test set was taken from digits obtained from a distinct population sector.

Multilayered Perceptual Neural Network

A multilayered perceptual model is a logistic regression where instead of feeding entry to a usual logistic regression layer, an intermediate layer named hidden layer having a non-linear activation function (in this case tangent hyperbolic). You can use as many hidden layers as you want by increasing the depth of the model architecture but with the risk of suffering the effect of gradient fading as explained above.

The topology used for the classification problem of the MNIST database was consisting of a single hidden layer, 500 units hidden by each layer, size of minibatch of 20 examples and learning rate of 0.01.

With this topology, we obtained a validation error of 1.79% and a test error of 1.92%, the execution time of the algorithm was 183 minutes. The dynamics of error follows similarly to the case of logistic regression with oscillatory behavior and downward trend because not all minibatches are representative examples of the probability distribution of the complete training set.

The validation and test errors have a downward trend without the oscillations presented by the training error, this is because the both sets of examples have representation over the totality of the distribution of training as opposed to the minibatches used during training. To itself, the test error again is greater than the validation error due to the similarities that each set has with the training set.

Noisy stacked Autoencoders

Noisy autoencoders are building blocks for an architecture deep. To test the performance of a deep model based on autoencoders noisy stacks, the MNIST classification problem was considered again that each autocoder will share with a hidden type layer sigmoid pertaining to a deep Perceptron. The exit layer was determined by a usual logistic regression algorithm, with 10 probability outputs, for each one of the categories to be classified whose maximum probability value constitutes the output prediction model.

A 3-layer hidden architecture with 500 units was chosen for test each, 20? Pre-training times for each layer and 1000. Supervised with learning rates of 0.1 and 0.01 respectively.

Since we work with noisy autoencoders, we consider corruption indices for each layer, in this problem we used indexes 0.1, 0.2 and 0.3 respectively for each of the layers of the model, it is appreciated that it is granted a greater degree of corruption to the layer closest to the exit, this is aimed at granting it greater capacity of representation to the complete model.

The cost function in the pre-training stage was the cross-entropy that had to be diminished by means of a downward gradient per ego layer for each one of the intermediate representations.

The mean validation error was 1.22% with a total test error of 1.34%. The dynamics of training error shows similar oscillatory behavior presented by the logistic regression models and multi-layer perceptual network having only the discrete values of 0 and 1 because minibatches were unitary, thus eliminating the variables of representativeness of the set of training.

The validation error and the test error again exhibit behaviors descendants due to the training algorithm used, being again the test error value slightly greater than the validation error for all periods of training.

CHAPTER 6

RESTRICTED LEARNING FROM BOLTZMANN IN THE IDENTIFICATION OF NON-LINEAR SYSTEMS

Ivan Stanimirović

RESTRICTED BOLTZMANN MACHINES

Energy-based models associate a scalar energy with each congruation of the set of interval variables. The learning process corresponds to modifying the form of this energy function such that it has desired properties, for example, in some cases it is desirable to have low energy congruations. The models based in probabilistic energies of a probability distribution through the function of energy such as:

$$p(x) = \frac{e^{-E(z)}}{Z}$$

The normalization factor Z is called the participation function by its analogy with systems physicists.

$$Z = \sum_{x} e^{-E(x)}$$

An energy-based model can learn by implementing a gradient process down-stochastic on the log-likelihood of negative empirical data training. As in the case of logistic regression, log-likelihood and after-effects of the cost function as its negative version with the goal of minimizing it:

$$\mathcal{L}(\Theta, D) = \frac{1}{N} \sum_{x^{(i)} \in D} \log p\left(x^{(i)}\right)$$

$$\ell(\Theta, D) = -\mathcal{L}(\Theta, D)$$

Power-Based Models with Hidden Units

In many cases of interest, it is impossible to observe the examples of output x in its totality, or you may want to introduce some unobserved variables to increase the expressive capacity of the model being designed. Let us consider first an observed part (a? A denoted by x) and a hidden part h. It can be written

$$P(x) = \sum_h P(x,h) = \sum_h \frac{e^{-E(x,h)}}{Z} \qquad (6.1.1)$$

In such cases, to bring this form to that presented for fully visible systems, introduces the concept of free energy, de nida as:

$$F(x) = -\log \sum_h e^{-E(x,h)}$$

which allows us to reformulate

$$P(x) = \frac{e^{-F(x)}}{Z} \; con \; Z = \sum_x e^{-F(x)}$$

Thus, the negative log-likelihood gradient of the data has a divisible shape in two addends:

$$\frac{\partial \log p(x)}{\partial \Theta} = \frac{\partial_F(x)}{\partial \Theta} - \sum_{\tilde{x}} P(\tilde{x}) \frac{\partial_F(\tilde{x})}{\partial \Theta} \qquad (6.1.2)$$

Note that the gradient contains two terms, which are referred to as positive phases and negative. The positive and negative terms do not refer to positive signs and negatives of the equation, but the effect they have on the probability density of the model. The first term increases the probability of training data (reducing the corresponding free energy), while the second term decreases the probability of the samples generated by the model. It is usually difficult to determine an expression for this gradient analytically, because it involves the calculation of EP. This expression represents hope mathematics on all possible combinations of the input x (under the distribution P formed by the model)

The first step in the realization of this viable calculation is to estimate hope using number of samples of the model. The samples used for the estimation of the negative phase of the gradient are known as negative particles. The gradient is then written as

$$-\frac{\partial \log p(x)}{\partial \Theta} \approx \frac{\partial_F(x)}{\partial \Theta} - \frac{1}{|N|}\sum_{\tilde{x} \in N} \frac{\partial_F(\tilde{x})}{\partial \Theta} \qquad (6.1.3)$$

where it is desired that ideally elements are sampled according to P (for instance MonteCarlo algorithm). With the previous formula, one has Algorithm for the training of an EBM, the only missing element is the choice of these negative particles. The literature [40] suggests that methods such as MonteCarlo Markov chains work especially well for models such as the restricted Boltzmann machines which form the basis of a network of deep belief.

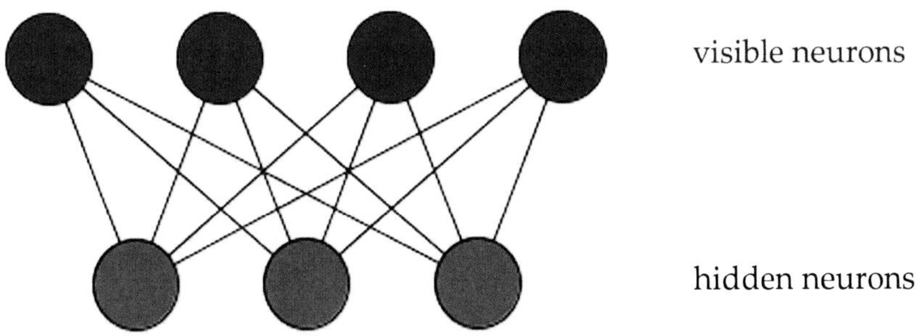

Figure 6.1.1

Boltzmann's machines are a particular form of the so-called random field of Markov in its linear logarithmic version, i.e. for which the power function is linear in its parameters. To make this model so powerful enough to represent complicated distributions (i.e. go from a parametric configuration limited to a non-parametric one), we will consider that some of the variables that describe representations are never observed (hidden calls). Having more variables (or hidden units), the modeling

(But also increases its computational cost). The Boltzmann (RBM) restricted machines only take into account those models in which there are no visible-visible and hidden-hidden connections (see Figure 6.1.1).

The energy function E (v, h) of a RBM is defined as:

$$E(v,h) = -b^T v - c^T h - h^T W_v \tag{6.1.4}$$

where W represents the weights connecting the hidden layer to the visible layer and b, c are the or sets of the units visible and hidden respectively. This translates directly into the formula for free energy:

$$F(v) = -b^T v - \sum_t \log \sum_{h_t} e^{ht(e_t + w_t v)} \tag{6.1.5}$$

Because of the specific structure of the RBMs, the hidden and visible units are conditionally independent given one or the other. Using this property one can write:

$$p(h \mid v) = \prod_t p(h_t \mid v)$$

$$p(v \mid h) = \prod_j p(v_j \mid h)$$

RBMS WITH BINARY INPUTS

In the simplified case of using binary units (where vj and hi 2 f0; 1g), we obtain probabilistic versions of the usual sigmoid-type neuronal activation function:

$$P(h_t = 1 / v) = sigm(c_t + W_t v)$$

$$P(v_j = 1 \mid h) = sigm(b_j + W_j^T h)$$

Thus, the free energy of a RBM is simplified to:

$$F(v) = -b^T - \sum_t \log\left(1 + e^{(e_t + W_t v)}\right) \tag{6.1.6}$$

Equations of Learning and Updating with Units

If a learning method based on the descending gradient is applied for example if a learning method based on the descending gradient is applied for example the likelihood between the patterns and the outputs, we obtain the equations of update [34] for the matrix of free weights W and bias b, c.

$$-\frac{\partial \log p(v)}{\partial W_{ij}} = E_v\left[p(h_t \mid v) - v_j\right] - v_j^{(t)} \bullet sigm\left(W_t \bullet v^{(t)} + e_t\right)$$

$$-\frac{\partial \log(v)}{\partial e_t} = E_v\left[p(h_t \mid v)\right] - sigm\left(W_t \bullet v^{(t)}\right)$$

$$-\frac{\partial \log p(v)}{\partial b_j} = E_v\left[p(v_j \mid h) \bullet v_j\right] - v_j^{(t)}$$

However, these formulas were not used during the implementation of the algorithm of learning in the present study because of the use of management of symbolic algebra to obtain the value of the gradient over the training.

Sampling in a RBM

The samples of p(x) can be obtained by implementing a Markov chain for convergence, using Gibbs sampling as the transitional operator.

The Gibbs sampling of the union of N random variables $S = (S_1; \ldots, S_N)$ is performed at traversal of a sequence of N sub-steps of sampling the form $S_i \sim p(S_i I S_{i-1})$ where S_{-1} contains the other N - 1 random variables in S excluding S_i.

For RBM, S consists of a set of visible and hidden units. Nevertheless, because these units are conditionally independent, we can simplify the process of sampling Gibbs to one carried in blocks. With these characteristics, visible units are simultaneously sampled from unit values hidden. Similarly, hidden units are simultaneously sampled data values for the visible units. A step in the Markov chain is then made as:

$$h^{(n+1)} \sim sigm\left(W^T v^{(n)} + e\right)$$

$$v^{(n+1)} \sim sigm\left(Wh^{(n+1)} + b\right)$$

where $h^{(n)}$ refers to the set of all the units hidden in step n of the Markov chain. What this means is that, for example, it is randomly chosen as 1 (or 0) with sigmoid probability, And similarly, $v_j^{(n+1)}$ is randomly chosen as 1 (or 0) with sigmoid probability

As t tends to positive infinity, the samples $(v^{(t)},h^{(t)})$ appropriately represent examples of .

With these considerations, each pair is updated during the learning process it would require running a particular string to converge. This is, however, immensely expensive from the computational point of view and therefore have been developed algorithms for the training of RBMs in order to perform less expensive and costly sampling process of the probability p(v, h) during the training process.

Contrastive Divergence (CD-k)

The contrastive divergence uses 2 events to accelerate the sampling process:

As what we want is that eventually $p(v) \approx t\ p_{train}(v)$ (which is the true form of data distribution), we initialize the Markov chain with an example of training (that is, of a distribution expected to be similar to p, such that the string is from the beginning close to converging to its distribution p)

- The CD does not wait until the string converges. Samples are obtained after of only k steps of the Gibbs process. In practice, k=1 is quite a choice common.

Persistent CD

Persistent contrastive divergence uses another approximation to perform the samples of p(v, h). This lies on a Markov chain, which has a persistent state (That is, not restarting one string for each observed instance). For each pair updated, we extract new examples by simply running the string for k steps. The state of the string is then preserved for subsequent updates.

The heuristic behind this is that if the updates of the parameters are your - relatively small compared to the mixing rate of the Markov chain, then the chain should be able to “catch” these small changes in the model

Monitoring and Evaluation of Training Progress

RBMs require stochastic processes during their training. This is because due to the partition function Z, it is not possible to estimate the log-likelihood log(P(x)) during the training. There is no direct way of obtaining a metric suitable for the choice of the hyperparameters of the network.

Some solutions are as follows:

- Inspection of negative samples. Negative samples obtained during training can be visualized. As the training progresses, we have established that the nest model by the RBM is progressively closer to the true probability distribution underlying, $p_{train}(x)$. Thus, the negative samples should resemble the examples taken from the training set, if this is not met you should choose a new set of hyperparameters.
- Visual inspection of the layers. The factors learned by the model can be deployed. That involves recording the weights of each unit as an image in grayscale (once changed its shape to a square matrix). The layers must catch important data characteristics, this is problematic because for an arbitrary set of training data there is no known of how these characteristics should be interpreted; However, in problem of classification of patterns in which data from images, layers behave as detectors of maximum changes in gradient.

Plausibility Approximations

These methods of choosing the best hyperparameters depend to a great extent on the analytical capacity of the person who is implementing the algorithm so that it does not recommended in all cases. During the training of a RBM with PCD, Pseudo-likelihood can be used as an approximation. The pseudo-likelihood (PL) is much less expensive to calculate, because it assumes that all bits of information are independent.

$$PL(x) = \prod_t P(x_t \mid x_{-t})$$

$$\log PL(x) = \sum_t \log P(x_t \mid x_{-t})$$

Here X_{-I} denotes the set of all bits of x with the exception of bit i. The log-PL is then the sum of the log-probabilities of each bit xi, conditioned to the state of the rest of the bits. In the case of the classic problem of classification of MNIST will entail adding about 784 entries, which despite previous considerations it's an expensive one. For this, a stochastic approximation of log-PL is introduced:

$$g = N \bullet \log P\left(x_t \mid x_{-t}\right), donde\, i \sim U\left(0, N\right),$$

$$E \mid g \mid = \log PL\left(x\right)$$

where mathematical hope is taken over the uniformly random choice of I, and N is the number of visible units. In order to work with binary units, enter the notation

$$\log PL\left(x\right) \approx N \log \frac{e^{-FE(x)}}{e^{-FE(x)} + e^{-FE(\tilde{x}_t)}}$$

$$\approx N \bullet \log\left[sigm\left(FE\left(\tilde{x}_t\right) - FE\left(x\right)\right)\right]$$

Then the log-PL for an RBM with binary units is written as:

$$\log PL\left(x\right) \approx N \log \frac{e^{FE(x)}}{e^{-FE(x)} + e^{-FE(\tilde{x}_t)}}$$

$$\approx N \bullet \log\left[sigm\left(FE\left(\tilde{x}_t\right) - FE\left(x\right)\right)\right]$$

During the implementation of the calculation of this cost, it should be returned the same as the corresponding parametric updates. It should be emphasized that the update dictionary cache for all parameters should be modified to increase the bit rate i. This translates into the bit i moving on all the possible values from one update to another.

Note that for training with contrastive divergence the entropy cost function between the entrance and the reconstruction (the same function used in a self-censoring noisy) is more aware than the pseudo-log-likelihood [3].

CONTINUOUS RESTRICTED BOLTZMANN MACHINES

The literature [6, 7, 9] adequately describes the behavior of restricted machines of Boltzmann when they are exposed to binary inputs but lack of extrapolation to the continuous case that is necessary for the implementation of techniques of deep learning for the resolution of the problem of systems identification. For this reason, the following Section describes some possibilities for solving the problem of considering a RBM with continuous inputs.

Hidden and Visible Binary Units

In the case of the RBMs discussed in Section 6.1, the simplification of consider the input to the model as binary, that is, where the incoming values to visible units can only take values of 0 and 1, taking this in consideration is how the conditional probability formulas were calculated used during the Gibbs sampling process obtained for classification purposes. Obtaining the probability formulas is explained below:

From equation 6.1.1 we know that the conditional probability is given by:

$$P(v)=\sum_{h}P(v,h)=\sum_{h}\frac{e^{-E(v,h)}}{Z}$$

The probability of an event h happening is represented by a similar expression:

$$P(h)=\sum_{v}\frac{e^{-E(v,h)}}{Z}$$

While the joint probability P (v,h) is determined by:

$$P(v,h)=\frac{e^{-E(v,h)}}{Z}$$

where v represents the values taken by the visible units and h the values taken by the hidden units. From the probability theory it is known that the probability conditional can be obtained by the quotient of the joint probability of events and the probability value of the conditioning event, as well as the conditional probability is calculated as:

$$P(v\mid h)=\frac{P(v,h)}{P(h)}=\frac{\frac{e^{-E9v,h}}{Z}}{\sum_{v}\frac{e^{-E(v,h)}}{Z}}=\frac{e^{-E(v,h)}}{\sum_{v}e^{-E(v,h)}}$$

In addition, it is known from equation 6.1.4 that the energy for a specific configuration is given by:

$$E(v,h) = -b^T v - c^T h - h^T W_v$$

Substituting in the expression for the conditional probability gives:

$$P(v \mid h) = \frac{e^{b^T v + e^T h + h^T W v}}{\sum_v e^{b^T v + e^T h + h W v}} \tag{6.2.1}$$

In the case of the case studied that corresponds to a restricted machine of Boltzmann there is no direct dependence between the values taken by the visible units regarding to the other visible units so the dependence is direct with the units hidden (the opposite also happens). We can write:

$$p(h \mid v) = \prod_i p(h_i \mid v)$$

$$p(v \mid h) = \prod_j p(v_j \mid h)$$

According to this last expression and associated with the fact that the probability conditional of the complete vector of visible units is the product of the probabilities conditionals of each of the elements can be made the following equality:

$$p(v_j \mid h) = \frac{e^{b_j v_j + h^T W_j v_j}}{\sum_v e^{b_j v_j + h^T W_j v_j}} \tag{6.2.2}$$

In the denominator it is observed that we must calculate the value of the summation by traversing all possible values of v, in the case of binary units these values are only 0 and 1, so we can calculate the conditional probability p(v_j=1 I h) as:

$$p(v_j = 1 \mid h) = \frac{e^{b_j v_j + h^T W_j v_j}}{\sum_v e^{b_j v_j + h^T W_j v_j}} \mid v_j = 1 \frac{e^{b_j + h^T W_j}}{1 + e^{b_j + h^T W_j}}$$

$$p(v_j = 1 \mid h) = sigm(b_j + h^T W_j) \tag{6.2.3}$$

A similar analysis can be made to obtain the conditional probability of the values of the hidden units with respect to values given of the visible units with which we obtain:

$$P(h_i = 1 \mid v) = sigm(c_i + W_i v) \tag{6.2.4}$$

$$P(v_i = 1 \mid v) = sigm(b_i + h^T W_j) \tag{6.2.5}$$

In spite of the limitations that are had considering the binary visible units, has been functioning properly even with systems with inputs (such as grayscale images) [40], this is because the input is assimilated as a probability distribution that has affected to the original entry.

Continuous Visible Units in the Interval [0,]

From equation 6.2.2 we can generalize the idea of conditional probability for units of continuous type transforming the summation into the visible domain.

$$p(v \mid h) = \frac{e^{(b_j + h^T W_j)v_j}}{\int_v e^{(b_j + h^T W_j)v_j dv_j}} \tag{6.2.6}$$

It can be shown that in general the case where lacks convergence in the integral term of the equation with which only the semi-nested case is analyzed positive [0,].

$$p(v_j \mid h) = \frac{e^{(b_j + h^T W_j)v_j}}{\int_\alpha^\infty e^{(b_j + h^T W_j)v_j dv_j}}$$

Let us make the following notation:

$$\alpha_j = \alpha_j(h) = b_j + h^T W_j$$

at the end that multiplies the value which takes visible unity, then we can develop the expression as:

$$p(v_j \mid h) = \frac{e^{\alpha_j v_j}}{\int_\alpha^\infty e^{\alpha_j v_j dv_j}} = \frac{e^{\alpha_j v_j}}{\frac{1}{\alpha_j} e^{\alpha_j v_j |_0^\infty}}$$

This last expression only has small values if the structure of the system guarantees that , for all , in the case that the condition is met, the distribution conditional probability:

$$p(v_j \mid h) = -\alpha_j e^{\alpha_j v_j} \tag{6.2.7}$$

where p is positive according to the condition of existence of the integral.

With this formula for the probability distribution we cannot directly perform the process of sampling since in general the computational tools cannot sample from a particular probability distribution, to solve this problem we proceed to calculate the cumulative probability distribution of the variable which will always have increasing values in the interval [0, 1]. Using the inverse cumulative probability function p_a^{-1} we could then sample from a distribution uniform. The calculation of pa is done in the usual way:

$$p\alpha\left(v_j \mid h\right) = \int_0^{v_j} p\left(v_j \mid h\right) dv_j = \int_0^{v_j} -\alpha_j e^{\alpha_j v_j dv_j}$$

$$P_a\left(v_j \mid h\right) = 1 - e^{a_j V_j} \qquad (6.2.8)$$

If we denote by U the result of a sampling process on a distribution uniform in the interval then we can relate this value U to the value of the accumulated density pa with which it would be left to find the value of the visible unit corresponding.

$$v_j = \frac{In\left(1-U\right)}{a_j} \qquad (6.2.9)$$

To the visible units that have as distribution of probability presented previously they are known as exponential units.

Continuous Units Visible in the Range [0, 1]

In many of the applications the interval I of possible values for the visible units is restricted (depends mainly on the dynamics and the source of the input to the model), the simplest case is where the input has normalized (either by natural conditions of the system or by some normalization process) where I = [0, 1]

When the calculation of the probability distribution is seen bounded as in this case we speak of truncated exponential visible units, from equation 6.2.6 we get:

$$p(v_j \mid h) = \frac{e^{a_j v_j}}{\int_a^1 e^{a_j v_j dv_j}} = \frac{e^{a_j v_j}}{\frac{1}{a_j} e^{a_j v_j}\Big|_0^1}$$

$$p(v_j \mid h) = \frac{a_j e^{a_j v_j}}{e^{a_j} - 1} \qquad (6.2.11)$$

In order to implement a Gibbs sampling process it is necessary to calculate of the conditional probability as follows:

$$p_a(v_j \mid h) = \int_0^{v_j} p(v_j \mid h) dv_j = \frac{a_j}{e^{a_j} - 1} \int_0^{v_j} e^{a_j v_j dv_j} \qquad (6.2.12)$$

Clearing the value of v_j is determined:

$$v_j = \frac{\log\left[1 + p_a\left(e^{a_j} - 1\right)\right]}{a_j}$$

Expression that can be used to calculate the new value of vj during a process of sampling by replacing the value of the variable pa by some distribution sample unit uniform U, substituting:

$$v_j = \frac{\log\left[1 + U\left(e^{a_j} - 1\right)\right]}{a_j} \qquad (6.13)$$

NETWORKS OF DEEP BELIEF

Geoffrey Hinton demonstrated that RBMs can be stacked and trained in an egoistic way to form what he called deep belief networks (DBN) [29]. Deep belief networks are graphic models that learn to extract a deep hierarchical representation of training data. These models characterize the joint distribution h^k between the observed vector x and the l hidden layers as:

$$P(x.h^l, \ldots, h^l) = \left(\prod_{k=0}^{l-2} P(h^k \mid h^{k+1})\right) P(h^{l-1}, h^l)$$

where $x = h^0$, $P(h^k \mid h^{k+1})$ Is a conditional distribution for the visible units limited on the hidden units belonging to the RBM at level k, and P $(h^{l-1},$

h^l) is the hidden-visible joint distribution in the upper or outbound RBM. This is illustrated by Figure 6.3.1.

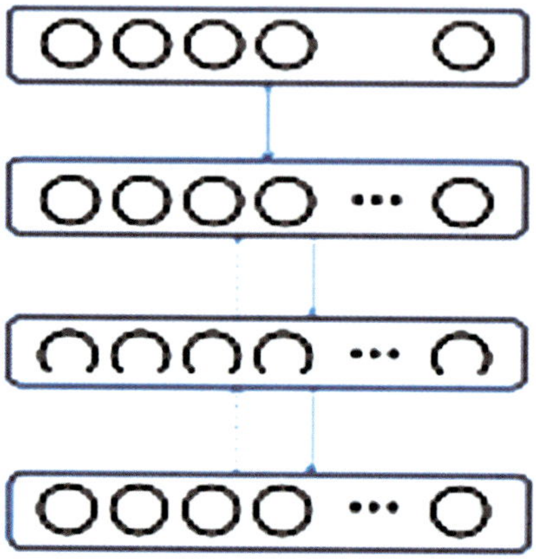

Figure 6.3.1

The principle of training can be applied to DBNs with RBMs following the concept of Hinton and Bengio [9,29]. The process is performed as follows:

1. The first layer is trained as an RBM that gives the original input $x = h^{(0)}$ as its visible layer.
2. Use this first layer to get a representation of the input that will be used as input data for the second layer, since a RBM does not contain in its structure an inherent output, we have two options for the assignment of this representation. The data set chosen is generally chosen as the mean activations $p\ (h^{(1)} = 1 \mid h^{(0)})$ or the samples of $p(h^{(1)} \mid h^{(0)})$.
3. Train the second layer as a separate RBM, taking the transformed data (samples or activations) as training examples (for layer visible from that RBM).
4. Repeat steps 2 and 3 according to the desired number of layers, propagating above either the samples or the activations.
5. Perform a procedure of fine-tuning all the parameters of this deep architecture with respect to a proxy function of the lognormality of the DBN, or with respect to a supervised learning criterion, for the latter, it is necessary to add extra machinery in the network architecture that converts the learned

representation into predictions supervised, for example a classifier or linear function.

Specifically, the logistic regression algorithm is used to classify the input x based on the output of the last hidden layer $H^{(l)}$ of the DBN. The fine-tuning is implemented by minimizing the negative log-likelihood function for classification purposes (in the MNIST problem for example) or the mean square error for function identification. Since the gradient used in the parameter update rule is only different from zero for weights and biases of the hidden sublayer of each layer of the DBN (i.e., it is null for biases of each RBM), this procedure is equivalent to the initialization of parameters of a deep architecture MLP with the values of the parameters obtained through an unsupervised training strategy.

In order to explain the effect of the ego-type pre-training, takes as example a shallow DBN of 2 layers with hidden layers $h^{(1)}$ and $h^{(2)}$ (with their respective variables containing $W^{(1)}$ and $W^{(2)}$ weights). Hinton and Bengio [9, 29] state that can be written in the form

$$\log(x) = KL\left(Q\left(h^{(1)} \mid x\right) \| \, p\left(h^{(1)} \mid x\right)\right) + H_{Q\left(h^{(1)} \mid x\right)}$$

$$+\sum_{H} Q\left(h^{(1)} \mid x\right)\left(lpgp\left(h^{(1)}\right) + \log p\left(x \mid h^{(1)}\right)\right)$$

where $KL(Q(h^{(1)} \mid x) \| p(h^{(1)} \mid x))$ represents the divergence KL between Q $(h^{(1)} \mid x)$ later of the first RBM if it worked on its own, and the probability $p(h^{(1)} \mid x)$ for the same layer but defined by the complete DBN (i.e., taking into account the $P(h^{(1)}, h^{(2)})$ Above defined by the RBM of the top layer). $HQ(h^{(1)} \mid x)$ is the distribution entropy $Q(h^{(1)} \mid x)$. It can be shown that if both hidden layers are initialized such that then the term of the divergence KL is zero. If then the RBM of the first level will be trained and its fixed $W^{(1)}$ parameters will be conserved, the optimization process with respect to $W^{(2)}$, could only increase the probability the likelihood p (x).

Optimizing this result with respect to $W^{(2)}$ is equivalent to training a RBM for the second step using the output of $Q(h^{(1)}|x)$ such as training distribution, when x is taken from a training distribution for the first RBM.

ALGORITHMS BASED ON RESTRICTED ENERGY MODELS

Boltzmann Restricted Machines

As with noisy auto-builders, the implementation of the RBM training process is done by constructing a class. This option becomes rather useless when a RBM is used as a layer pertaining to a deep architecture, in which case the matrix of p and those bias or cult are shared as parameters of a sigmoid layer of an MLP. The definition of the attributes that will have the RBM is important, should include W, B and c, since a weighted weight setting is being used. A usual way of initializing the attributes of that class would be:

W = random(size = [visible_units, hidden_units], min, max)

visible = numofzeros (size_visible_units])

hidden = numofzeros (size_hidden_units])

In this case there is no variable for the hidden weight matrix because a RBM it is constituted only by an interaction layer unlike a self-encoder that is formed by two layers. The minimum and maximum values correspond to the limits of the probability distribution function used to obtain the initial values from W.

min = -4 * sqrt (6 / (hidden_units + visible_units))

max= 4* sqrt (6 / (hidden_units + visible_units))

The next step is to define functions that build the relationships associated with conditional probabilities

As mentioned above, the calculation of these probabilities allows to construct a probability distribution from which the sampling of new units both hidden and visible. This process is necessary for the realization of a step of the Gibbs sampling. The process of sampling the visible units conditioned to the value of the hidden units (or vice versa) is done by assigning a probability.

hidden_samples = binomial(size = hidden_units, 0 success = sigmoid (visible hidden prob))

prob_visible_given_hidden = transposed (W) * hidden_ samples + visible_bias visible_samples = binomial (size = visible_units, success = sigmoid (occult _hidden_prob))

The Gibbs sampling process is not complete until new values are obtained for the visible units so the samples obtained from the units are used hidden to generate a new probability distribution that allows the creation of these new values.

The above series of steps constitutes a Gibbs stage of a “vhv” (visible-hidden-visible) that starts data initial values of the visible units and which is necessary to obtain samples from the model RBM. In some cases, the implementation of a “hvh” type Gibbs step (Hidden - visible - hidden) that starts from information in the hidden units and is used for the start of updates for CD and PCD.

In general, these procedures should be included as RBM class methods to facilitate the training of a hypothetical DBN. The calculation is also necessary of the free energy which will be denoted by free energy according to Equation 6.1.4. The next step is to generate the necessary gradients for CD updates and PCD, a sample string must be calculated. You can use the same a code procedure that allows the choice between CD and PCD processes using the persistent variable.

The hidden samples are calculated from the training input to the model using Gibbs half-step as described above. If the variable persistent is null, the Gibbs string is initialized with the hidden samples generated which would involve the use of a CD process. Once the starting point of the chain, the sample obtained at the end of the Gibbs chain can be calculated. Is necessary for the calculation of the optimization gradient (see equation 6.2.6). For the calculation of this sample, an iterative process during k steps of the process of sampling is performed, where each step consists of an hvh-type Gibbs process.

The sample that one can consider as the end of the chain is the sample of the units visible from the last step of the process (final sample), with the value of this sample is then calculated to calculate the free energy corresponding to the negative phase of the likelihood gradient with which it is convenient to first calculate the function the function of the cost of the training process, which is the difference between the positive and the negative phase of the sampling process.

This cost function is dependent on all modifiable parameters of the model so that a learning process would use the gradient with respect to the weights and biases during the parametric model update for each iteration of training. Let be a gradient (cost, parameter) a function that calculates the gradient of cost with respect to parameter, the new values of the parameters characterized by some learning rate are calculated as:

Once the parameters have been updated, the cost of these updates on the general behavior of the model, the definition of the form of the cost function depends on the Gibbs sampling process being used (PC or PCD).

It is observed in the pseudocode that if CD is being used a reconstruction error defined as the cross-entropy of the model similar to that used during the optimization of noisy autoencoders, the problem lies in defining which ones will be the values that we will take as predictions of the model and that will be compared with the original input, [31] proposes to use as data of reconstruction the data belonging to the vector of the last step of the chain of Gibbs. On the other hand, if PCD is used it is necessary to use pseudo-likelihood, which is calculated using the stochastic procedure ending in Equation 6.1.8. Having defined these parameters within the class of type RBM an algorithm for the training can be easily implemented.

It is important to initialize the Gibbs chain using the persistent variables to guarantee the use of PCD, in this case a stop algorithm is not implemented early because a non-supervised solution is being used.

Deep Belief Networks (DBN)

The process of implementing a DBN is very similar to that used in an SDA because both involve the principle of unsupervised pre-training of layer type ego followed by a supervised fine adjustment similar to that of an MLP. The difference is that RBM type layers are used instead of autoencoders. Like with the autoencoders stacked, the parameters of the RBM layers will be linked with the weights and biases of the sigmoid layers that construct an MLP.

During the training of loud stacked autoencoders, the parameters of the decoding step with their corresponding variables of the sigmoid layers (and thus only serve for the pre-drilling stage), visible bias of the

RBM is not subject to such a link being taken only as parameters shared the matrix of weights and hidden bias.

The rest of the training algorithm is carried out similarly to the case where autoencoders are used but considering that the pre-training process is notoriously more extensive and costs. During the pre-adjustment, account must be taken of the following considerations:

Gibbs sampling process is implemented using constrastive divergence (CD), ie, the persistent variable will be set to zero.

The recommended values for k (number of Gibbs steps) are between 1 and 10 [31].

APPLICATION IN CLASSIFICATION OF MNIST CODES

A deep belief network is obtained by stacking several Boltzmann restricted machines. When used for classification, a DBN is treated as if it were an MLP by adding a logistic regression layer as the output layer. For the problem of MNIST classification discussed in section 5.4, we opted for a deep architecture composed of 3 hidden layers with 500 units each (similar to the topology used with noisy autoencoders). The reconstruction error to be minimized was the cross-entropy when working with contrastive divergence (CD), the learning rates were 0.1 and 0.01 for the fine tuning and pre-training stages, respectively. The number k of Gibbs steps is set to 1 while the number of examples per minibatch was 10.

Finally, the number of supervised adjustment maintains in 1000 with 20 pre-training periods for each of the 3 layers of the model. Defining the topology to be used, we proceeded with the ego-type layering pre-drilling where cross-entropy was used as the error function to be minimized by the learning of each of the RBMs, the sampling process was carried out by medium of a unitary Gibbs process.

The mean final validation error was 1.52% with a total test error of 1.61%. The dynamics of the training error shows minor oscillations magnitude as presented by the logistic regression models and multilayer perceptron due to the use of minibatches with a smaller number of examples. Eliminates in magnitude the variance with respect to the training set.

The validation error and the test error again present minimizing computes due to the descending gradient algorithm used (it is sought to maximize the log-likelihood of the model), it is observed that the error value of test is greater than the validation error during all training periods.

APPLICATION IN THE IDENTIFICATION OF NONLINEAR SYSTEMS

First Order Nonlinear System

The problem of identification of nonlinear systems was attacked using as an example the system presented by [41] described by equation 6.6.1.

$$y(k+1) = \frac{y(k)}{1+y^2(k)} + u^3(k) \quad (6.6.1)$$

where u(k) is a discrete periodic input that will change according to the stage or process learning algorithm in which the training algorithm is, the input of the system is described by equation 6.6.2

$$u(k) = A\sin\left(\frac{\pi k}{50}\right) + B\sin\left(\frac{\pi k}{20}\right) \quad (6.6.2)$$

where parameters A and B change according to the purpose assigned to it to the dynamics of the system.

In accordance with Section 6.1, the pre-selection process (according to the limits of the convergence) requires a set of system values in the interval [0, 1], so the inputs and outputs must be normalized to this interval of the system, the update to the new values is given by equation 6.6.3.

$$NewValue = (old\ value - \min imum) / (\max imum - \min imum) \quad (6.6.3)$$

The training data set consists of the dynamics of the system along of the first 2000 iterations while the validation and test sets contain only 400 examples.

The training algorithm was implemented using the random selection presented in previous sections to form a choice for the number of units and number of layers of the model. The pre-training algorithm used was that of a DBN architecture using one Gibbs pass with fine tuning or average down gradient and early stop with threshold improvement of 5%. For both stages were used only one training period. Valuable values were considered for the units visible in the continuous interval of [0, 1].

The plot of the mean squared error dynamics on the test set regarding to the variation of the hyperparametres is observed in Figure 6.6.1.

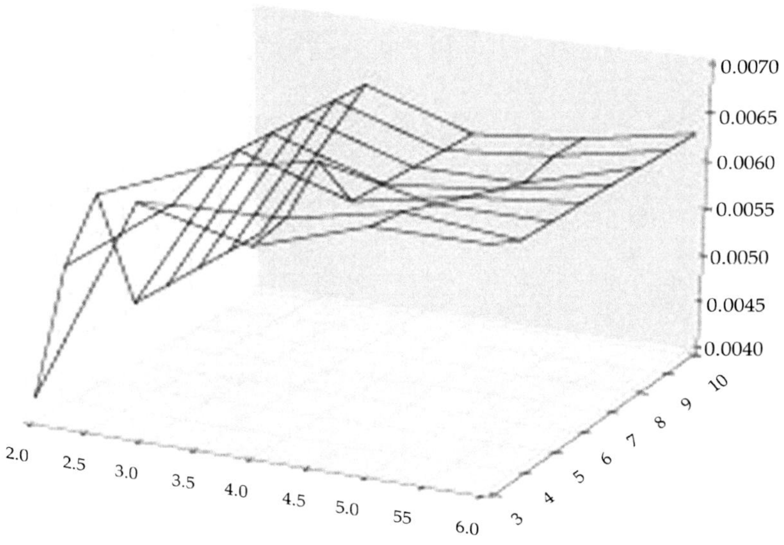

Figure 6.6.1

It is observed that using only two layers of identification there is a minimum test error. However, it is also observed that by increasing the number of layers and units, a new local minimum corresponds to an optimum configuration. Of the model within an appropriate neighborhood of the hyper-parameter space (ideal depth). This configuration is determined by a 4-layer architecture with 5 units per layer, which will be described by means of the notation 5, 5, 5, 5.

Using this architecture, the DBNs outputs are compared by varying the visible units (continuous in [0, 1], [0, ∞] and discrete in (0, 1)) with the results obtained by means of an SDA architecture, an MLP of the same dimensions but without pre-concentration and an MLP with architecture 20, 10 [41].

To obtain a visual orientation about the true behavior of the model, it is observed that none of the models has a complete output overlapping with the original but the DBN (architecture 5, 5, 5, 5) is more similar, the output of MLP with architecture 20, 10 is barely slightly less satisfactory, but it must be taken into account that this architecture has a smaller number of layers with it which avoids the problem of fading of the gradient signal during tuning fine. This fading problem is solved in the DBN by means of pre-training, the latter also being observed in the outflow of the MLP without pre-training with architecture 5, 5, 5, 5 which has been stagnant at a local minimum, poor performance compared to its pre-trained counterpart (DBN).

CHAPTER 7

ADVANCES IN SUPPORT VECTOR REGRESSION METHOD

Ivan Stanimirović

Support vector regression (SVR) is a machine learning method used for creating unique patterns between different descriptors. SVR uses the structural risk minimization principle and the insensitive loss function ε in order to search for a global optimum and estimate errors in training set. The difference between SVR and the support vector machine for classification problems lies in the usage of kernels.

Consider the set of descriptors $x = \{a, b, c\}$ which should be mapped to a high dimensional space. The decision function can be defined as follows:

$$f(w, x) = \langle w, x \rangle + z,$$

where $w \in N$ and $z \in R$ and $\langle w, x \rangle$ denotes the inner product.

In order to guarantee the flatness of the generated function, SVR uses the minimization of the Euclidean norm $\frac{1}{2} ||w||^2$, considering the Curie temperature T_{C_i} in the conditions. The minimization problem is given as follows:

$$\min \quad \frac{1}{2}||w||^2, \quad (1)$$

$$such\,that \quad T_{C_i} - \langle w, x_i \rangle - z \le \varepsilon \quad (2)$$

$$\langle w, x_i \rangle + z - T_{C_i} \le \varepsilon$$

Next, SVR incorporates the slack variables (θ_i, θ_i^*) for creating space for infeasible constraint in the optimization problem:

$$\min \quad \frac{1}{2}||w||^2 + C\sum_{i=1}^{n}(\theta_i, +\theta_i^*), \quad (3)$$

$$such\,that \quad T_{C_i} - \langle w, x_i \rangle - z \le \varepsilon + \theta_i \quad (4)$$

$$\langle w, x_i \rangle + z - T_{C_i} \le \varepsilon + \theta_i^*$$

$$\theta_i, \theta_i^* \ge 0$$

where n represents the number of data points and C is a regularization factor.

Lagrange multipliers are often use to cope with optimization problems (Owolabi, et al., "Development and validation of surface energies estimator (SEE) using computational intelligence technique," Comput. Mater. Sci. 101, 143–151 (2015)).

Therefore, if α and α^* are Lagrange multipliers, then the kernel function $\varphi(x_i, x)$ can be included in the decision function as follows:

$$f(x) = \sum_{i=1}^{n}(\alpha_i - \alpha_i^*)\varphi(x_i, x) + z \quad (5)$$

The equation (5) represents the general form of the decision function. The exact value of the decision function may differ in different models.

It is obvious that the value of the so-called SVR hyper parameters highly influences the accuracy of models that include Support vector regression. These hyper parameters are the regularization factor C,

Gaussian kernel and the loss function ε. In order to optimize the hyper parameters of SVR, many kernel functions were tested and the optimal function that often provides the best performance is given as

$$\varphi\left(x_i, x_j\right) = e^{-\frac{1}{2}\left(\frac{\left\|x_i - x_j\right\|}{\sigma}\right)^2}$$

where σ denotes one of the kernel parameters that are used.

CHAPTER 8

A BALANCED MOTOR PRIMITIVE FRAMEWORK CAN SIMULTANEOUSLY EXPLAIN MOTOR LEARNING IN UNIMANUAL AND BIMANUAL MOVEMENTS

Ken Takiyama [a], Yutaka Sakai [b]

[a] Tokyo University of Agriculture and Technology, Department of Engineering, 2-24-16, Nakacho, Koganei, Tokyo 184-8588, Japan

[b] Brain Science Institute, Tamagawa University, 6-1-1, Tamagawa-gakuen, Machida, Tokyo 194-8610, Japan

ABSTRACT

Certain theoretical frameworks have successfully explained motor learning in either unimanual or bimanual movements. However, no single theoretical framework can comprehensively explain motor learning in both types of movement because the relationship between these two types of movement remains unclear. Although our recent model of a balanced motor primitive framework attempted to simultaneously explain motor learning in unimanual and bimanual movements, this model focused only on a limited subset of bimanual movements and therefore did not elucidate the relationships between unimanual movements and various bimanual movements. Here, we extend the balanced motor primitive framework to simultaneously explain motor learning in unimanual and various bimanual movements as well as the transfer of learning effects between unimanual and various bimanual movements; these phenomena can be simultaneously explained if the mean activity of each primitive for various unimanual movements is balanced with the corresponding mean activity for various bimanual movements. Using this balanced condition, we can reproduce the results of prior behavioral and neurophysiological experiments. Furthermore, we demonstrate that the balanced condition can be implemented in a simple neural network model.

INTRODUCTION

In our daily lives, we flexibly switch from unimanual to bimanual movements and vice versa (e.g., we unimanually manipulate a smartphone and bimanually manipulate a tablet). Although motor learning in unimanual and bimanual movements has been intensively investigated, distinct theoretical frameworks have been proposed for each type of movement. In the motor primitive framework (Donchin, Francis, & Shadmehr, 2003; Takiyama, 2015; Takiyama, Hirashima, & Nozaki, 2015; Thoroughman & Shadmehr, 2000; Yokoi, Hirashima, & Nozaki, 2011), a theoretical framework for motor learning, neural activities $A_i(\theta)$ are nonlinearly determined by the desired movement direction θ, and a linear combination of these activities determines motor command:

$x(\theta) = \sum_{i=1}^{N} W_i A_i(\theta)$ (where N is the number of neurons). In this framework, the compatibility of nonlinear motor commands appropriate for nonlinear upper limb dynamics and linear learning curves in motor learning experiments, a characteristic of motor learning, can be explained (Donchin et al., 2003; Takiyama, 2015; Thoroughman & Shadmehr, 2000). An original motor primitive framework successfully reproduced the basic pattern of trial-dependent changes in the movement error and how motor learning can be generalized under changing kinematics (e.g., alterations in movement direction) (Donchin et al., 2003; Thoroughman & Shadmehr, 2000). The transfer of learning effects from a trained movement to other untrained movements is referred to as generalization. In the framework, the activities of motor primitives determine motor commands and a recruitment pattern of motor primitives is determined by the desired movement direction. An extended framework of motor primitives was proposed to reproduce the generalization pattern in bimanual movements (Yokoi et al., 2011); after training left arm movements with bimanual movements, generalization to other bimanual movements is restricted to similar kinematics of the left (trained) arm but also spread among wide-range kinematics of the right (untrained) arm. However, the distinct modeling of unimanual and bimanual movements cannot explain the generalization between the two types of movements. Learning effects in bimanual movements toward a fixed target direction are "partially" generalized to unimanual movements (Nozaki, Kurtzer, & Scott, 2006; Wang, Lei, Xiong, & Marek, 2013). A state space model (a model for trialdependent changes of motor commands) can explain this partial generalization (Nozaki & Scott, 2009). This model abstractly assumed that there are three different neural populations: one for unimanual, one for bimanual, and one for both unimanual and bimanual movements (i.e., a neural population for unimanual movements overlaps with that for bimanual movements). Although the overlap model predicts that the generalization between the two types of movements is always partial, learning effects in bimanual reaching movements toward eight directions are "perfectly" generalized to unimanual movements (Wang et al., 2013). The difference exists in the number of target directions during the bimanual training phase, but it is not clear why the number of training target directions affects the generalization between bimanual and unimanual movements.

Our recent model (Takiyama & Sakai, 2016), a balanced motor primitive framework, suggested a novel relationship between unimanual and bimanual movements. In the model, each motor primitive shows a different activity pattern for unimanual and bimanual movements, but the averaged activity across various unimanual movements equals the averaged activity across bimanual movements, which was termed unimanual–bimanual balance. Our proposal is that unimanual–bimanual balance determines the relationship between the two types of movements. However, the balanced motor primitive model assumed only parallel bimanual reaching movements in which the target directions are the same for the left and right arms, which leaves the relationship between unimanual and bimanual reaching movements toward various patterns of target directions unresolved. Generalizations of bimanual movements from one pattern to another pattern of target directions have been investigated in detail (Yokoi et al., 2011), but our conventional balanced motor primitive model failed to explain those generalization patterns. When considered together, we have not yet identified a single framework that can concurrently explain motor learning effects in unimanual movements, those in bimanual movements, and the generalization between the two types of movements.

Here, we extended the balanced motor primitive model to simultaneously explain motor learning in unimanual and in various bimanual movements as well as the generalization between those movements in a unified manner (within a single framework with an identical set of parameters). Our model is proposed based on the experimental results of the perfect generalization from bimanual to unimanual movements when bimanual movements for training trials are parallel or symmetrical (the sign of the target direction is different for the left and right arms ($\theta^{L} = -\theta^{R}$)) (Wang et al., 2013), which yields an extended version of the unimanual–bimanual balanced condition. The extended version of the balanced motor primitive model can successfully explain not only the results of behavioral experiments, motor learning in unimanual and in various bimanual movements, as well as the generalization between those movements, but also the results of neurophysiological experiments (Donchin et al., 2002; Rokni, Steinberg, Vaadia, & Sompolinsky, 2003). Furthermore, we demonstrate that the extended version of unimanual–bimanual balance can be implemented in a simple biologically inspired neural network model.

RESULTS

General Framework

The present study focused on reaching movements toward radially distributed target directions: $\theta_1, \ldots, \theta_K$. The target direction was randomly sampled from the K target directions in each trial.

During each reaching movement, an unpredictable perturbation was given, such as a force field (Shadmehr & Mussa-Ivaldi, 1994), which yielded a movement error e perpendicular to the movement direction

(Fig. 1(a), (c)). The aim of the task was to accurately reach toward a given target by generating additional motor command x perpendicular to the movement direction to compensate for the movement error.

Following the original motor primitive model (Donchin et al., 2003; Thoroughman & Shadmehr, 2000), we assumed that the motor command x was a linear summation of motor primitive activities $A_1(\theta), \ldots, A_N(\theta)$ that were determined by a target direction θ (Fig. 1(a), (c)) (i.e. $x = \sum_{i=1}^{N} W_i A_i(\theta)$, where W_i determined how the ith primitive contributed to generate the motor command). Each weight W_i was modified by $-\frac{\eta}{2}\frac{\partial e^2}{\partial W_i}$ (gradient descent rule) in each trial to reduce the squared movement error e^2 (see Methods), where the positive constant η denoted the learning rate. This framework could explain trial-dependent changes of the movement error and the generalization effects on untrained movements (Donchin et al., 2003; Thoroughman & Shadmehr, 2000).

The framework of motor primitives could be applied for bimanually reaching movements (Takiyama & Sakai, 2016; Yokoi et al., 2011). Throughout this study, we supposed a perturbation imposed only on the left arm (the left arm was trained, and the right arm was untrained). Therefore, the additional motor command x should be considered only for left arm movements. We considered various types of bimanual movements in which target directions for the left and right arms were defined as θ^L and θ^R, respectively. The present study assumed that the activity pattern of each motor primitive for bimanual movements was determined by $\theta^L \in \{\theta_1, \ldots, \theta_K\}$ and $\theta^R \in \{\theta_1, \ldots, \theta_K\}$, and for unimanual movement, it was determined by θ^L, $A_i^{\mathrm{uni}}(\theta^L)$. The weight value W_i was assumed to be common for unimanual and bimanual movements. This assumption did not mean common contributions of a weight W_i to unimanual and bimanual movements. The contribution of W_i depended on the primitive activities $A_i^{\mathrm{uni}}(\theta^L)$ in unimanual movements and $A_i^{\mathrm{bi}}(\theta^L, \theta^R)$ in bimanual movements (i.e., each motor primitive contributed $W_i A_i^{\mathrm{uni}}(\theta^L)$ or $W_i A_i^{\mathrm{bi}}(\theta^L, \theta^R)$ to the generation of motor command).

Unimanual–Bimanual Balance

The learning effect trained with a parallel bimanual movements ($\theta^L = \theta^R = \theta_1$) toward a fixed target direction, K = 1, was partially generalized to unimanual movements with $\theta^L = \theta_1$ (Nozaki et al., 2006; Wang et al., 2013), whereas the learning effect for K = 8 with $\theta^L = \theta^R$ (parallel bimanual movements) or $\theta^L = -\theta^R$ (symmetric bimanual movements) was perfectly generalized to unimanual movements (Wang et al., 2013). When the generalization from bimanual to unimanual movements was partial, movement error increased at the trial when bimanual movements were switched to unimanual movements. In contrast, when the generalization from bimanual to unimanual movements was perfect, movement error did not change at the trial when bimanual movements were switched to unimanual movements. The goal in this section was to analytically derive the condition to reconcile the partial generalization when K = 1 and the perfect generalization when K = 8. The learning speed for unimanual movements was not significantly different from that for bimanual movements (Tcheang, Bays, Ingram, & Wolpert, 2007). Under the assumption of equivalent learning speeds, we analytically proved that the generalization was perfect if and only if $\sum_{k=1}^{K} A_i^{\mathrm{uni}}(\theta_k) = \sum_{k=1}^{K} A_i^{\mathrm{bi}}(\theta_k, \theta_k) = \sum_{k=1}^{K} A_i^{\mathrm{bi}}(\theta_k, -\theta_k)$ for all primitives and was partial otherwise (see Methods). The partial generalization for K = 1 was observed in the case of a certain target direction θ (Nozaki et al., 2006; Wang et al., 2013);hence, activity patterns of motor primitives for unimanual and bimanual movements toward θ were different, i.e.,

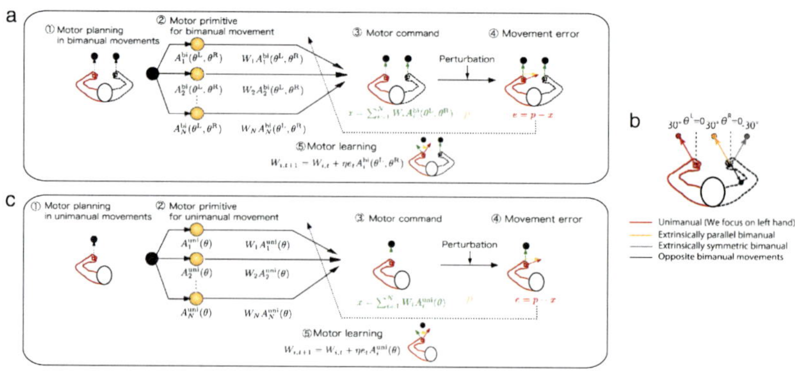

Figure. 1. Reaching tasks and motor primitive framework. (a) Motor primitive framework for bimanual movements. Target directions (desired movement

direction) θ^L for the left arm and θ^R for the right arm determine the activities of motor primitives $A_i(\theta^L, \theta^R)$ in motor planning, and the linear summation of primitive activities determines an additional motor command x to compensate for the movement error e induced by a given perturbation p. We focus on the case when the perturbation is applied only for left arms based on some behavioral experiments. The activities of motor primitives $A_i(\theta^L, \theta^R)$, as discussed in the current study, contribute to generate a motor command for the left arm. (b) In the unimanually reaching task, a target is given (red circle) for the left arm in the direction θ^L (red arrow). In the bimanually reaching task, two targets are given for the left and right arms in the directions θ^L and θ^R, respectively. The targets of parallel bimanual movements are given in the same direction (red and orange arrows, $\theta^L = \theta^R$), and the targets of symmetric bimanual movements are given in the symmetric directions (red and gray arrows, $\theta^L = -\theta^R$). (c): Motor primitive framework for unimanual movements. We focus on left arm movements, similar to some behavioral experiments. (For interpretation of the references to color in this figure legend, the reader is referred to the web version of this article.)

$$A_i^{\text{uni}}(\theta) \neq A_i^{\text{bi}}(\theta, \theta) \neq A_i^{\text{bi}}(\theta, -\theta), \tag{1}$$

for some primitives. The perfect generalization for K = 8 was observed in the case of equally distributed targets, $\theta_k = 2\pi k/8$ (Wang et al., 2013). Hence, the condition for perfect generalization implied that the integrals of tuning curves might be balanced,

$$\begin{aligned}\frac{1}{2\pi}\int_{-\pi}^{\pi} A_i^{\text{uni}}(\theta)d\theta &= \frac{1}{2\pi}\int_{-\pi}^{\pi} A_i^{\text{bi}}(\theta, \theta)d\theta \\ &= \frac{1}{2\pi}\int_{-\pi}^{\pi} A_i^{\text{bi}}(\theta, -\theta)d\theta,\end{aligned} \tag{2}$$

for all primitives i = 1, . . . , N. This unimanual–bimanual balance represented that, for unimanual and bimanual movements, tuning patterns for each motor primitive might be different, but the integrals of tuning curves should be balanced. Notably, in our previous study, unimanual–bimanual balance was considered only for unimanual and parallel bimanual movements. Here, we extended the framework to include symmetrical bimanual movements, which enabled us to construct $A_i^{\text{bi}}(\theta^L, \theta^R)$ and $A_i^{\text{uni}}(\theta^L)$ A bi i (θ L , θ R) and A uni i (θ L) to satisfy the balanced condition (see below and Methods). In other words, the extended version of unimanual–bimanual balance allowed us to consider motor primitive activities in various bimanual movements. Eqs. (1) and (2) were the only plausible candidates for reconciling the partial generalization when K = 1, the perfect generalization when K = 8, and the equivalent learning speeds. Additionally, weight values for unimanual and bimanual movements needed to be common

$$W_i^{\mathrm{uni}} = W_i^{\mathrm{bi}} = W_i \tag{3}$$

for all primitives to reconcile those phenomena. Although these conclusions were obtained by assuming Eq. (3), we could achieve the same conclusions (i.e., Eqs. (1), (2), and (3)) if we started the analysis with assuming Eq. (2) and without assuming Eq. (3) (see our previous publication Takiyama & Sakai, 2016). Therefore, the three previous equations were theoretically validated conditions to reproduce the partial transfer, the perfect transfer, and the equivalent learning speed. Inspired mainly by Eq. (2), we referred to these conditions as extended unimanual–bimanual balance in the following.

Activities of each Motor Primitive to Satisfy the Extended Unimanual–Bimanual Balance

Our goal was to clarify the explicit forms of $A_i^{\mathrm{bi}}(\theta^{\mathrm{L}}, \theta^{\mathrm{R}})$ and $A_i^{\mathrm{uni}}(\theta^{\mathrm{L}})$ to satisfy the extended unimanual–bimanual balance. We analytically derived the explicit form of $A_i^{\mathrm{bi}}(\theta^{\mathrm{L}}, \theta^{\mathrm{R}})$ to satisfy the extended unimanual–bimanual balance; the condition $\frac{1}{2\pi}\int_{-\pi}^{\pi} A_i^{\mathrm{bi}}(\theta, \theta)d\theta = \frac{1}{2\pi}\int_{-\pi}^{\pi} A_i^{\mathrm{bi}}(\theta, -\theta)d\theta$ could be satisfied when

$$A^{\mathrm{bi}}(\theta^{\mathrm{L}}, \theta^{\mathrm{R}}) = h(\theta^{\mathrm{L}}, \theta^{\mathrm{R}}) - \frac{1}{2\pi}\int_{-\pi}^{\pi} h(\theta^{\mathrm{L}} - z, \theta^{\mathrm{R}} - z)dz - \frac{1}{2\pi}\int_{-\pi}^{\pi} h(\theta^{\mathrm{L}} - z, \theta^{\mathrm{R}} + z)dz, \tag{4}$$

where $h(\theta^{\mathrm{L}}, \theta^{\mathrm{R}})$ is an arbitrary two-dimensional periodic function $h(\theta^{\mathrm{L}}, \theta^{\mathrm{R}}) = h(\theta^{\mathrm{L}} + 2\pi, \theta^{\mathrm{R}}) = h(\theta^{\mathrm{L}}, \theta^{\mathrm{R}} + 2\pi)$ (proof was given in Methods section). In the current study, we constructed an example solution $A_i^{\mathrm{bi}}(\theta^{\mathrm{L}}, \theta^{\mathrm{R}})$ when $h(\theta^{\mathrm{L}}, \theta^{\mathrm{R}})$ was a multiplicative Gaussian tuning curve, which could be well fit to the experimental observation in a previous study (Yokoi et al., 2011). Eq. (4) indicated that standard functions (e.g., a two-dimensional multiplicative Gaussian) could not satisfy Eq. (2), and some modification terms were necessary. The colormap in Fig. 2(a) shows an example of motor primitive activities.

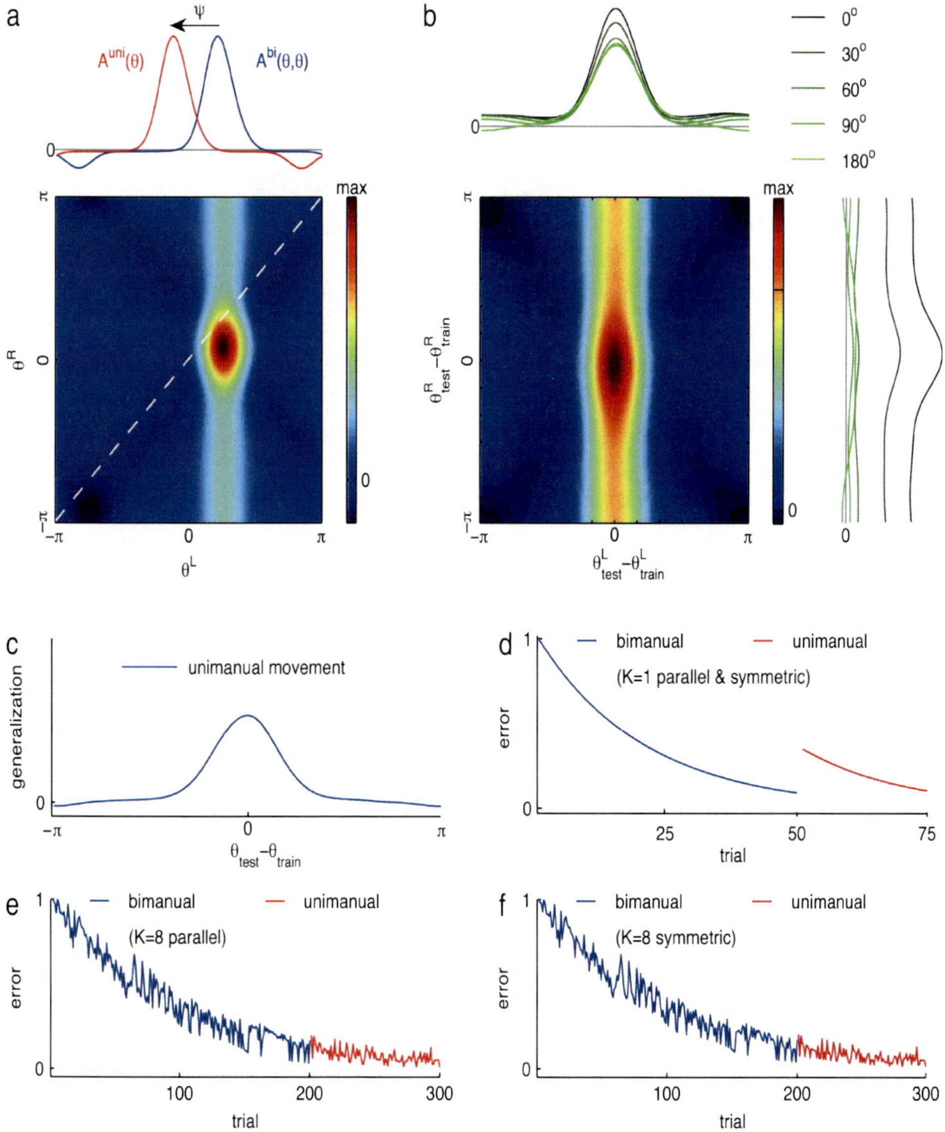

Figure. 2. Properties of extended unimanual–bimanual balance. (a) Examples of a two-dimensional tuning curve for bimanual movements $A_i(\theta^L, \theta^R)$ (pseudo color plot normalized from the minimum to the maximum) and a tuning curve for unimanual movements $A_i^{uni}(\theta^L)$ (red line in the upper plot) that satisfy the extended unimanual–bimanual balance. The cross section

through the diagonal line (white dashed) that corresponds parallel bimanual movements $(\theta^{L} = \theta^{R}), A_i^{bi}(\theta, \theta),$, was shown as the blue line in the upper plot. The unimanual tuning curve $A_i^{uni}(\theta)$ (red line) was obtained by shifting $A_i^{bi}(\theta, \theta)$ with ψ_i. (b) Generalization from a bimanual movement toward θ^{L}_{train} and θ^{R}_{train} (set as $\theta^{L}_{train} = \theta^{R}_{train} = 0$) to various patterns of bimanual movements toward θ^{L}_{test} and θ^{R}_{test}. Cross sections at 0°, 30°, 60°, 90° and 180° for $\theta^{L}_{test} - \theta^{L}_{train}$ and for $\theta^{R}_{test} - \theta^{R}_{train}$ train are shown in the right and upper plots, respectively. (c) Generalization effects from a unimanual movement toward θ^{L}_{train} (set as $\theta^{L}_{train} = 0$) to various patterns of unimanual movements toward θ^{L}_{test}. (d) Trial-dependent changes of movement errors of bimanual movements toward a fixed target direction (K = 1 and $\theta_1 = 0$), which is both parallel and symmetric. After 50 trials of the bimanual task (blue line), the task switched to the unimanual task (red line). Generalization was partial. (e, f) Trial-dependent changes of movement errors of parallel (e) and symmetric (f) bimanual movements toward eight target directions (K = 8). After 200 trials of the bimanual task (blue lines), the task switched to the unimanual task (red lines). Generalization was perfect in both cases. (For interpretation of the references to color in this figure legend, the reader is referred to the web version of this article.)

The extended unimanual–bimanual balance could determine the relationship between unimanual and bimanual movements based on the equations $\frac{1}{2\pi}\int_{-\pi}^{\pi} A_i^{uni}(\theta)d\theta = \frac{1}{2\pi}\int_{-\pi}^{\pi} A_i^{bi}(\theta, \theta)d\theta$ for parallel bimanual movements and $\frac{1}{2\pi}\int_{-\pi}^{\pi} A_i^{uni}(\theta)d\theta = \frac{1}{2\pi}\int_{-\pi}^{\pi} A_i^{bi}(\theta, -\theta)d\theta$ for symmetric bimanual movements. Here, we focused on the case when the activity of each primitive was described with a common tuning curve g(·) for unimanual and parallel bimanual movements: $A_i^{uni}(\theta) = \alpha_i^{uni} g(\theta - \varphi_i^{uni}),$ and $A_i^{bi}(\theta, \theta) = \alpha_i^{bi} g(\theta - \varphi_i^{bi}),$ where the amplitude α_i and preferred direction (PD) ϕ_i are parameters for the ith primitive, and g(·) is a periodic function, $g(\theta) = g(\theta + 2\pi)$. Under the assumption of a common tuning curve, unimanual–bimanual balance was equivalent to the conditions $\alpha_i^{uni} = \alpha_i^{bi}$ for all primitives and $\varphi_i^{uni} \neq \varphi_i^{bi}$ for some primitives (Takiyama & Sakai, 2016). Namely, the PD, not the amplitude, was modulated by the opposite arm movements.

We referred to this condition as PD modulation. Notably, these analytical results were independent of the shape of the tuning curve and the distribution of PDs and difference of PDs in unimanual and bimanual movements.

Under the assumption of a common tuning curve, the relationship between unimanual and bimanual movements could be determined by PD modulation. In parallel bimanual movements, motor primitive activities were determined by $A_i^{\mathrm{bi}}(\theta,\theta)$) in Eq. (4). These activities could be fit well by using a tuning curve g: $A_i^{\mathrm{bi}}(\theta,\theta) \simeq \alpha_i g(\theta-\varphi_i^{\mathrm{bi}})$ (the white dotted line in Fig. 2(a)). After this fit, the PD modulation determined the motor primitive activities in unimanual movements; $A_i^{\mathrm{uni}}(\theta) = \alpha_i g(\theta - \varphi_i^{\mathrm{uni}}) = \alpha_i g(\theta - (\varphi_i^{\mathrm{bi}} + \psi_i))$ (the blue and red lines in the top panel of Fig. 2(a)). We conducted numerical simulations by using $A_i^{\mathrm{bi}}(\theta^{\mathrm{L}},\theta^{\mathrm{R}})$ and $A_i^{\mathrm{uni}}(\theta^{\mathrm{L}})$ as noted above.

Reproduction of the Generalization between Unimanual and Bimanual Movements

To validate the theoretical analysis, we simulated the generalization from bimanual to unimanual movements. After task switching from the bimanual task (blue lines in Fig. 2(d), (e), (f)) to the unimanual task (red lines in Fig. 2(d), (e), (f)), the errors at the initial unimanual movements showed smaller values than did those at the initial bimanual movements. The learning effects of bimanually training were generalized to untrained unimanual movements. Under the condition of the fixed target directions (K = 1), the error in the first trial of the unimanual task was larger than the error in the last trial of the bimanual task (Fig. 2(d)). This meant that the generalization effects were partial. Thus, the extended unimanual–bimanual balance reproduced the experimental results under the condition of the fixed target directions (K = 1) (Nozaki et al., 2006; Wang et al., 2013). Under the condition that a target direction in each trial was sampled from the eight radial directions (K = 8), the generalization from parallel and symmetrical bimanual to unimanual movements was perfect (Fig. 2(e), (f)). Our previous balanced motor primitive (Takiyama & Sakai, 2016) could not reproduce the perfect generalization from symmetrical bimanual to unimanual movements. We confirmed that our extended unimanual–bimanual balance could concurrently reproduce partial generalization when K = 1 (Nozaki et al., 2006; Wang et al., 2013) and perfect generalization for K = 8 with parallel or symmetric bimanual movements (Wang et al., 2013); these findings validated our analytical results.

Reproduction of Motor Learning in Various Bimanual Movements

To confirm whether Eq. (4) could reproduce generalization from one bimanual movement to various other bimanual movements, we numerically calculated a generalization from the bimanual movement with $(\theta^L, \theta^R) = (\theta^L_{train}, \theta^R_{train}) = (0, 0)$ to other bimanual movements with $(\theta^L, \theta^R) = (\theta^L_{test}, \theta^R_{test})$ (Fig. 2(b)). In this section, the degree of generalization is denoted as $f((\theta^L_{test}, \theta^R_{test}), (0, 0))$. The generalization pattern with fixed θ^R and varied θ^L, such as $(\theta^L, \theta^R) = (\theta^L_{test}, 0), (\theta^L_{test}, 30), \ldots$ is shown in the cross sections in the upper portion of Fig. 2(b). In this case, the generalization pattern showed multiplicative modulation, i.e., $f((\theta^L_{test}, 30), (0, 0)) = C_{30} f((\theta^L_{test}, 0), (0, 0))$ $f((\theta^L_{test}, 60), (0, 0)) = C_{60} f((\theta^L_{test}, 0), (0, 0)), \ldots,$ where $C_{30}, C_{60}.$ are constants. In contrast, the generalization pattern with fixed θ^L and varied θ^R, such as $(\theta^L, \theta^R) = (0, \theta^R_{test}), (30, \theta^R_{test}), \ldots,$ is shown in the cross sections on the right side of Fig. 2(b), and the generalization pattern showed both multiplicative and additive modulations, i.e., $f((30, \theta^R_{test}), (0, 0)) = D_{30} f((0, \theta^R_{test}), (0, 0)) + E_{30}$, $f((60, \theta^L_{test}), (0, 0)) = D_{60} f((0, \theta^L_{test}), (0, 0)) + E_{60}$, $D_{30}, D_{60}, \ldots$ and $E_{30}, E_{60}, \ldots$ are constants. These were qualitatively equivalent to the observation reported in the previous study (Yokoi et al., 2011). Therefore, unimanual–bimanual balance could reproduce the qualitative characteristics of the generalization pattern within various bimanual movements.

Reproduction of Motor Learning in Unimanual Movements

PD modulation could reproduce the generalization from one unimanual movements with $\theta^L = \theta_{train}$ to other unimanual movements with $\theta^L = \theta_{test}$. The generalization function within unimanual movements was frequently approximated by a Gaussian

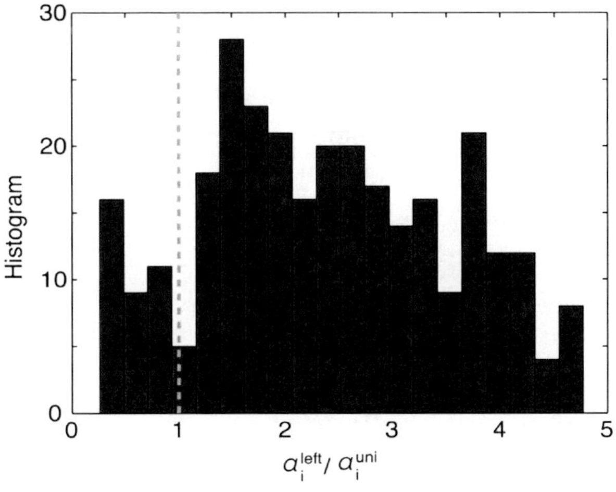

Figure. 3. The ratio of α_i^{left} to α_i^{uni}. If the amplitude was not estimated to be modulated between unimanual and bimanual movements, the ratio should be 1 (dotted vertical line). Amplitudes α_i^{left} estimated by fitting a statistical model α left $\alpha_i^{\text{left}} g(\theta_l - \varphi_i) + \alpha_i^{\text{right}} g(\theta_r - \phi_i)$ to the activities of motor primitives shown in Fig. 2(a) when $|\theta_l - \theta_r| = 0$ (parallel bimanual movements) and $|\theta_l - \theta_r| = \pi$ (opposite bimanual movements). The tuning curve g in the statistical model was determined based on the activities of motor primitives in parallel bimanual movements (Fig. 2(a)). Because the primitive activities in unimanual movements were $A_i^{\text{uni}}(\theta) = \alpha_i g(\theta - \varphi_i^{\text{uni}})$ and those in parallel bimanual movements were $A_i^{\text{bi}}(\theta, \theta) = \alpha_i g(\theta - \varphi_i^{\text{bi}})$., the amplitudes were not modulated. However, the amplitudes were estimated to be modulated between unimanual and bimanual movements.

(Donchin et al., 2003; Thoroughman & Shadmehr, 2000; Yokoi et al., 2011), and the reproduced generalization function by our model could also be approximated by a Gaussian (Fig. 3(c)). The extended unimanual–bimanual balance reproduced motor learning within unimanual movements.

Relation to Neurophysiological Results

Although unimanual–bimanual balance, or PD modulation, could reproduce various types of results in behavioral experiments, it remained unclear whether unimanual–bimanual balance was consistent with neurophysiological results. In particular, a prior study reported that both

PD and amplitude were modulated between unimanual and bimanual movements (Rokni et al., 2003); at a glance, this finding appears to be inconsistent with our notion of PD modulation. In this previously published study, neural activities were investigated in parallel ($|\theta_l - \theta_r| = 0°$) and opposite ($|\theta_l - \theta_r| = 180°$) (Fig. 1(b)) bimanual movements. To confirm whether this prior report actually contradicts PD modulation as defined in the present study, we applied the same procedure used in this previous investigation to our model and examined whether we also reached the conclusion that amplitude was modulated. Additionally, the statistical models $A_i^{\text{bi,stat}}(\theta_l, \theta_r) = \alpha_i^{\text{left}} g(\theta_l - \varphi_i^{\text{left}}) + \alpha_i^{\text{right}} g(\theta_r - \varphi_i^{\text{right}})$ in bimanual movements and $A_i^{\text{uni,stat}}(\theta) = \alpha_i^{\text{uni}} g(\theta - \varphi_i^{\text{uni}})$ in unimanual movements were fit to recorded neural activities, where g is a tuning curve common to unimanual and bimanual movements; $\alpha_i^{\text{left}}, \alpha_i^{\text{right}}$, and α_i^{uni} are amplitudes for left arm movements, right arm movements, and unimanual movements, respectively; and $\varphi_i^{\text{left}}, \varphi_i^{\text{right}}$, and φ_i^{uni} are the PDs for left arm movements, right arm movements, and unimanual movements, respectively. Similarly, we fit the same statistical model to $A_i^{\text{bi}}(\theta, \theta)$ and $A_i^{\text{bi}}(\theta, \theta + \pi)$ as well as $A_i^{\text{uni}}(\theta)$, , as determined by PD modulation (Fig. 2(a)). In this fitting, the tuning curve g was determined by $A_i^{\text{uni}}(\theta)$.

Fig. 3 represents the histogram of $\alpha_i^{\text{left}}/\alpha_i^{\text{uni}}$ estimated using the same statistical model. We confirmed that the ratios of estimated amplitudes were not concentrated around 1 ($\alpha_i^{\text{left}} \simeq \alpha_i^{\text{uni}}$) but were instead widely distributed. This finding indicates that amplitudes of many primitives were estimated to be modulated. Although PD modulation determined a relationship between unimanual and bimanual movements (i.e., amplitude was held constant between unimanual and bimanual movements), amplitude was also predicted to be modulated between unimanual and bimanual movements when the same procedure used in previous neurophysiological experiments was applied (Rokni et al., 2003) (Fig. 3). Thus, we confirmed that the extended unimanual–bimanual balance, or PD modulation, found in our model does not conflict with results from the aforementioned prior study.

Neural Implementation of Unimanual–Bimanual Balance

Unimanual–bimanual balance enabled us to simultaneously reproduce results of behavioral and neurophysiological experiments (Figs. 2 and 3).

However, it remained unclear whether unimanual–bimanual balance was biologically plausible. Here, we attempted to confirm the plausibility by constructing a biologically inspired neural network model that could achieve extended unimanual–bimanual balance. Extended unimanual–bimanual balance could be achieved by interaction between the left and right hemispheres through the corpus callosum. In a primary motor cortex, a neuron received input from a small number of neurons through the corpus callosum, or the corpus callosum connectivities were sparse in a primary motor cortex (Rouiller et al., 1994). We supposed a simple neural network model that consisted of two neural populations to control left and right arm movements in the right and left motor cortices that interact with each other through sparse connections (Fig. 4(a), see Methods for details). The firing rate of each neuron was determined by the membrane potential through a standard rate model described as a sigmoid function. Each neuron received feedforward input that was determined by a given target direction, θ L or θ R .When no target direction was given, the feedforward input was zero (e.g. in the case of a unimanual movement by the left arm) and the neural population to control the right arm received no feedforward input. The neurons in the right and left cortices interacted through sparse connections in which the excitation and inhibition were balanced. For a given pattern of target directions, the dynamics of the firing rates of the neurons approached asymptotic values (Fig. 4(b)). By using these values, we obtained tuning curves of the neurons to various patterns of target directions. As shown in Fig. 4(c), tuning curves of neurons were modulated by the opposite arm target. To validate the unimanual–bimanual balance defined in Eq. (2), we compared the mean activity of each neuron for unimanual movements with the mean activities for various bimanual movements: parallel (Fig. 4(d)) and symmetric (Fig. 4(e)) (see Methods). These activities were all approximately balanced. Based on the primitive activities $A_i^{uni}(\theta^L)$ and $A_i^{bi}(\theta^L, \theta^R)$ determined by the firing rates of the neurons in the neural network model, we demonstrated the behavioral reaching tasks in the same manner as Fig. 2. The neural network model also reproduced the partial generalization when K = 1 (Fig. 4(f)) and the perfect generalization when K = 8 (Fig. 4(g), (h)).

DISCUSSION

We successfully extended our conventional balanced motor primitive framework (Takiyama & Sakai, 2016) to reconcile the partial generalization from bimanual to unimanual movements (Fig. 2(d)) when K = 1 (Nozaki et al., 2006; Wang et al., 2013) and the perfect generalization from parallel and symmetric bimanual to unimanual movements when K = 8 (Wang et al., 2013) (Fig. 2(e) and (f)). The balanced motor primitive could reproduce not only the generalizations between unimanual and bimanual movements but also motor learning in unimanual movements (i.e., the generalization from one unimanual movement to other unimanual movements (Donchin et al., 2003; Thoroughman & Shadmehr, 2000), Fig. 2(c)), motor learning in bimanual movements (i.e., the generalization from one bimanual movement to other various bimanual movements (Yokoi et al., 2011), Fig. 2(a)). The extended unimanual–bimanual balance could determine the relationship between unimanual and various bimanual movements based on PD modulation (Fig. 2(b)). Furthermore, the balanced motor primitive that satisfies the extended unimanual–bimanual balance (Eq. (4)) could reproduce the result of a neurophysiological experiment (Rokni et al., 2003) (Fig. 3). Unimanual–bimanual balance could be implemented in a simple neural network model in which the left and right cortices interact with each other (Fig. 4). The simple neural network model could reproduce the learning effects and generalization of unimanual and bimanual movements.

Comparison with Conventional Models

Conventional frameworks of motor primitive successfully reproduced the generalization pattern when the kinematics of reaching movements (e.g., target direction) changed in unimanual (Donchin et al., 2003; Takiyama, 2015; Thoroughman & Shadmehr, 2000) or bimanual movements (Yokoi et al., 2011). However, those models could explain the generalization pattern only in unimanual movements or those only in bimanual movements. In the model for unimanual movements, some motor primitives were activated when the desired movement direction is 0°, but how these primitives were activated in bimanual movements remained unclear. In the model for bimanual movements, some motor primitives were activated when the desired movement direction for left and right arms were 0°, but it remained unknown how these primitives were activated in unimanual movements. A candidate for the answers was an overlap: partial generalization between unimanual and bimanual movements could be reproduced by an overlap of a neural population for unimanual movements and a neural population for bimanual movements (Nozaki & Scott, 2009). In that model, learning effects were embedded into a non-overlapped and an overlapped neural population. Learning effects in the overlapped neural population were shared between unimanual and bimanual movements, thereby resulting in partial generalization between those movements. Nevertheless, based on the overlap model, the generalization was always partial independent of the number of targets, which contradicts to the results of behavioral experiments (Takiyama & Sakai, 2016; Wang et al., 2013). To reconcile the partial and perfect generalizations, we have proposed a previous version of balanced motor primitive. However, the previous framework considered only parallel bimanual movements in which $\theta^{L} = \theta^{R}$, and it was impossible to simulate the generalization from symmetric bimanual to unimanual movements and motor learning in various bimanual movements. When considered together, the extended unimanual–bimanual balance was the first framework that could simultaneously explain all of these phenomena in a unified manner.

Relation to Neural Activities

PD modulation appeared contradictory to a previous neurophysiological result, wherein both the PD and amplitude were modulated between unimanual and bimanual movements (Rokni et al., 2003). However, PD modulation was a simple solution of unimanual–bimanual balance assuming a common tuning curve. As long as it is balanced, the amplitude modulation was also allowed

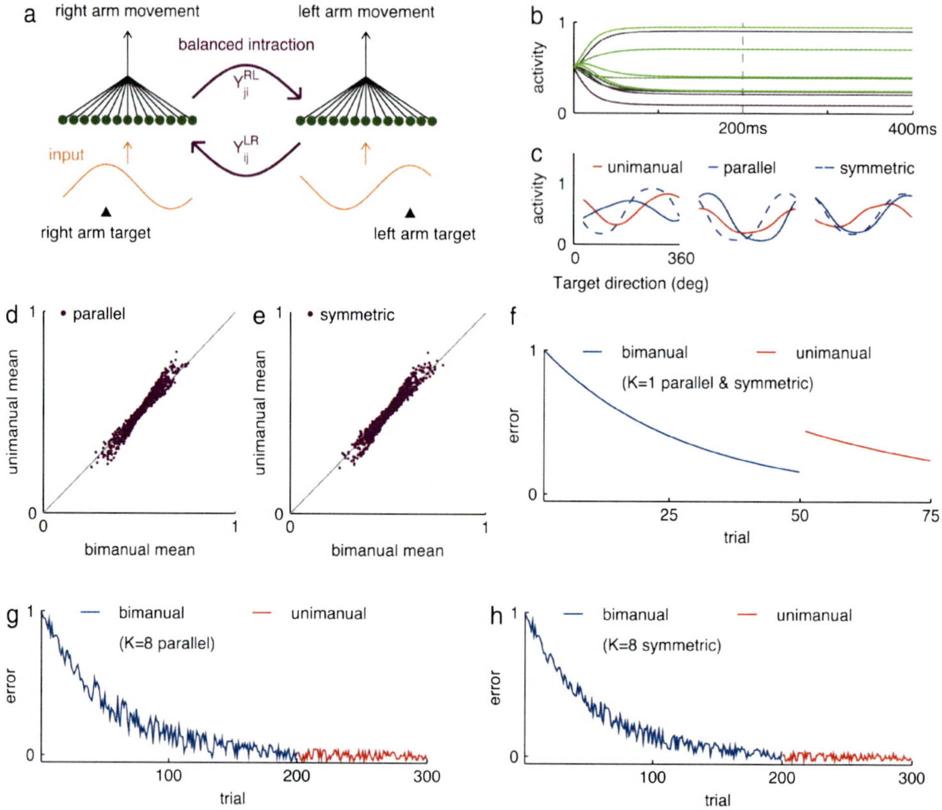

Figure. 4. Neural implementation of unimanual–bimanual balance. (a) Schematic diagram of a neural network model. Two neural populations (N = 1000 for each, green circles) to control left and right arm movements in right and left motor cortices were assumed. Each neuron received feedforward input (orange curves) determined by a given target direction, θ^L or θ^R (triangles). The neurons in the right and left cortices interacted with the sparse connections Y_{ij}^{LR} and Y_{ji}^{RL} (purple arrows), respectively. (b) Time courses of the firing rates of 10 example neurons (solid curves). The values approached asymptotic values in approximately 100 ms. The values at 200 ms (vertical dashed line) were used as the activities of the neurons. (c) Tuning curves of three example neurons to target directions for unimanual movements (red lines), parallel bimanual movements (blue solid lines), and symmetric bimanual movements (blue dashed lines). (d, e) Unimanual–bimanual balance in the neural network model. The mean activity of each neuron for unimanual movements was plotted to the mean for various distributions of bimanual movements: parallel (d) and symmetric (e) bimanual movements. The dots were around the diagonal line (gray lines). Therefore, the unimanual–bimanual balance was approximately satisfied for various distribu-

tions of bimanual target directions. ((f)–(h)) Trial-dependent changes of movement errors and generalization effects in the behavioral reaching task based on the primitive activities determined by the neuronal activities in the neural network model. The task procedures and analyses were the same as those in Fig. 2(d)–(f). (For interpretation of the references to color in this figure legend, the reader is referred to the web version of this article.)

if the shape of tuning curve was also modulated. Moreover, we applied the model fitting procedure in the previous study (Rokni et al., 2003) to $A_i^{bi}(\theta^L, \theta^R)$ in Fig. 2 and found that amplitude was also estimated to be modulated (Fig. 3). Hence, the unimanual–bimanual balance was consistent with the neurophysiological result. In particular, the PD modulation proposed by previous studies (Rokni et al., 2003; Takiyama & Okada, 2012) was consistent with the neurophysiological result.

Acquiring the Unimanual–Bimanual Balance

We demonstrated that unimanual–bimanual balance could be implemented in a simple neural network model. However, how the neural network could acquire unimanual–bimanual balance remained unclear. In other words, what type of learning rules enable to achieve unimanual–bimanual balance remained unclear. We perform various bimanual movements in our daily life. Bimanually reaching or manipulating an extrinsic object requires extrinsically cooperative movements, whereas maintaining body balance requires intrinsically cooperative movements. Through various movements in our daily life, each motor primitive may keep the balance of mean activity. It remains unclear what type of synaptic plasticity can achieve unimanual–bimanual balance. Notably, dense and Hebbian learned callosal connections are unable to satisfy unimanual–bimanual balance (Takiyama, Naruse, & Okada, 2009), which suggests that sparsity of callosal connections is likely required. We must further investigate how to acquire unimanual–bimanual balance in a future project.

Inter-Limb Transfer

Motor learning involving both the right and left arms has been investigated. A phenomenon that has been intensively investigated is inter-limb transfer: The training of a new motor task with one arm can improve the performance of this task with the other hand (Sainburg&Wang, 2002). Our balanced motor primitive framework can be expected to be applicable to a transfer if the number of targets affects the degree of transfer. However, a larger number of training targets does not affect transfers (Wang & Sainburg, 2004). Other theoretical frameworks should therefore be proposed to explain inter-limb transfer.

METHODS

Learning Rule and Generalization Function

The movement error feedback e perpendicular to the movement direction was supposed to be proportional to the difference between the given perturbation p (p was set to 1 or -1 in this study) and the generated motor command x, $e \propto p - x$. The motor command x generated for a target direction θ was described as $x(\theta) = \sum_{i=1}^{N} W_i A_i(\theta) = \boldsymbol{W} \cdot \boldsymbol{A}(\theta)$, where the vectors represented $\boldsymbol{W} = (W_1, \ldots, W_N)$ and $\boldsymbol{A}(\theta) = (A_1(\theta), \ldots, A_N(\theta))$ respectively, and the dot · denoted the inner product of vectors. The gradient of the square error e^2 for a weight Wi was obtained as $\frac{\partial e^2}{\partial W_i} \propto -2eA_i(\theta)$. Thus, the gradient learning rule in trial t was obtained as $W_{t+1} = W_t + \eta e A(\theta)$, where error feedback e and target direction θ were given in trial t. A positive constant η represents the learning rate. This learning rule enabled us to predict the availability of learning effects from the tth trial with θ during the (t + 1)th trial with θ ′, where θ ′ (which may differ from θ) is a target direction to be presented at the trial. Multiplying A(θ′) to the learning rule, we obtained $x_{t+1}(\theta') = x_t(\theta') + \eta e_t \boldsymbol{A}(\theta') \cdot \boldsymbol{A}(\theta)$, where $x_t(\theta')$ is the motor command to be generated for a target θ ′. Because the motor command $x_t(\theta')$ was unobservable, it indicated a predicted value. When $A(\theta) \cdot A(\theta') = 0$, $x_{t+1}(\theta') = x_t(\theta')$ indicated that learning effects from the tth trial with θ were not available during the next trial θ ′. In contrast, when $A(\theta) \cdot A(\theta') = A(\theta) \cdot A(\theta)$, $x_{t+1}(\theta') - x_t(\theta') = x_{t+1}(\theta) - x_t(\theta)$ indicated that learning effects from the tth trial with θ were completely available during the next trial with θ ′. We referred to this perfect availability of learning effects as perfect generalization. The case when θ ′ = θ was a self-evident case of perfect generalization. Furthermore, when $A(\theta) \cdot A(\theta') < A(\theta) \cdot A(\theta)$, $x_{t+1}(\theta') - xt(\theta') < x_{t+1}(\theta) - x_t(\theta)$ indicated that learning effects from the tth trial with θ were partially available during the next trial with θ ′. We referred to this partial availability of learning effects as partial generalization. Thus, the generalization of learning was determined by the factor A(θ′)·A(θ). The factor for the same direction A(θ) · A(θ) = |A(θ)| 2 determines the learning speed. This could be applied for generalization between unimanual and bimanual movements and that from one bimanual movement to other various bimanual movements (i.e., the generalization from one unimanual movement to bimanual movements) was determined by the factor $\boldsymbol{A}^{\text{bi}}(\theta^{\text{L}}, \theta^{\text{R}})\cdot\boldsymbol{A}^{\text{uni}}(\theta)$ and the generalization from one bimanual with $(\theta^{\text{L}}, \theta^{\text{R}}) = (\theta^{\text{L}}_{\text{train}}, \theta^{\text{R}}_{\text{train}})$ is determined by $\boldsymbol{A}^{\text{bi}}(\theta^{\text{L}}_{\text{test}}, \theta^{\text{R}}_{\text{test}})\ \boldsymbol{A}^{\text{bi}}(\theta^{\text{L}}_{\text{train}}, \theta^{\text{R}}_{\text{train}})$

Equivalent Condition to Perfect or Partial Generalization

The generalization of learning effects per trial in bimanual movements toward multiple training directions was determined by the average over all training target directions, $\frac{1}{K}\sum_{k=1}^{K}\boldsymbol{A}\cdot\boldsymbol{A}^{\text{bi}}(\theta_k, \theta_k)$. The learning effect and generalization effect were experimentally evaluated as the average over the test target directions which were common to the training directions (Wang et al., 2013). Therefore, the average effects of bimanual learning and generalization to unimanual movements were determined by $\frac{1}{K^2}\sum_{k'=1}^{K}\sum_{k=1}^{K}\boldsymbol{A}^{\text{bi}}(\theta_{k'}, \theta_{k'})\cdot\boldsymbol{A}^{\text{bi}}(\theta_k, \theta_k)$ and $\frac{1}{K^2}\sum_{k'=1}^{K}\sum_{k=1}^{K}\boldsymbol{A}^{\text{uni}}(\theta_{k'})\cdot\boldsymbol{A}^{\text{bi}}(\theta_k, \theta_k)$, respectively. Comparisons between these two values determined whether the generalization from bimanual to unimanual movements was perfect or partial. Under the assumption of equivalent learning speeds,

$\left|\sum_{k=1}^{K}\boldsymbol{A}^{\text{uni}}(\theta_k)\right|^2 = \left|\sum_{k=1}^{K}\boldsymbol{A}^{\text{bi}}(\theta_k, \theta_k)\right|^2$, we obtained

$$\sum_{k'=1}^{K}\boldsymbol{A}^{\text{uni}}(\theta_{k'})\cdot\sum_{k=1}^{K}\boldsymbol{A}^{\text{bi}}(\theta_k, \theta_k) \le \left|\sum_{k'=1}^{K}\boldsymbol{A}^{\text{uni}}(\theta_{k'})\right|\left|\sum_{k=1}^{K}\boldsymbol{A}^{\text{bi}}(\theta_k, \theta_k)\right|$$
$$= \left|\sum_{k=1}^{K}\boldsymbol{A}^{\text{bi}}(\theta_k)\right|^2,$$

by using the Cauchy–Schwarz inequality. The equality condition of the Cauchy–Schwarz inequality was $\sum_{k'=1}^{K}\boldsymbol{A}^{\text{uni}}(\theta_{k'}) = \sum_{k=1}^{K}\boldsymbol{A}^{\text{bi}}(\theta_k, \theta_k)$. Thus, under the assumption of equivalent learning speeds, the perfect generalization occurred if and only if

$$\sum_{k'=1}^{K} \boldsymbol{A}^{\mathrm{uni}}(\theta_{k'}) = \sum_{k=1}^{K} \boldsymbol{A}^{\mathrm{bi}}(\theta_k, \theta_k), \tag{5}$$

and otherwise the partial generalization occurred. Similarly, the perfect generalization from symmetric bimanual to unimanual movements occurred if and only if

$$\sum_{k'=1}^{K} \boldsymbol{A}^{\mathrm{uni}}(\theta_{k'}) = \sum_{k=1}^{K} \boldsymbol{A}^{\mathrm{bi}}(\theta_k, -\theta_k). \tag{6}$$

Generalized Bimanual–Unimanual Balance

Unless only the parallel and symmetric bimanual movements were the special cases, unimanual–bimanual balance might hold true for various patterns of reaching movements. We extended unimanual–bimanual balance for extrinsically cooperative movements, including parallel movements, such that the probability distribution of target directions (θ^{L}, θ^{R}) was determined by the difference $\theta^{\mathrm{L}} - \theta^{\mathrm{R}}$ and for intrinsically cooperative movements, including symmetric movements, such that the probability distribution of target directions (θ^{L}, θ^{R}) was determined by the sum $\theta^{\mathrm{L}} + \theta^{\mathrm{R}}$. The extended unimanual–bimanual balance is described as the condition

$$\begin{aligned}\frac{1}{2\pi}\int_{-\pi}^{\pi} A_i^{\mathrm{uni}}(\theta^{\mathrm{L}})d\theta^{\mathrm{L}} &= \int_{-\pi}^{\pi}\int_{-\pi}^{\pi} A_i^{\mathrm{bi}}(\theta^{\mathrm{L}}, \theta^{\mathrm{R}})p(\theta^{\mathrm{L}} - \theta^{\mathrm{R}})d\theta^{\mathrm{R}}d\theta^{\mathrm{L}} \\ &= \int_{-\pi}^{\pi}\int_{-\pi}^{\pi} A_i^{\mathrm{bi}}(\theta^{\mathrm{L}}, \theta^{\mathrm{R}})q(\theta^{\mathrm{L}} + \theta^{\mathrm{R}})d\theta^{\mathrm{R}}d\theta^{\mathrm{L}}\end{aligned} \tag{7}$$

for arbitrary probability distributions $p(\theta^{\mathrm{L}} - \theta^{\mathrm{R}})$ and $q(\theta^{\mathrm{L}} + \theta^{\mathrm{R}})$. Condition (7) indicates that the mean primitive activities for extrinsically cooperative movements, intrinsically cooperative movements, and unimanual movements were all balanced with each other.

Two-Dimensional Tuning Curves to Bimanual Movement Directions

The explicit form of the solution to satisfy the extended unimanual–bimanual balance (7) could be described as

$$A^{\text{bi}}(\theta^{\text{L}},\theta^{\text{R}}) = h(\theta^{\text{L}},\theta^{\text{R}}) - \frac{1}{2\pi}\int_{-\pi}^{\pi} h(\theta^{\text{L}}-z,\theta^{\text{R}}-z)dz - \frac{1}{2\pi}\int_{-\pi}^{\pi} h(\theta^{\text{L}}-z,\theta^{\text{R}}+z)dz, \tag{8}$$

where $h(\theta^{\text{L}}, \theta^{\text{R}})$ is an arbitrary two-dimensional periodic function $h(\theta^{\text{L}},\theta^{\text{R}}) = h(\theta^{\text{L}}+2\pi,\theta^{\text{R}}) = h(\theta^{\text{L}},\theta^{\text{R}}+2\pi)$. Below, we prove that this solution (8) satisfies the aforementioned condition (7).

For a periodic function $g(\theta) = g(\theta+2\pi)$, the integrals $\int_{-\pi}^{\pi} d\theta g(\theta) = \int_{-\pi}^{\pi} d\theta g(\theta+\varphi) = \int_{-\pi}^{\pi} d\theta g(2\theta)$ for arbitrary ϕ. Due to the periodicity of the function $h(\theta^{\text{L}},\theta^{\text{R}})$ and probability distribution $p(\theta^{\text{L}}-\theta^{\text{R}})$, $\int_{-\pi}^{\pi} d\theta^{\text{L}} p(\theta^{\text{L}}-\theta^{\text{R}})h(\theta^{\text{L}}-z,\theta^{\text{R}}\pm z) = \int_{-\pi}^{\pi} d\theta^{\text{L}} p(\theta^{\text{L}}+z-\theta^{\text{R}})h(\theta^{\text{L}},\theta^{\text{R}}\pm z)$, $\int_{-\pi}^{\pi} d\theta^{\text{R}} p(\theta^{\text{L}}+z-\theta^{\text{R}})h(\theta^{\text{L}},\theta^{\text{R}}\pm z) = \int_{-\pi}^{\pi} d\theta^{\text{R}} p(\theta^{\text{L}}+z-\theta^{\text{R}}\pm z)h(\theta^{\text{L}},\theta^{\text{R}})$ and $\int_{-\pi}^{\pi} dz p(\theta^{\text{L}}-\theta^{\text{R}}+2z) = \int_{-\pi}^{\pi} dz p(2z) = \int_{-\pi}^{\pi} dz p(y) = \frac{1}{2\pi}$ because the probability distribution satisfies $\int_{-\pi}^{\pi} d\theta^{\text{L}} \int_{-\pi}^{\pi} d\theta^{\text{R}} p(\theta^{\text{L}}-\theta^{\text{R}})A^{\text{bi}}(\theta^{\text{L}},\theta^{\text{R}}) = \int_{-\pi}^{\pi} d\theta^{\text{L}} \int_{-\pi}^{\pi} d\theta^{\text{R}} q(\theta^{\text{L}}+\theta^{\text{R}})A^{\text{bi}}(\theta^{\text{L}},\theta^{\text{R}})$ for arbitrary probability distributions $p(\theta^{\text{L}}-\theta^{\text{R}})$ and $q(\theta^{\text{L}}+\theta^{\text{R}})$.

The shift-and-scale transformation $\alpha A^{\text{bi}}(\theta^{\text{L}}-\varphi^{\text{L}},\theta^{\text{R}}-\varphi^{\text{R}})$ is also described as the form (8) for the arbitrary function replaced with $\alpha h(\theta^{\text{L}}-\varphi^{\text{L}},\theta^{\text{R}}-\varphi^{\text{R}})$ Therefore, primitives with various PDs and amplitudes were allowed as solutions of the extended unimanual–bimanual balance (7). The activity pattern for unimanual movements $A^{\text{uni}}(\theta)$ such that $\int_{-\pi}^{\pi} d\theta A^{\text{uni}}(\theta) = -\frac{1}{2\pi}\int_{-\pi}^{\pi} d\theta^{\text{L}} \int_{-\pi}^{\pi} d\theta^{\text{R}} h(\theta^{\text{L}},\theta^{\text{R}})$ was allowed for a solution of the unimanual–bimanual balance (7). Example solutions for $A_i^{\text{uni}}(\theta)$A uni i (θ) could be described as $A_i^{\text{uni}}(\theta) = A_i^{\text{bi}}(\theta-\psi_i, \pm\theta-\phi_i)$ by PD shift ψ_i and ϕ_i.

Procedure to Demonstrate the Behavioral Task

The perturbation imposed on the left arm was set as a constant (p = 1 or −1, assumed as clockwise perturbation) for the figures to show the movement error (Figs. 2(d)–(f) and 4(f)–(h)). In the simulations for K = 1 (Figs. 2(d) and 4(f)), the fixed target direction was set as $\theta^{\text{L}} = \theta^{\text{R}} = \theta_1 = 0$, which was both parallel and symmetric. After 50 trials of the bimanually reaching task, the task switched to the unimanual task. The perturbation was still consistent in the unimanual task.

In the simulations for K = 8 (Figs. 2(e), (f) and 4(g), (h)), a target direction was sampled independently from $\theta_k = \pi k/4 - \pi$ (k = 1, . . . , 8) in each trial, and the sampled direction θ_k determined the target directions, $\theta^L = \theta_k$ in the unimanually reaching task, $\theta^L = \theta^R = \theta_k$ in the parallel reaching task, and $\theta^L = -\theta^R = \theta_k$ in the symmetrically reaching task. After 200 trials of the bimanually reaching task (parallel or symmetric), the task switched to the unimanual task. The perturbation was still consistent in the unimanual task.

Motor Primitives for Fig. 2

We considered a product of truncated Gaussian tuning curves to θ^L and θ^R as an arbitrary function h in Eq. (8),

$$h(\theta^L, \theta^R) = \left(\exp\left(-\frac{\|\theta^L\|^2}{2\sigma_L^2}\right) + b^L\right)\left(\exp\left(-\frac{\|\theta^R\|^2}{2\sigma_R^2}\right) + b^R\right) - c, \quad (9)$$

where parameters $b^L = 0.0158$, $b^R = 0.318$, $\sigma_L = 21.1°$ and $\sigma_R = 27.2°$ were set to be equal to the parameters of the multiplicative model fitted to the experimental results in a previous study (Yokoi et al., 2011). The parameter c = 0.1651 was set such that the

integral of $A^{bi}(\theta, \theta')$ was equal to the integral of the multiplicative model (Yokoi et al., 2011). By discretizing θ^L and θ^R at 240 sample points (i.e., $\theta^L = \theta^R = \theta_k = \pi k/120 - \pi$ $(k = 1, \ldots, 240)$), we approximately calculated the second and third terms in Eq. (8) by diagonally averaging $h(\theta_{k+j}, \theta_{k'\pm j})$ for j, and we obtained a base tuning curve $A^{bi}(\theta_k, \theta_{k'})$. The preferred directions φ_i^L and φ_i^R were sampled independently from discretized directions θ_k $(k = 1, \ldots, 240)$ with an even probability of 1/240. The tuning curve of each primitive was obtained by periodically shifting the base tuning curve $A^{bi}(\theta_k, \theta_{k'})$ with $(\varphi_i^L, \varphi_i^R)$. The unimanual tuning curve $A_i^{uni}(\theta^L)$ was obtained by periodically shifting the tuning curve of parallel bimanual movements, $A_i^{bi}(\theta_k, \theta_k)$, with ψ_i

sampled from a Gaussian distribution $P(\psi_i) \propto \exp(-\frac{1}{2\sigma_p^2}\|\psi_i\|^2)$ (i.e., $A_i^{\text{uni}}(\theta) = A_i^{\text{bi}}(\theta - \psi_i, \theta - \psi_i)$). The number of primitive was N = 1000, as in Fig. 2. The learning rate was set as η = 0.0015 such that the net learning speed might be approximately equivalent to those in Fig. 2.

Simulations for Fig. 3

A statistical model $\alpha_{l,i} g(\theta_{l,k} - \varphi_i) + \alpha_{r,i} g(\theta_{r,k'} - \phi_i)$ was fit for $A_i^{\text{bi}}(\theta_k, \theta_k)$ (parallel bimanual movements) and $A_i^{\text{bi}}(\theta_k, \theta_{k'} + \pi)$ (opposite bimanual movements). $(\varphi_i(m), \phi_i(n))$ were sampled as $(-\pi + \pi \times \frac{m}{128}, -\pi + \pi \times \frac{n}{128})\ (m, n = 1, \ldots, 128)$. For each pair (m, n), the best $\alpha_{l,i}(m, n)$ and $\alpha_{r,i}(m, n)$ were calculated to minimize the squared error $E_i(m, n) = (A_i^{\text{bi}}(\theta_{l,k}, \theta_{r,k}) - \alpha_{l,i}(m, n) g(\theta_{l,k} - \varphi_i(m)) - \alpha_{r,i}(m, n) g(\theta_{r,k'} - \phi_i(n)))^2 + (A_i^{\text{bi}}(\theta_{l,k}, \theta_{r,k} + \pi) - \alpha_{l,i}(m, n) g(\theta_{l,k} - \varphi_i(m)) - \alpha_{r,i}(m, n) g(\theta_{r,k'} + \pi - \phi_i(n)))^2$. Similar procedures were conducted for $A_i^{\text{uni}}(\theta_{l,k})$ by fitting the statistical model $\alpha_i^{\text{uni}} g(\theta_k - \varphi_i^{\text{uni}})$. Each $\alpha_{l,i}(m', n')$ and $\alpha_i^{\text{uni}}(m')$ in Fig. 3 represented values obtained when $(m', n') = \arg\min_{m,n} E_i(m, n)$ and $(m') = \arg\min_m E_i(m)$

Neural Implementation of Unimanual–Bimanual Balance

We supposed two neural populations (N = 1000 for each) to control left and right arm movements in the right and left motor cortices, respectively (Fig. 4(a)). The firing rate y of each neuron was determined with a rate model as a function of the membrane potential v, i.e., $y_i^{\text{L}} = f(v_i^{\text{L}})$ for neuron i to control the left arm (in the right cortex) and $y_j^{\text{R}} = f(v_j^{\text{R}})$ for neuron j to control the right arm. The rate function was set as a sigmoid function, f(x) = (tanh(x) + 1)/2. The dynamics of the membrane potentials v L i and v_j^{R} were described using differential equations,

$$\tau \frac{dv_i^{\text{L}}}{dt} = I_i^{\text{L}} - v_i^{\text{L}} + \sum_j Y_{ij}^{\text{LR}} f(v_j^{\text{R}}),$$

$$\tau \frac{dv_j^{\text{R}}}{dt} = I_j^{\text{R}} - v_j^{\text{R}} + \sum_i Y_{ji}^{\text{RL}} f(v_i^{\text{L}}), \qquad (10)$$

where τ represents the membrane time constant, set as τ = 20 ms in the simulation. The feedforward inputs I_i^{L} and I_j^{R} were determined with the PDs of the neurons, φ_i^{L} and φ_j^{R}, as functions of the respective target directions, $I_i^{\mathrm{L}}(\theta^{\mathrm{L}}) = 0.5\cos(\theta^{\mathrm{L}} - \varphi_i^{\mathrm{L}})$ and $I_j^{\mathrm{R}}(\theta^{\mathrm{R}}) = 0.5\cos(\theta^{\mathrm{R}} - \varphi_j^{\mathrm{R}})$. When no target direction was provided (e.g. for the opposite arm of the unimanual movement), the input I = 0. The interactions between the right and left cortices, Y_{ij}^{LR} and Y_{ji}^{RL} were assumed to be sparse, and the connecting rate was set as 0.2. If a connection existed, the connection strength was to be set as $Y_{ij}^{\mathrm{LR}} = 0.05\cos(\psi_i^{\mathrm{L}} - \varphi_j^{\mathrm{R}})$ or $Y_{ji}^{\mathrm{RL}} = 0.05\cos(\psi_j^{\mathrm{R}} - \varphi_i^{\mathrm{L}})$. Hence, the excitatory and inhibitory connections that neuron i received approximately balanced, $\sum_j Y_{ij}^{\mathrm{LR}} \simeq 0$, which depended on the sampling of sparse connections. Neuronal parameters φ_i^{L}, ψ_i^{L}, φ_j^{R} and ψ_j^{R} were independently sampled from a uniform distribution in the range [−π , π].

After the input patterns $\{I_i^{\mathrm{L}}\}$ and $\{I_j^{\mathrm{R}}\}$ were changed for a given pattern of target directions, $(\theta^{\mathrm{L}}, \theta^{\mathrm{R}})$ for bimanual or θ^{L} for unimanual movements, the firing rates y_i^{L} and y_j^{R} approached asymptotic values in approximately 100 ms (Fig. 4(b)). The values at 200 ms were used as the activities of the neurons for Fig. 4. The target directions θ^{L} and θ^{R} were discretized at 32 sample points (i.e., $\theta^{\mathrm{L}} = \theta^{\mathrm{R}} = \theta_k = \pi k/16 - \pi$ $(k = 1, \ldots, 32)$), and the activity at 200 ms of each neuron in the right cortex was obtained as a function $y_i^{\mathrm{L}}(\theta^{\mathrm{L}}, \theta^{\mathrm{R}})$ and denoted for unimanual movements as $y_i^{\mathrm{L}}(\theta^{\mathrm{L}})$

.

To validate the unimanual–bimanual balance defined in Eq. (7), we supposed five different probability distributions of bimanual target directions: $p(\theta^{\mathrm{L}} - \theta^{\mathrm{R}}) = \frac{1}{2\pi}\delta(\theta^{\mathrm{L}} - \theta^{\mathrm{R}})$(parallel movements), $q(\theta^{\mathrm{L}} + \theta^{\mathrm{R}}) = \frac{1}{2\pi}\delta(\theta^{\mathrm{L}} + \theta^{\mathrm{R}})$ (symmetric movements), $p(\theta^{\mathrm{L}} - \theta^{\mathrm{R}}) = q(\theta^{\mathrm{L}} + \theta^{\mathrm{R}}) = \frac{1}{4\pi^2}$ (independent movements), $p(\theta^{\mathrm{L}} - \theta^{\mathrm{R}}) \propto \exp(-2\|\theta^{\mathrm{L}} - \theta^{\mathrm{R}}\|^2/\pi^2)$ (an example distribution of extrinsically cooperative movements), and $q(\theta^{\mathrm{L}} + \theta^{\mathrm{R}}) \propto \exp(-2\|\theta^{\mathrm{L}} + \theta^{\mathrm{R}}\|^2/\pi^2)$(an example distribution of intrinsically cooperative movements), where δ(θ) is Dirac's delta function. The integrals were approximated by the summations at the discretized sample points θ_k(k = 1, . . . , 32) (e.g., the means for unimanual, parallel, symmetric and independent movements) were calculated as $\frac{1}{32}\sum_{k=1}^{32} y_i^{\mathrm{L}}(\theta_k)$, $\frac{1}{32}\sum_{k=1}^{32} y_i^{\mathrm{L}}(\theta_k, \theta_k)$, $\frac{1}{32}\sum_{k=1}^{32} y_i^{\mathrm{L}}(\theta_k, -\theta_k)$, $\frac{1}{32^2}\sum_{k=1}^{32}\sum_{k'=1}^{32} y_i^{\mathrm{L}}(\theta_k, \theta_{k'})$, respectively.

The present neuron model exhibited spontaneous activity. Hence, we adopted an offset to determine the motor primitive activity used for behavioral demonstrations as $A_i^{uni}(\theta^L) = y_i^L(\theta^L) - 0.2$ for unimanual movements, and $A_i^{bi}(\theta^L, \theta^R) = y_i^L(\theta^L, \theta^R) - 0.2$ for bimanual movements. The learning rate was set as $\eta = 0.0002$ such that the net learning speed might be approximately equivalent to those in Figs. 2 and 3.

ACKNOWLEDGMENTS

This work was supported by a Grant-in-Aid for Young Scientists B (16K16122) to KT and a grant to YS from the MEXTSupported Program for the Strategic Research Foundation at Private Universities, 2013–2017.

REFERENCES

1. Donchin, O., Francis, J. T., & Shadmehr, R. (2003). Quantifying generalization from trial-by-trial behavior of adaptive systems that learn with basis functions: theory and experiments in human motor control. The Journal of Neuroscience, 23(27), 9032–9045.
2. Donchin, O., Gribova, A., Steinberg, O., Mitz, A. R., Bergman, H., & Vaadia, E. (2002).
3. Single-unit activity related to bimanual arm movements in the primary and supplementary motor cortices. Journal of Neurophysiology, 88(6), 3498–3517.
4. Nozaki, D., Kurtzer, I., & Scott, S. H. (2006). Limited transfer of learning between unimanual and bimanual skills within the same limb. Nature Neuroscience, 9(11), 1364–1366.
5. Nozaki, D., & Scott, S. H. (2009). Multi-compartment model can explain partial transfer of learning within the same limb between unimanual and bimanual reaching. Experimental Brain Research, 194(3), 451–463.
6. Rokni, U., Steinberg, O., Vaadia, E., & Sompolinsky, H.(2003). Cortical representation of bimanual movements. The Journal of Neuroscience, 23(37), 11577–11586.
7. Rouiller, E. M., Babalian, A., Kazennikov, O., Moret, V., Yu, X. H., & Wiesendanger, M. (1994). Transcallosal connections of the distal

forelimb representations of the primary and supplementary motor cortical areas in macaque monkeys. Experimental Brain Research, 102(2), 227–243.

8. Sainburg, R. L., & Wang, J. (2002). Interlimb transfer of visuomotor rotations: independence of direction and final position information. Experimental Brain Research, 145(4), 437–447.
9. Shadmehr, R., & Mussa-Ivaldi, F. A. (1994). Adaptive representation of dynamics during learning of a motor task. The Journal of Neuroscience, 14(5 Pt 2), 3208–3224.
10. Takiyama, K. (2015). Context-dependent memory decay is evidence of effort minimization in motor learning: a computational study. Frontiers in Computational Neuroscience, 9(4), 1–10.
11. Takiyama, K., Hirashima, M., & Nozaki, D. (2015). Prospective errors determine motor learning. Nature Communications, 6, 5925.
12. Takiyama, K., Naruse, Y., & Okada, M. (2009). Statistical mechanics of mexicanhat-type horizontal connection. Journal of the Physical Society of Japan, 78(6), 064002.
13. Takiyama, K., & Okada, M. (2012). Recovery in stroke rehabilitation through the rotation of preferred directions induced by bimanual movements: a computational study. PLoS One, 7(5), e37594.
14. Takiyama, K., & Sakai, Y. (2016). Balanced motor primitive can explain generalization of motor learning effects between unimanual and bimanual movements. Scientific Reports, 6, 23331.
15. Tcheang, L., Bays, P. M., Ingram, J. N., & Wolpert, D. M. (2007). Simultaneous bimanual dynamics are learned without interference. Experimental Brain Research, 183(1), 17–25.
16. Thoroughman, K. A., & Shadmehr, R. (2000). Learning of action through adaptive combination of motor primitives. Nature, 407(6805), 742–747.
17. Wang, J., Lei, Y., Xiong, K., & Marek, K. (2013). Substantial generalization of sensorimotor learning from bilateral to unilateral movement conditions. PLoS One, 8(3), e58495.
18. Wang, J., & Sainburg, R. L. (2004). Limitations in interlimb transfer of visuomotor rotations. Experimental Brain Research, 155(1), 1–8.

19. Yokoi, A., Hirashima, M., & Nozaki, D. (2011). Gain field encoding of the kinematics of both arms in the internal model enables flexible bimanual action. The Journal of Neuroscience, 31(47), 17058–17068.

CHAPTER 9

A NEW METHOD FOR QUANTIFYING THE PERFORMANCE OF EEG BLIND SOURCE SEPARATION ALGORITHMS BY REFERENCING A SIMULTANEOUSLY RECORDED ECOG SIGNAL

Naoya Oosugi [a,b], Keiichi Kitajo[c], Naomi Hasegawa [a], Yasuo Nagasaka [a], Kazuo Okanoya [b], Naotaka Fujii [a]

[a] Laboratory for Adaptive Intelligence, BSI, RIKEN, Saitama, Japan

[b] Graduate School of Arts and Sciences, University of Tokyo, Tokyo, Japan

[c] Laboratory for Advanced Brain Signal Processing, RIKEN Brain Science Institute, Japan

ABSTRACT

Blind source separation (BSS) algorithms extract neural signals from electroencephalography (EEG) data. However, it is difficult to quantify source separation performance because there is no criterion to dissociate neural signals and noise in EEG signals. This study develops a method for evaluating BSS performance. The idea is neural signals in EEG can be estimated by comparison with simultaneously measured electrocorticography (ECoG). Because the ECoG electrodes cover the majority of the lateral cortical surface and should capture most of the original neural sources in the EEG signals. We measured real EEG and ECoG data and developed an algorithm for evaluating BSS performance. First, EEG signals are separated into EEG components using the BSS algorithm. Second, the EEG components are ranked using the correlation coefficients of the ECoG regression and the components are grouped into subsets based on their ranks. Third, canonical correlation analysis estimates how much information is shared between the subsets of the EEG components and the ECoG signals. We used our algorithm to compare the performance of BSS algorithms (PCA, AMUSE, SOBI, JADE, fastICA) via the EEG and ECoG data of anesthetized nonhuman primates. The results (Best case >JADE = fastICA >AMUSE = SOBI ≥ PCA (Best case >JADE =fastICA >AMUSE = SOBI ≥ PCA >random separation) were common to the two subjects. To encourage the further development of better BSS algorithms, our EEG and ECoG data are available on our Web site (http://neurotycho.org/) as a common testing platform.

INTRODUCTION

Goal of this Study

Electroencephalography (EEG) measures electric potentials via electrodes placed on the scalp. The electrodes are isolated from the neural sources in the brain by multiple layers consisting of dura mater, skull, muscle, and skin. These layers cause attenuation of the neural signals propagating from the source. In addition, signal contamination by nonneural artifacts is caused by the distance between the electrode and the brain because of these layers. Thus, extracting the desired neural signals from the EEG signal with a low signal-to-noise ratio is an important problem (Daly, Pichiorri, Faller, Kaiser, & Kreilinger, 2012). To address this problem, various multivariate analyses have been used (Sanei & Chambers, 2008), including artifact reduction (Soong & Koles, 1995; Vigário, Sarela, Jousmiki, Hamalainen, & Oja, 2000), source localization (Valdés-Sosa, Vega-Hernández, Sánchez-Bornot, Martínez-Montes, & Bobes, 2009; Zhukov, Weinstein, & Johnson, 2000), and solving the colinearity problem (Besthorn, Zerfass, Geiger-Kabisch, Sattel, & Daniel, 1997; Ombao & Ringo Ho, 2006). Blind source separation (BSS) is one method for artifact reduction (Choi, Cichocki, Park, & Lee, 2005; Vorobyov &

Cichocki, 2002).

There are many BSS algorithms with different criteria for source separation. They are categorized into two groups: one is based on second-order statistics (SOS) and the other is based on higher-order statistics (HOS). Typical SOS-based BSSs include: principal component analysis (PCA) using a covariance matrix; algorithms for multiple unknown signals extraction (AMUSE) using a covariance matrix and a single time-delayed covariance matrix (Tong, Soon, Huang, & Liu, 1990); and second-order blind identification (SOBI) using multiple time-delayed covariance matrices (Belouchrani, Abed-Meraim, Cardoso, & Moulines, 1997). Typical HOS-based BSSs are joint approximate diagonalizations of eigenmatrices (JADE) using cross-cumulants (Cardoso & Souloumiac, 1993) and fast independent component analyses (fastICA) using negentropy (Hyvarinen, 1999). These major algorithms compute weights for the separation. The BSS algorithms per se simply perform the separation and do not indicate whether the separated components are neural signals or noise. Identification of the separated EEG components is thus another important problem in EEG analysis. Identification problems are often solved using decisions by human experts (Choi et al., 2005) or automatic identification techniques (Mognon, Jovicich, Bruzzone, & Buiatti, 2011; Zou, Hart, & Jafari, 2012).

The use of artificial data (Delorme, Sejnowski, & Makeig, 2007) is one way to validate BSS algorithms because noise and neural signals are completely dissociable from the data. However, these model-based validations are limited because they do not probe the validity of real EEG signals. Other studies have shown that BSS algorithms could extract task-specific neural information by implementing brain–computer interface (BCI) technology (Nicolas-Alonso & Gomez-Gil, 2012; Winkler, Haufe, & Tangermann, 2011) or event-related potentials (Makeig, Westerfield, Jung, Enghoff, & Townsend, 2002; Mennes, Wouters, Vanrumste, Lagae, & Stiers, 2010). The results of these studies suggest that BSS algorithms are useful for neural source estimation when associated with task-specific parameters. Other studies used real EEG data and simultaneously measured eye movements and eye blinks (Jung, Makeig, Humphries, Lee, & McKeown, 2000; Makeig, Bell, Jung, & Sejnowski, 1996; Mennes et al., 2010; Romero, Mañanas & Barbanoj, 2008) to remove eye-related artifacts. Similarly, artifact reduction using muscle activity has also been studied (Muthukumaraswamy, 2013).

Some studies tried to find the best BSS algorithms for EEG data, but there is no consensus. For example, Crespo-Garcia et al. compared BSS algorithms and concluded that AMUSE, SOBI, and infoMAX were better than JADE by showing how well the BSS techniques separated electromyographic (EMG) components from human sleep EEGs by simultaneously measuring the EMGs (Crespo-Garcia, Atienza, & Cantero, 2008). Romero et al. compared various BSS algorithms by simultaneously measuring EEG and electrooculography (EOG) and concluded that AMUSE and SOBI were better than PCA, JADE, infoMAX, and fastICA (Romero et al., 2008). Kachenoura et al. also compared BSS algorithms via simulated EEGs for BCI and concluded that JADE and contrast function optimization algorithms were better than SOBI, infoMAX, and fastICA (Kachenoura, Albera, Senhadji, & Comon, 2008). However, these assessments cannot be compared directly because they used different datasets and criteria.

We propose a new criterion for evaluating the performance of BSS algorithms using simultaneously recorded EEG and electrocorticography (ECoG). We measured real EEG and ECoG signals and tested well-known BSS algorithms using the datasets and our new criterion. We have already showed EEG signals could be reconstructed from our ECoG signals sufficiently and shared the EEG and ECoG datasets (Oosugi & Fujii, 2011). In this study, we showed more practical result and richer dataset than previous datasets.

ECoG Signals for Evaluation of BSS Algorithms

The main problem with BSS studies for real EEG signals is distinguishing neural signals from nonneural signals. To solve this problem, we recorded ECoG activity to estimate the neural components in EEG signals as a reference. ECoG electrodes cover the majority of the lateral cortex and can capture the raw neural components of an EEG. Propagation of the raw neural signals is considered spatially continuous; therefore, most of the neural signals in the EEG signal must be attributed to the ECoG signals. Of course, the ECoG signal is not a pure neural signal because it also is contaminated by artifact noise. However, we assumed that ECoG signals would be useful for evaluating BSS algorithms for EEG signals because the signal-to-noise ratios of ECoG signals are higher than those of EEG signals (Ball, Kern, Mutschler, Aertsen, & Schulze-Bonhage, 2009), and the non-ECoG-correlated components in the EEG signals must be artifacts. Thus, extraction of a BSS component similar to the ECoG closely equates to extraction of neural components from the EEG signals.

METHOD

Recording Experiment

Ethics

All of the experimental and surgical procedures were performed in accordance with the experimental protocols (No. H24-2-203(4)) approved by the RIKEN ethics committee and the recommendations of the Weatherall report, "The use of nonhuman primates in research". Implantation surgery was performed under sodium pentobarbital anesthesia, and all efforts were made to minimize suffering. No animals were sacrificed in this study. Overall care was managed by the Division of Research Resource Center at the RIKEN Brain Science Institute. The animals were housed in large individual enclosures with other animals visible in the room, and maintained on a 12/12-h light/dark cycle. The animals were given food (PS-A; Oriental Yeast Co., Ltd., Tokyo, Japan) and water ad libitum, and daily fruit/dry treats as a means of enrichment and novelty. The animals were occasionally given toys in the cage. The in-house veterinarian checked the animals frequently and adjusted their daily feedings to maintain their weight. We have attempted to maintain as humane a treatment of our subjects as possible.

ECoG Recording

A customized multichannel ECoG electrode array (Unique Medical, Tokyo, Japan) was implanted in the subdural space between the arachnoid membrane and dura in two adult macaque monkeys (M1 was *Macaca mulatta* and M2 was *M. fuscata*). The interelectrode distance was 5 mm. The electrode was a platinum disk 3 mm in diameter insulated by a silicone sheet with a tiny hole that was 0.8 mm in diameter at its center. A 128-channel ECoG array was implanted in the left hemispheres of both M1 and M2. The electrode array completely covered the frontal, parietal, temporal, and occipital lobes (Fig. 1). The right hemisphere of M1 was also covered with a 128-channel ECoG array in the same manner. The reference electrode consisted of rectangular platinum plates placed in the subdural space between the ECoG array and the dura mater. A platinum ground electrode was placed in the epidural space (see Nagasaka, Shimoda, & Fujii, 2011 for details) between the dura and the skull. We recorded ECoG data via these electrodes at a sampling rate of 1000 Hz.

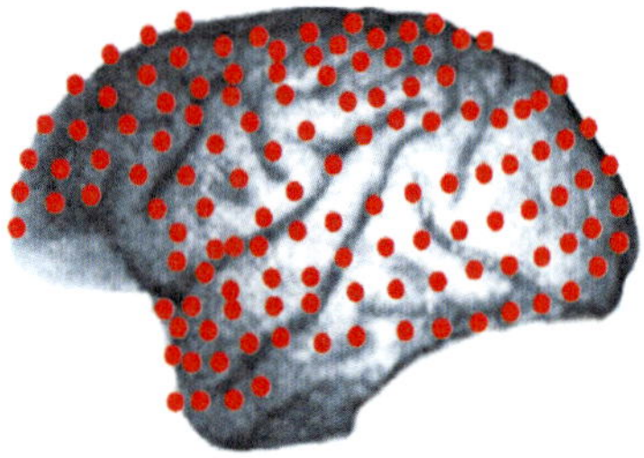

M1

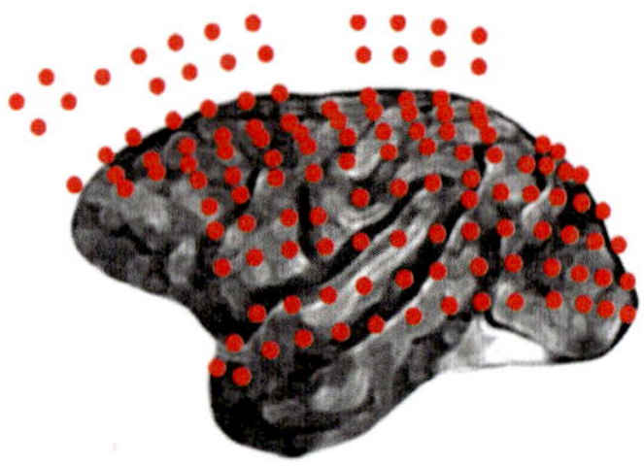

M2

Figure. 1. Location of ECoG electrodes on the monkeys' left hemispheres. Locations of the 128 ECoG electrodes on the left hemisphere are illustrated by red circles on magnetic resonance brain images (left: M1; right: M2). Red circles shown outside the brain in M2 illustrate the unfolded set of electrodes that covered the medial wall. In M1, the same number of ECoG electrodes was implanted on the right hemisphere with a symmetric pattern.

EEG Recording

For the EEG recordings, 16 or 17 Ag/AgCl electrodes at a sampling rate of 4096 Hz via NEURO PRAX (neuroConn GmbH, Ilmenau, Germany) were used. The electrodes were placed according to the international 10–20 system adapted to the monkey head using a custom-made plastic EEG cap. Some electrodes in the 10–20 system could not be placed (M1: Cz, and Pz; M2: Cz, Pz, and P3) because the ECoG connector occupied the space. In addition, in M1, the signal quality of channel P8 was not adequate and the channel was discarded. All signal channels were referenced to an activity that was averaged between the signals recorded from the bilateral mastoids. The hair on the head and earlobes was shaved. The EEG and ECoG recordings were synchronized by a common external trigger. The EEG signals were down-sampled to 1000 Hz in offline analysis.

Anesthesia Experiment

EEG and ECoG signals were recorded after injection of an anesthetic agent and an antagonist. The subject was seated in a primate chair with the movements of both arms and the head restricted. Each subject's heart rate and respiration were monitored carefully during the entire experiment. The anesthetic agent was a combination of ketamine (8.21 mg/kg for M1 and 5.00 mg for M2) and medetomidine (0.05 mg/kg for M1 and 0.02 mg/kg for M2).

We collected five minutes of data for the following analyses to control any individual variation of the effect of the anesthetic agents and the antagonist. These five-minute data blocks were called the "low-anesthetic", "deep-anesthetic", and "recovery" states. In the low-anesthetic state, 5 min of EEG and ECoG data was recorded shortly after the injection. In the deep-anesthetic state, 5 min of EEG and ECoG data was recorded when sneeze, grasp, and blink reflexes had vanished. In the recovery state, 5 min of EEG and ECoG data was recorded shortly after injection of the antagonist (atipamezole, 0.21 mg/kg for M1 and 0.09 mg/kg for M2).

Evaluation of BSS Algorithms

ECoG electrodes capture neural activity from the cortical surface that propagates to the scalp surface, while EEG electrodes capture the propagated activity as well as nonneural signals from the scalp surface. BSS algorithms are often used to isolate the nonneural signals from neural signals in EEGs; therefore, detecting ECoG source signals from the EEG signal must be a good index to quantify BSS algorithms. We developed an evaluation algorithm with five steps: (1) preprocessing, (2) BSS, (3) evaluation of each EEG component, (4) evaluation of a subset of EEG components, and (5) final evaluation of the BSS algorithm. Steps 3 and 4 are the core idea of our proposed method.

Step 1: In the preprocessing, fourth-order band-pass filtering was performed to divide the EEG and ECoG signals into specific frequency bands. Seven frequency bands were used (delta + theta: 0.3–7, alpha: 8–13, low beta:14–20, high beta: 21–30, low gamma: 31–48, middle gamma: 52–60, and high gamma: 61–100 Hz). There were four reasons why the EEG signals were divided into these frequency bands. First, artifact noise in the EEG signals could be characterized by its frequency band. For example, EMG, EOG and power supply noise are mainly distributed above 20 Hz, below 10 Hz, and at 50 Hz, respectively (Daly et al., 2012). Second, the frequency characteristics of neural signals are also characterized by their frequency bands, such as alpha (8–13 Hz) and beta (14–30 Hz) waves. Third, some BSS algorithms (in this research, AMUSE and SOBI) based on the assumption that a source signal does not change drastically in the time direction (Choi et al., 2005). Fourth, previous research showed frequency dependencies of the transmission of the EEG from cortex to scalp (Pfurtscheller & Cooper, 1975).

Step 2: In the BSS step, five well-known BSS algorithms were applied to the preprocessed EEG signals. Three SOS-based BSS algorithms (PCA, AMUSE, and SOBI) and two HOS-based BSS algorithms (JADE and fastICA) were selected. We used open-source Matlab codes: that for AMUSE was coded by A. Cichocki and R. Szupiluk; SOBI was coded by A. Belouchrani and A. Cichocki; JADE was coded by J.-F. Cardoso; and ICASSO, which is a wrapper for fastICA, was coded by Himberg and Hyvarinen (2003). In this paper, we called ICASSO "fastICA" to simplify the technical terms. These codes are included in FieldTrip (Oostenveld, Fries, Maris, & Schoffelen, 2011) and the SOM toolbox (Vesanto, Himberg, Alhoniemi, & Parhankangas, 1999). As references, two more types of linear separation methods were applied. One was unblind separation, canonical correlation analysis (CCA) (Hotelling,

1936), to define the upper limit of BSS performance. We used Matlab codes for CCA that were coded by M. Borga. The other was random separation, which estimates chance levels of BSS performance for our EEG and ECoG data. Random separation was defined as a linear combination by random weight vectors that were created by resampling from a uniform distribution between −1 and 1. Results of CCA separation and random separation were later used in Step 5. For the BSS analysis, default parameters of these codes were used. Each BSS algorithm (including CCA and random separation) separated 16 EEG components from the preprocessed EEG signals for each frequency band and subject.

Step 3: In the evaluation of the EEG components of each BSS, each EEG component was evaluated by measuring how much information the component shared with the ECoG signals. If the BSS algorithm worked well for EEG signals, then the EEG component is either very highly correlated or absolutely uncorrelated with the ECoG signals. One naïve idea for calculating the evaluation index is to measure the regression of a single EEG signal from all of the ECoG signals. This approach reveals how much information is shared between the single EEG signal and the ECoG signals. However, this approach does not work well when comparing multiple EEG signals because the overlap of information among EEG signals is not considered (the overlap problem). This is a critical problem when evaluating BSS algorithms: if the BSS algorithm fails to separate the original EEG signals and if some of the EEG components share the same sources, we cannot tell whether the BSS extracts neural components or components mixing neural and nonneural sources. To overcome the overlap problem, CCA was performed to evaluate how much information was shared between EEG components and the ECoG signals. To explain why the CCA approach is better than the simple linear regression approach, two extreme cases are helpful. If two EEG components overlap completely, they are the same and completely share information with the ECoG signals. Then, the correlation coefficients of these components are both one via the simple linear regression approach, but the canonical correlation coefficients are one and zero via the CCA approach. This extreme example shows that the CCA approach reduces the effect of the overlap problem. If BSS works well for EEG signals, then we can divide the EEG components into two clusters: one of the clusters has very high canonical correlations and the other has very low canonical correlations with the ECoG signals. The next question was how to generate the EEG component clusters. There are $2^{16} - 1 = 65,535$216−1=65,535 possible combinations of all EEG components, so we must confine the combinations used. Our idea was to define the order of EEG components and create subsets of them via the order recursively as EEG component clusters for the CCA evaluation. The order was defined by how much information was shared with the ECoG signals. We therefore calculated correlation coefficients between an EEG component and the ECoG signals using simple linear regression.

Step 4: We sorted the EEG components by descending order of the correlation coefficient of each component, and created 16 subsets of EEG components by picking the components from the top to the Nth (1/16th) component. For each subset of EEG components of each BSS, CCA was again applied and the sum of canonical correlation coefficients was calculated for each frequency, state, and subject; these are called evaluated values (EVs). Fig. 3 shows an example of EVs from CCA separation, PCA and random separation. If the BSS algorithms worked well, then EVs tended to be high except for a few EEG components and the EVs do not increase when these EEG components are added, because these EEG components must be nonneural signals. Thus, the area under the EV curve tends to be higher, for the better BSS algorithms.

Step 5: Finally, the area under the EV curve was measured as an index of the BSS algorithm performance. To compare the performance between BSSs, we used the EVs of CCA and random separation for normalization. The EV of random separation was computed 50 times and averaged. The area under the EV curve was normalized by a linear transformation, in which zero corresponded to the averaged areas under the EV curves of random separation (chance level) and one corresponded to the area under the EV curve of CCA separation (theoretical upper limit).

RESULTS

We applied our evaluation algorithm to six sets of EEG and ECoG data. These sets consisted of the data recorded from the two monkeys M1 and M2 under three conditions (low-anesthetic, deep-anesthetic, and "recovery" states). Five BSS algorithms (PCA, AMUSE, SOBI, JADE, and fastICA) were evaluated in seven frequency bands (Hz). Examples of interim results of the evaluation steps are shown in Figs. 2 and 3.

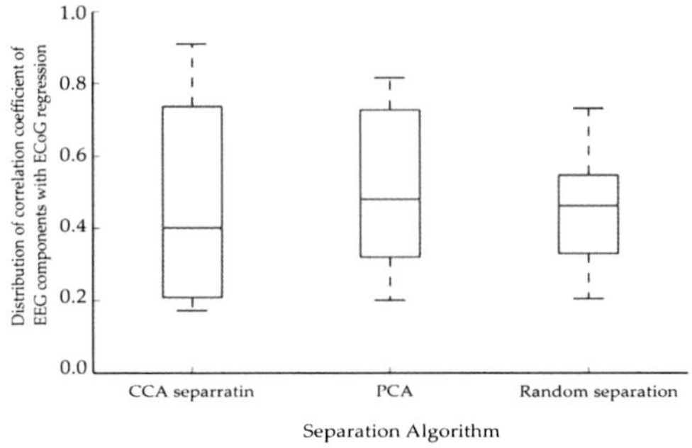

Figure. 2. An example of the distribution of shared information between an EEG component and ECoG signals (0.3–7 Hz, M1, low-anesthetic state). This figure shows the distribution of the correlation coefficients of EEG components with ECoG regression by box and whisker plots, showing minimum, lower quartile, median, upper quartile, and maximum values in the EEG components. The x-axis indicates the BSS methods. The y-axis indicates the correlation coefficient distribution of each EEG component with ECoG regression. The variance of each separation was CCA separation > PCA > random separation. These differences were significantly different ($p<0.05$). These results cannot be used directly for the evaluation of BSS performance because of the overlap problem (for details, see the "Evaluation of BSS algorithms" section).

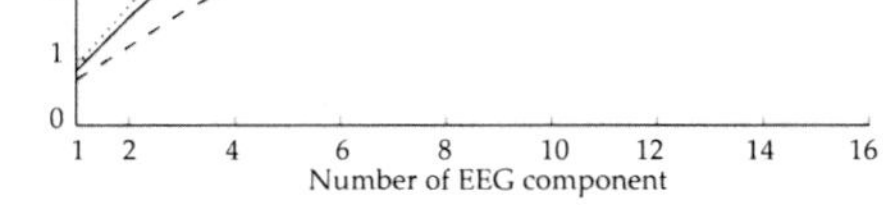

Figure. 3. An example of EV curves for CCA separation, PCA and random separation (0.3–7 Hz, M1, deep-anesthetic state). This figure shows the EV curve of PCA (black line), CCA separation (dotted line), and the averaged EV curve of random separation (dashed line) in the 0.3–7 Hz band, M1, and deep-anesthetic state. The x-axis indicates the number of EEG components used for the subset.

The y-axis indicates the sum of correlation coefficients of CCA from the subset of the EEG component and the whole ECoG signal. The figure indicates that the EV of CCA separation is larger than that for PCA or random separation. For comparison between BSS algorithm performances, the EVs were normalized by using the EV of CCA separation (theoretical upper limit as one) and averaged EV of random separation (chance level as zero).

Fig. 2 illustrates the distributions of correlation coefficients of EEG components, which were separated by PCA, CCA, and random separation for M1, low-anesthetic states, and the 0.3–7 Hz band. Two findings are illustrated in Fig. 2. First, EEG data and ECoG data shared rich information because correlation coefficients were quite high, even with random separation. The EEG signals were therefore not independent and identically distributed. Second, the BSS algorithms could separate ECoG-correlated and uncorrelated components better than random separation could.

Next, we measured how well whole ECoG signals could be reconstructed from subsets of EEG components by CCA (see the Method section). For each subset, a sum of canonical correlation coefficients was calculated that was called an EV. Fig. 3illustrates an example EV curve showing the relationships between the number of EEG components of the subset and the EVs of PCA at 0.3–7 Hz in M1 in the deep-anesthetic state. An EV curve was plotted for each BSS, frequency band, state, and subject. EV curves for CCA separation and random separation were also generated and used for normalizing EV values.

To compare the separation performance of BSSs, normalized areas under the EV curves of five BSS algorithm, five frequency bands, two subjects, and three experimental states were calculated (Fig. 4). Fig. 4 illustrates the merged results of the different states. The results show four tendencies: (1) the results were similar between subjects; (2) PCA tended to show poorer separation performance than the other BSS algorithms; (3) the areas under the EV curves of JADE and fastICA were larger than those for the other BSS algorithms, which suggests that HOS-based BSS has better separation performance than SOS-based BSS; and (4) almost all BSS performances were worse than CCA separation's performance and better than random separation's performance.

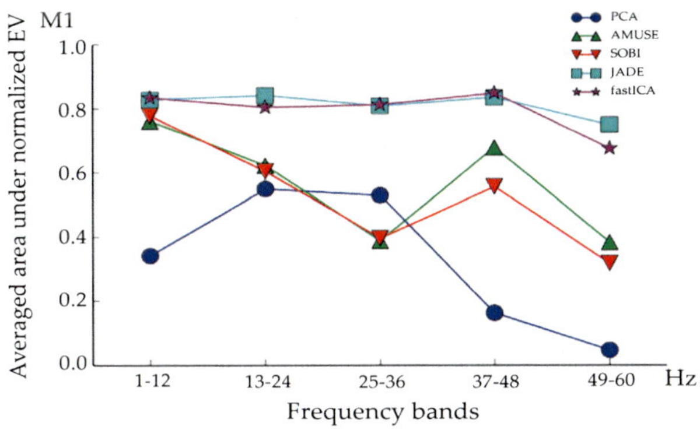

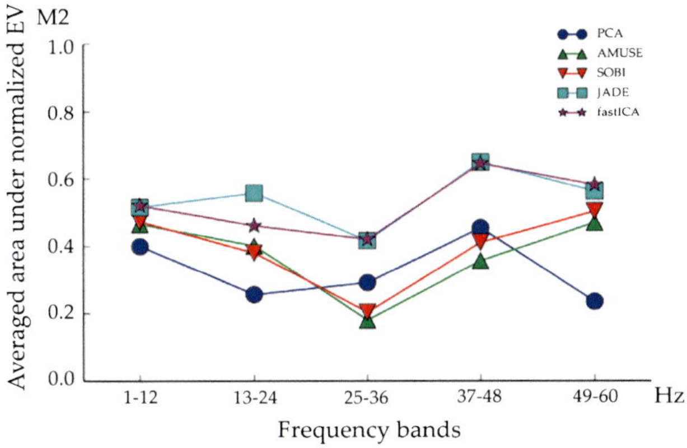

Figure. 4. The normalized area under the EV curves of each BSS, frequency band, and subject. One hundred fifty (5 BSS × 7 frequency bands × 2 subjects × 3 states) normalized areas under the EV curve were calculated and summarized by combining the three states. The figures show averaged normalized EVs for each subject, frequency band, and BSS algorithm. The upper figure shows results for M1 and the bottom shows results for M2. The x-axis indicates frequency band. The y-axis indicates the averaged normalized area under the EV curve. Each line represents one of the BSS algorithms, as indicated in the figures.

We confirmed the statistical significance of these findings via nonparametric statistical tests. The null hypotheses were: (1) "the area under the EV curve of PCA is equal to the areas under the EV curves of AMUSE, SOBI, JADE, or fastICA;" (2) "the areas under the EV curve of SOS-based BSS algorithms are equal to the areas under the EV curve of HOS-based BSS algorithms;" (3) "the area under the EV curve of BSS is equal to the areas under the EV curves of CCA separation;" and (4) "the area under the EV curve of BSS is equal to the areas under the EV curves of random separation".

The Mann–Whitney test was applied to find the significant differences in the areas under the EV curves of each BSS algorithm. We found that: (1) the area under the EV curve of PCA was significantly smaller than that for AMUSE and SOBI for M1 ($p<0.01$) and smaller than JADE and fastICA for M1 and M2; (2) the areas under the EV curves of HOS (JADE and fastICA) were significantly larger than those for SOS (AMUSE and SOBI) ($p<0.01$) for M1 and M2; (3) the area under the EV curve of BSS (PCA, SOBI, AMUSE, JADE and fastICA) was significantly smaller than that for CCA separation for M1 and M2; and (4) the area under the EV curve of BSS (PCA, SOBI, AMUSE, JADE and fastICA) was significantly larger than that for random separation for M1 and M2.

CONCLUSIONS

We have shown that the performance of HOS-based BSS algorithms was better than that of SOS-based algorithms for extracting neural components from EEG data. We also confirmed that PCA was not adequate for BSS in EEG studies. The performance rank of BSS separation was

(JADE=fastICA)>(SOBI=AMUSE)≥PCA.

We compared the different algorithms using the same dataset, but there should be room for further improvement of separation performance because even JADE and fastICA were still only about 60%–80% of their theoretically best performance (CCA separation). This suggests that there could be a better BSS algorithm for these datasets.

For the development of better BSS algorithms, sharing a common dataset must be helpful. We have shared our datasets and Matlab codes on our Web site (neurotycho.org) as a common testing platform. If a novel BSS algorithm is developed, the data and codes are freely available to anyone who wants to test their new algorithm. The shared testing platform may accelerate the speed of development, which will enhance the research field.

REFERENCES

1. Ball, T., Kern, M., Mutschler, I., Aertsen, A., & Schulze-Bonhage, A. (2009). Signal quality of simultaneously recorded invasive and non-invasive EEG. Neuroimage, 46, 708–716.
2. Belouchrani, A., Abed-Meraim, K., Cardoso, J.-F., & Moulines, E. (1997). A blind source separation technique using second-order statistics. IEEE Transactions on Signal Processing, 45, 434–444.
3. Besthorn, C., Zerfass, R., Geiger-Kabisch, C., Sattel, H., Daniel, S., et al. (1997).
4. Discrimination of Alzheimer's disease and normal ageing by EEG data. Electroencephalography and Clinical Neurophysiology, 103, 241–248.
5. Cardoso, J.-F., & Souloumiac, A. (1993). Blind beamforming for non-Gaussian signals. In IEE proceedings F (radar and signal processing), Vol. 140 (pp. 362–370). IET.
6. Choi, S., Cichocki, A., Park, H.-M., & Lee, S.-Y. (2005). Blind source separation and independent component analysis: A review. Neural Information Processing Review, 6, 1–57.
7. Crespo-Garcia, M., Atienza, M., & Cantero, J. L. (2008). Muscle artifact removal from human sleep EEG by using independent component analysis. Annals of Biomedical Engineering, 36, 467–475.
8. Daly, I., Pichiorri, F., Faller, J., Kaiser, V., & Kreilinger, A. et al. (2012). What does a clean EEG look like? In Conf proc IEEE eng med biol soc 2012 (pp. 3963–3966).
9. Delorme, A., Sejnowski, T., & Makeig, S. (2007). Enhanced detection of artifacts in EEG data using higher-order statistics and independent component analysis. Neuroimage, 34, 1443–1449.
10. Himberg, J., & Hyvarinen, A. (2003). Icasso: software for investigating the reliability of ICA estimates by clustering and visualization. In 2003 IEEE 13th workshop on neural networks for signal processing, 2003. NNSP'03 (pp. 259–268).
11. Hotelling, H. (1936). Relations between two sets of variates. Biometrika, 28, 321–377.

12. Hyvarinen, A. (1999). Fast and robust fixed-point algorithms for independent component analysis. IEEE Transactions on Neural Networks, 10, 626–634.
13. Jung, T., Makeig, S., Humphries, C., Lee, T., McKeown, M. J., et al. (2000). Removing electroencephalographic artifacts by blind source separation. Psychophysiology, 37, 163–178.
14. Kachenoura, A., Albera, L., Senhadji, L., & Comon, P. (2008). ICA: a potential tool for BCI systems. IEEE Signal Processing Magazine, 25, 57–68.
15. Makeig, S., Bell, A. J., Jung, T.-P., & Sejnowski, T. J. (1996). Independent component analysis of electroencephalographic data. In Advances in neural information process system, (pp. 145–151). MA: MIT Press.
16. Makeig, S., Westerfield, M., Jung, T.-P., Enghoff, S., Townsend, J., et al. (2002).
17. Dynamic brain sources of visual evoked responses. Science, 295, 690–694.
18. Mennes, M., Wouters, H., Vanrumste, B., Lagae, L., & Stiers, P. (2010). Validation of ICA as a tool to remove eye movement artifacts from EEG/ERP. Psychophysiology, 47, 1142–1150.
19. Mognon, A., Jovicich, J., Bruzzone, L., & Buiatti, M. (2011). ADJUST: An automatic EEG artifact detector based on the joint use of spatial and temporal features. Psychophysiology, 48, 229–240.
20. Muthukumaraswamy, S. D. (2013). High-frequency brain activity and muscle artifacts in MEG/EEG: a review and recommendations. Frontiers in Human Neuroscience, 7, 1–11.
21. Nagasaka, Y., Shimoda, K., & Fujii, N. (2011). Multidimensional recording (MDR) and data sharing: an ecological open research and educational platform for neuroscience. PLoS One, 6, e22561.
22. Nicolas-Alonso, L. F., & Gomez-Gil, J. (2012). Brain computer interfaces, a review. Sensors (Basel), 12, 1211–1279.
23. Ombao, H., & Ringo Ho, M. (2006). Time-dependent frequency domain principal components analysis of multichannel non-stationary signals. Computational Statistics and Data Analysis, 50, 2339–2360.

24. Oostenveld, R., Fries, P., Maris, E., & Schoffelen, J.-M. (2011). FieldTrip: open source software for advanced analysis of MEG, EEG, and invasive electrophysiological data. Computational Intelligence and Neuroscience, 1.
25. Oosugi, N., & Fujii, N. (2011). Neurotycho, http://neurotycho.org/eeg-and-ecogsimultaneous-recording.
26. Pfurtscheller, G., & Cooper, R. (1975). Frequency dependence of the transmission of the EEG from cortex to scalp. Electroencephalography and Clinical Neurophysiology, 38, 93–96.
27. Romero, S., Mañanas, M. A., & Barbanoj, M. J. (2008). A comparative study of automatic techniques for ocular artifact reduction in spontaneous EEG signals based on clinical target variables: a simulation case. Computers in Biology and Medicine, 38, 348–360.
28. Sanei, S., & Chambers, J. A. (2008). EEG signal processing. Southern Gate, Chichester: John Wiley & Sons.
29. Soong, A. C. K., & Koles, Z. J. (1995). Principal-component localization of the sources of the background EEG. IEEE Transactions on Biomedical Engineering, 42, 59–67.
30. Tong, L., Soon, V.C., Huang, Y.F., & Liu, R. (1990). AMUSE: a new blind identification algorithm. In Proceedings of IEEE ISCAS, 1990 (pp. 1784–1787).
31. Valdés-Sosa, P. A., Vega-Hernández, M., Sánchez-Bornot, J. M., Martínez-Montes, E., & Bobes, M. A. (2009). EEG source imaging with spatio-temporal tomographic nonnegative independent component analysis. Human Brain Mapping, 30, 1898–1910.
32. Vesanto, J., Himberg, J., Alhoniemi, E., & Parhankangas, J. (1999). Self-organizing maps in Matlab: the SOM toolbox. In Proceedings of the Matlab DSP conference. Vol. 99 (pp. 16–17).
33. Vigário, R., Sarela, J., Jousmiki, V., Hamalainen, M., & Oja, E. (2000). Independent component approach to the analysis of EEG and MEG recordings. IEEE Transactions on Biomedical Engineering, 47, 589–593.
34. Vorobyov, S., & Cichocki, A. (2002). Blind noise reduction for multisensory signals using ICA and subspace filtering, with application to EEG analysis. Biological Cybernetics, 86, 293–303.

35. Winkler, I., Haufe, S., & Tangermann, M. (2011). Automatic classification of artifactual ICA-components for artifact removal in EEG signals. Behavior and Brain Functions, 7, 30.
36. Zhukov, L., Weinstein, D., & Johnson, C. (2000). Independent component analysis for EEG source localization. IEEE Engineering in Medicine and Biology Magazine, 19, 87–96.
37. Zou, Y., Hart, J., & Jafari, R. (2012). Automatic EEG artifact removal based on ICA and hierarchical clustering. In IEEE int conf acoustics, speech and signal process, ICASSP, 2012 (pp. 649–652).

CHAPTER 10

SMART SAMPLING AND INCREMENTAL FUNCTION LEARNING FOR VERY LARGE HIGH DIMENSIONAL DATA

Diego G. Loyola R, Mattia Pedergnana, Sebastián Gimeno García

Aerospace Center (DLR), Oberpfaffenhofen, 82234 Wessling, Germany

ABSTRACT

Very large high dimensional data are common nowadays and they impose new challenges to data-driven and data-intensive algorithms. Computational Intelligence techniques have the potential to provide powerful tools for addressing these challenges, but the current literature focuses mainly on handling scalability issues related to data volume in terms of sample size for classification tasks.

This work presents a systematic and comprehensive approach for optimally handling regression tasks with very large high dimensional data. The proposed approach is based on smart sampling techniques for minimizing the number of samples to be generated by using an iterative approach that creates new sample sets until the input and output space of the function to be approximated are optimally covered. Incremental function learning takes place in each sampling iteration, the new samples are used to fine tune the regression results of the function learning algorithm. The accuracy and confidence levels of the resulting approximation function are assessed using the probably approximately correct computation framework.

The smart sampling and incremental function learning techniques can be easily used in practical applications and scale well in the case of extremely large data. The feasibility and good results of the proposed techniques are demonstrated using benchmark functions as well as functions from real-world problems.

INTRODUCTION

Computer-based simulations of tremendously complex mathematical systems describing multifaceted physical, chemical, dynamical and engineering models are usually associated with very expensive costs in terms of processing time and storage. Complex mathematical models are present in a wide variety of scientific areas such as the simulation of atmospheric processes in numerical weather prediction (Han & Pan, 2011; Hsieh & Tang, 1998; Lynch, 2006; Morcrette, 1991), climate modeling (Flato et al., 2013), (Gordon et al., 2000), chemical transport (Grell et al., 2005), (Menut et al., 2013), radiative transfer (Gimeno García, Trautmann, & Venema, 2012) and large eddy simulations (Sagaut, 2006). Other scientific disciplines such as genetics, aerodynamics, or statistical mechanics also make use of highly complex models. The input space of these models can be of high dimensionality with hundreds or more components. The usage of more realistic models usually introduces new dimensions leading to an exponential increase in volume, i.e. "Big Data" (Hilbert & López, 2011; Lynch, 2008).

The speeding-up of such models is a crucial problem in practical applications like weather forecasting, remote sensing, and climate modeling among others. These complex models can be approximated using for example neural networks (Haupt, Pasini, & Marzban, 2009; Loyola, 2006; Qazi & Linshu, 2006), support vector machines (Tripathi,

Srinivas, & Nanjundiah, 2006), kernel smoothing models (Cervellera & Macciò, 2013), and the resulting computational intelligence systems are being deployed in operational environments, see for example Krasnopolsky and Chevallier (2003) and Loyola (2006).

Related work found in the literature usually addresses function approximations from complex mathematical models in the general framework of statistical machine learning where it is assumed that the training samples are created independently according to an unknown probability density function (Krasnopolsky & Schiller, 2007). However, having a mathematical model to parameterize, we can freely choose the sample points that better characterize the input and output spaces of the model. The selection of samples to be used in function approximation problems is less explored in the literature, sometimes it is called design of experiments (Sacks, Welch, Mitchell, & Wynn, 1989) or active learning (Enăchescu, 2013).

In this study we develop a general function approximation framework in which the choice of the samples describing the function is an integral part of the learning problem. The goal is to find a regression model that accurately approximates a function $f : X \rightarrow Y$ with input $X \subset \mathbb{R}^n$ and output $Y \subset \mathbb{R}^m$ from a set of training samples $T = \{x_i, y_i\}$ for $i = 1$ to s, subject to the constraint of minimizing the number of calls to the target function f, i.e. minimizing the number of samples s. The dimensionality of the data needed to solve this problem is characterized by the number of samples s and the dimensionality of the mapping represented by f with n inputs and m outputs.

On one side, the complexity of a function does not necessarily increase for a high dimensional input space; moreover the limiting factor for an accurate function approximation is the intrinsic complexity of the target function and not the dimension of the input data space (Kolmogorov, 1957). Accuracy of approximation can be achieved in high dimensional cases for target functions with lower complexity (Gnecco, Kůrková, & Sanguineti, 2011). On the other side, systematically sampling the input space with k values per dimension will require in total $s=k^n$ calls to f, therefore the number of samples s for mappings with high dimensional input size grows exponentially with the input dimension n.

These difficulties are usually referred as "the curse of dimensionality" problem (Vapnik, 2006) and it has severe consequences not only for the time needed to create the training dataset but also for the algorithms needed to solve function approximation problems that must deal with very large number of samples of high dimensionality, i.e. "Big Data".

The data volume component of Big Data is a hot topic in machine learning, see for example Jin and Hammer (2014) and O'Leary (2013) and the references therein. But the main focus on the current literature is the sample size and not the sample dimension problem. Only a few studies address the big dimension problem such as for example theoretical investigations of computational models efficiency in high dimensional context (Kainen, Kůrková, & Sanguineti, 2012) and for classification tasks wherein the explosion of features brings about new challenges to computational intelligence (Yiteng, Yew-Soon, & Tsang, 2014).

In this article we focus on the under-explored topic of big dimensionality for regression tasks. The paper is organized as follows: Section 2 gives an overview of data sampling methods and discrepancy measurements. Section 3 shows a comparison of sampling methods and evaluates their performance for high dimension input problems. Section 4 presents the smart sampling and incremental function learning (SSIFL) algorithm that optimally solves this kind of function approximation problems and Section 5shows the results of applying SSIFL to a number of benchmark and real-world functions. Finally, the conclusions are given in Section 6.

DATA SAMPLING METHODS

Generally speaking a good sampling method should create training data that accurately represents the underlying function preserving the statistical characteristics of the complete dataset. This section presents first a survey of sampling methods grouped in four categories that can be applied to generating high dimensional sample data.

Stochastic Methods

Pseudo-Random number generators (PR) (Marsaglia & Tsang, 2000; Matsumoto & Nishimura, 1998) are commonly used to create samples in a given range. PR methods are fast and simple to use, but they are not distributed uniformly enough especially for the cases of low number of sampling points and/or large number of dimensions. PR sampling is sometimes called "pure" or "plain" Monte Carlo (Swiler, Slepoy, & Giunta, 2006).

Deterministic Methods

Uniform populations can be created using quasi-random or sub-random sequences that cover the input space quickly and evenly, the uniformity and coverage improves continually as more data points are added to the sequence. Deterministic methods can be divided into two subcategories (Kazimipour, Li, & Qin, 2013) described in the next subsections.

Low Discrepancy Methods

Low Discrepancy methods have the support of theoretical upper-bounds on discrepancy. Halton, Sobol, Niederreiter, Hammersley, and Faure are well known sequences from this category.

In this work we use Halton sequences, shortened as (HA), which are constructed according to a deterministic algorithm that uses prime numbers as bases for each dimension (Halton, 1960). Halton sequences work well in low dimensionality, but they lose uniformity in high dimensions. Workaround solutions such as using big prime numbers, setting leap values, scrambling and shuffling improve the sampling uniformity in such cases. The HA method is computationally efficient even for very high dimensional spaces.

Experimental Design Methods

Experimental Design methods are commonly used for initializing the population of evolutionary algorithms in order to accelerate convergence speed and improve stability. The most representative methods in this category are:

- Uniform Design: a space-filling method based on prime numbers that generates points uniformly scattered on the input domain (Peng, Wang, Dai, & Cao, 2012).
- Orthogonal Design: based on Latin squares for creating orthogonal arrays (Leung & Wang, 2001).

Geometrical Methods

Uniform Grid

Uniform grid (UG) is the simplest sampling method in which the samples are created using node points at fixed intervals uniformly distributed for every dimension. UG is commonly used for creating look-up tables to accelerate complex model computations (Perkins et al., 2012; Richter, Heege, Kiselev, & Schläpfer, 2014).

Using a uniform grid is convenient for storing the look-up tables in multi-dimensional arrays. As we will show in Sections 3.2 and 3.3, the coverage of the input space using UG is very poor for low number of sampling points and high dimensions.

Latin Hypercube

Latin hypercube sampling (McKay, Beckman, & Conover, 1979) partitions the input space into bins of equal probability and distributes the samples in such a way that only one sample is located in each axis-aligned hyperplane. This method and variations like nearly-orthogonal Latin hypercube (Cioppa, 2002) and Distributed Hypercube and Improved Hypercube sampling (Beachkofski & Grandhi, 2002) are very popular in computer model problems. A Particle Swarm Optimization algorithm for solving large-scale Latin hypercube design problems was proposed recently (Aziza & Tayarani-N., 2014).

Latin hypercube sampling is useful when the underlying function has a low order distribution but this method produces clustering of sampling points at high dimensions.

Centroidal Voronoi Tessellation

Centroidal Voronoi Tessellation (CVT) generates sample points located at the center of mass of each Voronoi cell covering the input space. The CVT algorithm starts with an initial partition and then iteratively updates the estimate of the centroids of the corresponding Voronoi sub regions. Recent advances in mathematical and computational studies and in practical applications of CVT are presented in Du, Gunzburger, and Ju (2010).

One important advantages of CVT is that sampling points may be easily distributed according to a prescribed density function. One drawback of the CVT method lies in being computationally demanding for high dimensional spaces.

Opposite Methods

The Opposition Based Learning initialization is a well-known method used to initialize the population in evolutionary algorithms (Rahnamayan, Tizhoosh, & Salama, 2008). The sequence of opposite points (complimentary point with respect to given bounds) is created from the original sequence, then both sequences are merged and a fitness function is used to select the best samples. The Quasi-opposition Based Learning (Rahnamayan, Tizhoosh, & Salama, 2007) is an improvement on the previous method that increases the sequence uniformity by creating opposite points using a random range.

The opposite sampling methods are computationally very effective and straightforward to use.

Hybrid Methods

Hybrid methods combine different sampling methods with the aim of improving the sampling coverage and uniformity. A sampling method can be used as (a) pre-processing step, e.g. the computation of CVT samples can be initialized with stochastic or deterministic methods and (b) as post-processing, e.g. using geometrical methods to reorganize or create new samples.

In this work we use Latinized CVT (LCVT) (Romero, Burkardt, Gunzburger, & Peterson, 2006). We create first a Halton sequence as initial population, then the CVT sample is computed and finally the sequence is Latinized.

Sampling in High Dimensional Space

The sampling methods usually perform very well in low dimensions, the top panels of Fig. 1 contrast the 2-D input space sequences created using 25 and 100 points drawn from uniform grid, pseudo random, Halton, CVT, and LCVT methods. A visual inspection of the sampling sequences in a 2-D space gives a fast qualitative feedback on the degree of uniformity achieved and how it varies over the input space. In the 2-D case using a uniform grid sampling is very appealing, the points are well spread and the samples with 25 and 100 points are complementary. Note that the Halton sequences with 25 and 100 points overlap whereas the corresponding sequences created with the other sampling methods do not. A workaround solution for creating complementary Halton sequences is to use different prime numbers as basis of different sequences. The random sequence is less uniform than the other ones.

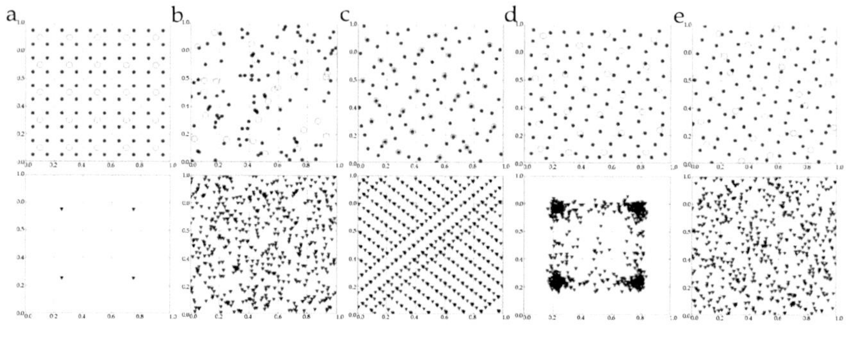

Figure. 1. Sampling sequences generated with (a) uniform grid, (b) pseudo-random, (c) Halton, (d) CVT, and (e) LCVT. The top panel shows the sequences corresponding to 2-D input space with a total of 25 (white circle) and 100 (black points) samples. The bottom panel shows plots of the same sampling sequences for a 9-D input space sampled with 2 points per dimension giving a total of 2^9=524 samples projected to the dimension 7 and 9.

Most of the sample strategies have problems with high dimensional sequences as can be appreciated in the bottom panels of Fig. 1 showing sampling sequences for a 9-D input space with 2 points per dimension given a total of 512 samples. The plots show the sampling points projected into dimensions 7 and 9. All the points with uniform grid sampling concentrate on only 4 places as this sampling algorithm allocate two fixed grid points per dimension. Halton has well known spurious correlation problems (Robinson & Atcitty, 1999) that leave large empty regions not covered, this problem can be at least partially solved using alternative algorithms like scrambled Halton. The points created with the CVT algorithm concentrate mainly in four regions reflecting the good volumetric but bad spatial uniformity of the samples when projected to single dimensions. Only LCVT behaves qualitatively better than pure random.

Developing algorithms for creating good sampling uniformity in high dimensional spaces with low computational cost remains an area of active research.

Sampling Non-Uniform Distributions

A sequence of uniformly distributed points can be mapped into another sequence that reflects a desired non-uniform join probability distribution function using different techniques like the rejection method or weighted sampling (Moskowitz & Caflisch, 1996). In this paper we use the more generic procedure described in Romero et al. (2006). The cumulative distribution function CDF(x) for a random variable x is

$$CDF(x) = \int_0^{x \leq 1} PDF(x')dx' \quad (1)$$

where PDF(x) is the probability density function of the random input x. Since the CDF is defined in the unit hypercube domain it is possible to inverse-map sequences uniformly distributed between 0 and 1 into sequences {xi} drawn from the density function PDF(x).

Importance Sampling

Importance sampling is a commonly used method for variance in Monte Carlo integration (Moskowitz & Caflisch, 1996) and system reliability analysis (Patelli, Pradlwarter, & Schuëller, 2011). An importance function h is introduced which mimics the behavior of the function f to approximate, sampling points are generated according to the distribution of h instead of being uniformly distributed. In this work we create importance sampling sequences following the distribution of h by first creating a uniformly distributed sequence and then computing the inverse-map using (1) and the normalized h (integral equals 1) as a density function.

The top panel of Fig. 2 depicts the sampling sequences corresponding to the top panel of Fig. 1 transformed using as important sampling function an exponential function. The middle panel of Fig. 2 illustrates the sampling sequences corresponding to the bottom panel of Fig. 1 transformed using as important sampling function a 2-D Gaussian probability distribution with standard deviation of 0.8 and 1.2. Finally the bottom panel of Fig. 2 shows histograms resulting from the sampling sequences of the middle panel. The problems of the sampling methods in this 9-D case are more evident in the histograms. All samples from uniform grid concentrate on 4 positions, similarly the samples from CVT concentrate on 4 regions.

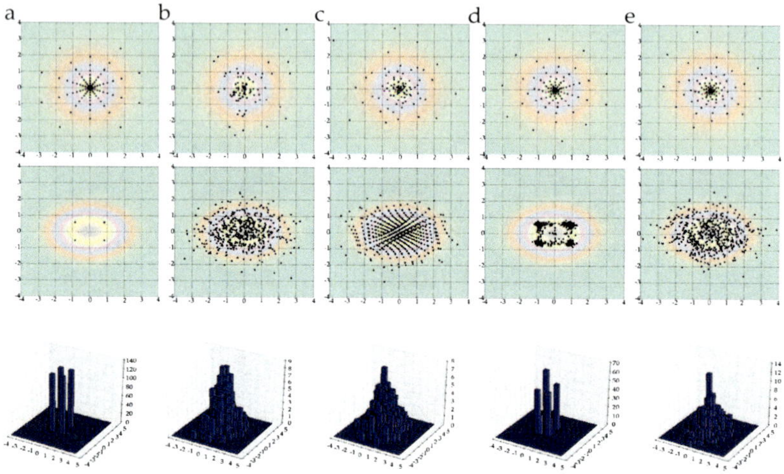

Figure. 2. Sampling sequences (black points) created from the sequences from Fig. 1 resampled using important sampling functions represented as color surface plot on the background. (top panel) 2-D input space with 100 samples and exponential important sampling function. (middle panel) A 9-D input space with 512 samples projected to the dimension 7 and 9 and resampled with a 2-D Gaussian probability distribution important sampling function. (bottom panel) Histogram showing the distribution of the samples shown in the middle panel. (For interpretation of the references to color in this figure legend, the reader is referred to the web version of this article.)

COMPARISON OF SAMPLING METHODS

This section presents a systematic comparison of selected sampling methods from the different categories described in the previous chapter. First the benchmark functions used for the comparison are presented, then we describe how to quantitatively assess the goodness of the sampling methods both in: the input space (uniformity measures) and in the output space (statistical measures). Finally the results of the different sampling methods applied to a number of benchmark function are discussed.

Benchmark Functions

Three benchmark functions commonly used to test the performance of global optimization algorithms are used in this work. Asymmetric benchmark functions containing multiple minima and maxima were selected to better evaluate the effects of the different sampling strategies.

Shifted Ackley Function

A number of unimodal and multimodal basic functions like Sphere, Rosenbrock, Rastrigin, Weierstrass, Griewank, and Ackley are commonly used for testing optimization problems. We selected the shifted Ackley's function defined in a multi-dimensional input space as

$$F_{SA}(x) = -20\exp\left(-0.2\sqrt{1/n\sum_{i=1}^{n} z_i^2}\right) - \exp\left(1/n\sum_{i=1}^{n}\cos(2\pi z_i)\right) + 20 + e + f\_bias_i \quad (2)$$

with z=(x−o) the shifted global optimum. This function is multi-modal, shifted, separable and scalable, see Tang et al. (2008).

Composition Function

More complex functions can be created by combining basic functions

$$F_{CO}(x) = \sum_{i=1}^{n} \{w_i * [f_i(x_i - o_i)/(\lambda_i * M_i) + bias_i] + f\_bias_i\}. \quad (3)$$

The various parameters used in this equation define the way the basic functions are combined, see CF5 in Liang, Suganthan, and Deb (2005) for details.

Fast Fractal Double Dip function

This multimodal function is defined as

$$F_{FD}(x) = \sum_{i=1}^{n} fractal(x_i + twist(x_{(i \bmod n)+1})). \quad (4)$$

Specifics about this function including the definition of *twist*() and *fractal*() can be found in Tang et al. (2008).

Fig. 3 shows 3-D plots of the functions Shifted Ackley, Composition, and Fast Fractal Double Dip for the case of 2-D input space.

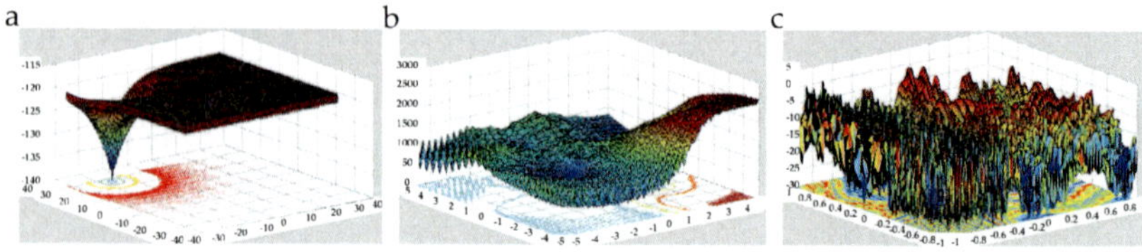

Figure. 3. Plots of the benchmark functions for a 2-D input space: (a) Shifted Ackley, (b) Composition, and (c) Fast Fractal Double Dip.

Discrepancy as Measure of Uniformity in the Input Space

Discrepancy is a quantity that measures the uniformity of a set of points in a hyper dimensional cube and it is usually employed in the integral approximation field. Given a dataset of s points $P = \{x_1, \ldots, x_s\}$ with $x \subset \mathbb{R}^n$ the n-dimensional unit cube In=$[0,1)^n$,n≥1, the infinite star discrepancy $D_s^*(P)$ is defined as

$$D_s^*(P) = \sup_{J \in I^*} \left| \frac{A(J, P)}{s} - V(J) \right| \tag{5}$$

where I^* is the family of all subintervals of $I^n=[0,1)^n$ of the form $\prod_{i=1}^{n} [0, J_i)$, $V(J)$ is the volume of a subinterval J, and A(J,P)is the number of points of P belonging to J (Niederreiter, 1992). The star discrepancy describes the relationship between the number of points in a subinterval and its volume. $D_s^*(P)$ varies between 0 and 1; low discrepancy values indicate that the points are well scattered over the input space. The optimal sampling methods are the ones that rapidly converge to a star discrepancy of 0 as the number of sample points increases.

The computation of $D_s^*(P)$ for $n \geq 2$ is a time consuming and not an easy task (Thiémard, 2001). Alternatives more suitable for high dimensional data are the L2-star discrepancy (Braaten & Weller, 1979)

$$D_s^{L_2}(P) = \sqrt{\frac{1}{s} \sum_{k=1}^{s} \left(\frac{A(J_{(k)}, P)}{s} - V(J_{(k)}) \right)^2} \tag{6}$$

and the modified L2-star discrepancy (Qazi & Linshu, 2006) defined as

$$D_s^{ML_2}(P) = \left(\frac{4}{3}\right)^n - \frac{2^{1-n}}{s}\sum_{d=1}^{s}\prod_{i=1}^{n}\left(3 - x_{di}^2\right) + \frac{1}{s^2}\sum_{d=1}^{s}\sum_{j=1}^{s}\prod_{i=1}^{n}\left(2 - \max\left(x_{di}, x_{ji}\right)\right). \quad (7)$$

The discrepancy of 5-D input space sequences as function of the number of samples created using different sampling methods is depicted in Fig. 4. The three plots correspond to discrepancy computed with (5), (6), and (7). The best sampling methods are HA and LCVT with similar convergence rates of discrepancy; both outperform pseudo random sequences by a factor of 10 regarding the number of samples needed to reach a given discrepancy. The worst sampling method is by far the uniform grid; it needs 10^5samples to reach similar discrepancy levels achieved with only 10^2samples created with HA or LCVT.

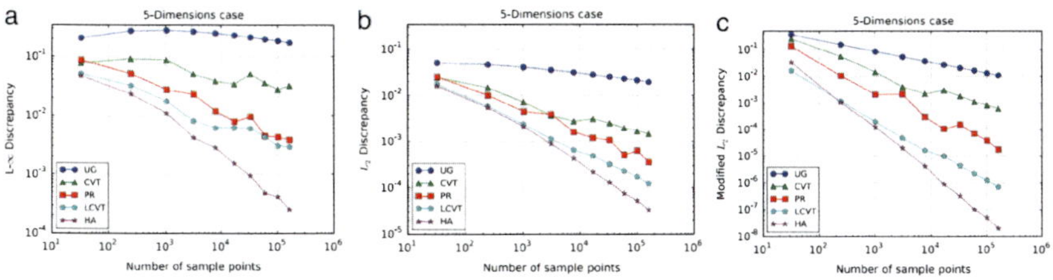

Figure. 4. Discrepancy measurements in (a) infinite, (b) L2 and (c) modified L2 star norms for 5-D input space sequences as function of the number of samples. The five curves per plot correspond to different sampling methods: uniform grid (UG), CVT, pseudo-random (PR), LCVT and Halton (HA).

It is important to note that discrepancy is calculated using the 5-D samples and not their 2-D projections depicted in the plots. Spurious correlation problems on samples projected into single dimensions, like the ones from HA shown in the bottom of Fig. 1, are not reflected in such discrepancy measurements. The left plot of Fig. 5 shows the modified L2-star discrepancy as function of the number of samples for a 9-D input space for LCVT and HA sampling sequences, the error bars correspond to the worst discrepancy computed in the 2-D projections. As it can be appreciated, LCVT outperforms HA when relative few samples are available but the situations change around 700 samples where HA has a lower discrepancy as LCVT. The right plot of Fig. 5 shows the inflection point between LCVT and HA modified L2-star discrepancy as function of the input space dimension and the number of samples.

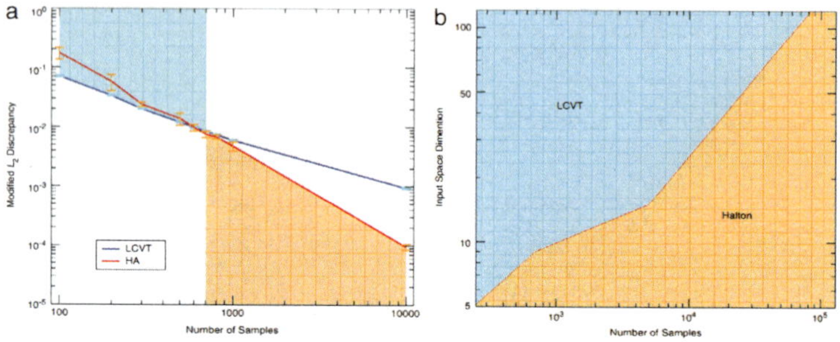

Figure. 5. (a) Modified L2-star discrepancy as function of the number of samples for 9-D input space for LCVT (cyan) and HA (orange) sampling sequences, LCVT performs better than Halton for sequences with less than 700 samples. (b) Sampling method with lower discrepancy as function of the input space dimension and the number of samples, LCVT outperforms Halton in case of low dimensions and relative few samples. (For interpretation of the references to color in this figure legend, the reader is referred to the web version of this article.)

Statistical Measures of the Response Function in the Output Space

The goodness of a sampling method with respect to the coverage in the output space of a response function f can be assessed by the multivariate function integration expressed as

$$I(f) = \int_{C^n} f(x)dx \tag{8}$$

with $C^n = [0, 1)^n$.. The sample mean gives an approximation to I(f) by

$$\hat{I}(f, P) = \frac{1}{s}\sum_{i=1}^{s} f(x_i) \tag{9}$$

where $P = \{x_i\}$ for $i = 1$ to s is a set of points in C^n. If the sampling points xi are i.i.d. uniformly distributed on Cn then it is known that the sample mean is unbiased. The Koksma–Hlawka inequality gives an upper bound for the approximation error

$$\left| I(f) - \hat{I}(f, P) \right| \leq D(P)\upsilon(f) \tag{10}$$

where D(P) is the discrepancy of P as defined in Section 3.2, and v(f) is a measure of the variation of f as defined in Fang, Ma, and Winker (2002). If the sampling method used to create Pconverges to a discrepancy of 0 as the number of sample points increases, then this inequality ensures that the sample mean converges to the multivariate function integral. Therefore the convergence rate of the sample mean from the output space of a response function depending on the number of sample points can be used to measure the goodness of a sampling method.

The efficacy of sampling methods is commonly compared using estimates of response function mean and standard deviation (Romero et al., 2006). In this work we propose to use not only the mean and standard deviation, but also the higher order statistics skewness (measure of the asymmetry) and the Kurtosis (measure of the peakedness) as they provide complementary information of f computed from the samples in P.

Fig. 6 depicts the statistical moments of the Fast Fractal Double Dip function for a 5-D input space and the Composition function for a 10-D input space as function of the number of samples. The uniform grid sampling for the case of 10 dimensions does not converge even with a large number of samples, the performance of CVT is slightly better, while PR, LCVT and HA nicely converge with increasing samples.

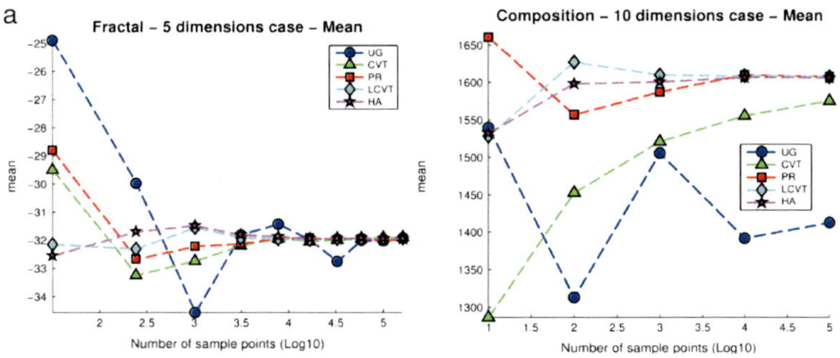

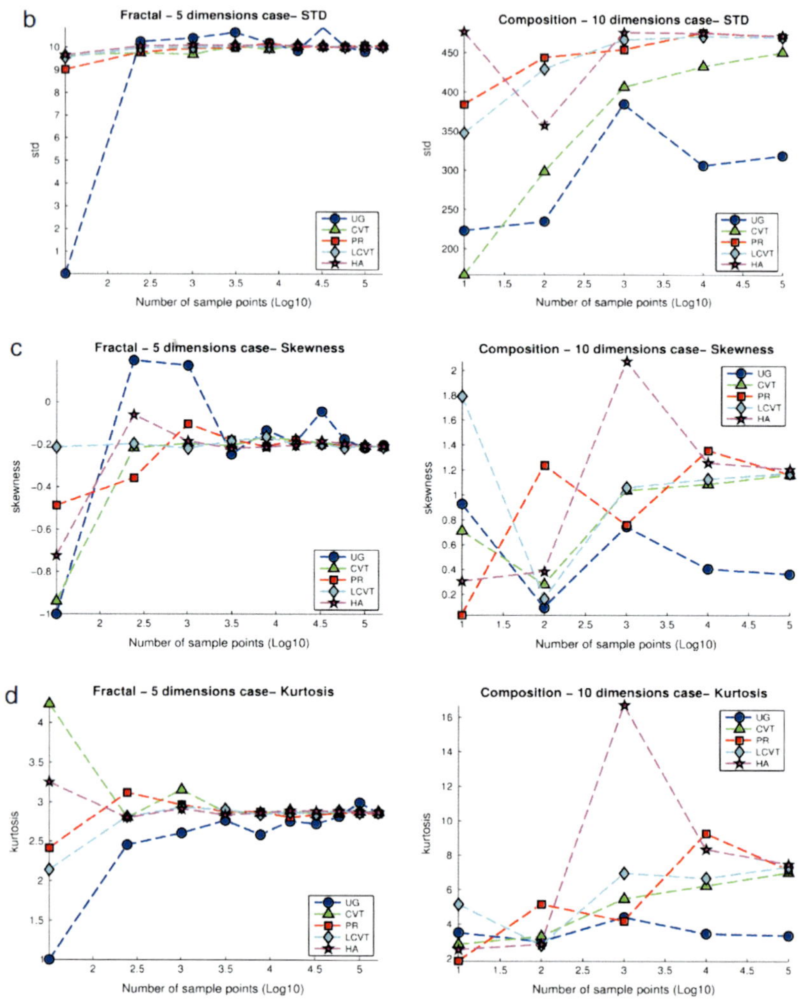

Figure. 6. Statistical moments (a) mean, (b) standard deviation, (c) skewness, and (d) kurtosis of the response function in the output space as function of the number of sample patterns for two of the benchmark functions depicted in Fig. 3. The left panels correspond to the Fast Fractal Double Dip function for a 5-D input space and the right panels correspond to the Composition function for a 10-D input space.

The convergence of the statistical moments on the response function can be used to take an informed decision on the adequate number of samples required to properly cover the output space.

SMART SAMPLING AND INCREMENTAL FUNCTION LEARNING ALGORITHM

In this section we present the Smart Sampling and Incremental Function Learning Algorithm (SSIFL) based on the results from the previous section. First the problem is formalized, then the algorithm SSIFL stages are listed and then finally the single steps are described in detail in the following subsections.

The goal of the SSIFL algorithm is to find a Probably Approximately Correct Computation (PACC) regression model f̂ that accurately approximates a Lebesgue measurable function $f:X \rightarrow Y$ with input $X \subset R^n$ and output $Y \subset R^m$ from a set of finite training samples $T=\{x_i,y_i\}$ for i=1 to s, subject to the constraint of minimizing the number of calls to the function fneeded to create the training samples s.

Following the PACC framework from Alippi (2014) and using as figure of merit $u(x) = \left(\left|f(x) - \hat{f}(x)\right|\right)$, we have that f̂ is a PACC of function f at accuracy τ and confidence η when

$$\Pr\left(\left|f(x) - \hat{f}(x)\right| \leq \tau\right) \geq \eta, \quad \forall x \in X. \tag{11}$$

In other terms, the function f̂ is approximately correct in the sense that it approximates $\hat{f}$ at level τ with probability η. If τ is small, then $\hat{f}$ ^ provides a value which approximates the true f with high probability.

The PACC formalism allows a probabilistic performance estimation of τ and η from a finite dataset as described in Section 4.6. The minimum number of samples sPACC required for assessing inequality (11) can be calculated using the Chernoff bound (Chernoff, 1952) as

$$s_{PACC} \geq \frac{1}{2\varepsilon^2} \ln \frac{2}{\delta} \tag{12}$$

where ε∈(0,1) are the expected accuracy and confidence 1−δlevels, δ∈(0,1). According to this inequality, the accuracy is more demanding on the number of samples (quadratic bound) than the confidence (linear bound). It is important to note that the Chernoff bound does not depend on the input dimension of f, that means that the PACC framework allow us to escape the curse of dimensionality problem for assessing τ and η (Alippi, 2014).

The SSIFL algorithm is described in the following table. The algorithm numbered steps are described in more detail in the following subsections, the step numbers are the same as the corresponding subchapter numbers.

Algorithm 1: Smart Sampling and Incremental Function Learning Algorithm

Input:

Lebesge measurable target function $f: X \rightarrow Y$ with $X \subset \mathbb{R}^n$ and $Y \subset \mathbb{R}^m$

Output:

Approximator function $\hat{f}$ with accuracy τ and confidence η

Begin

1. *// Initialization phase:*
   Select accuracy ε and confidence δ
   Compute s_{PACC} using (12)
   Set s_{val} such that $s(n, s_{val} - 1) \leq s_{PACC} \leq s(n, s_{val})$
   $iteration = 0$
   Create function approximator $\hat{f}(\theta_0)$ using initial model parameters θ_0
2. *// Initialize test data set and create train dataset using smart sampling*
   $Test = \{\}$
   Train = Create input/output dataset using $s(n, s_{val})$ sampling points
   Repeat
3. *// Use previous train as test dataset and create new train dataset*
   Test = *Train*
   Train = Create input/output dataset using $s(n, s_{val}+iteration)$ points
4. *// Incremental function learning*
   $\hat{f}(\theta_{iteration-1}, Train, Val) \mapsto \hat{f}(\theta_{iteration})$
   $iteration = iteration + 1$
5. **Until** Convergence reached
6. *// Determine approximator accuracy and confidence levels*
   Val = Create input/output dataset using $s(n, s_{val})$ sampling points
   Compute τ and η of $\hat{f}$ using *Val* dataset

End

Initialization Phase

The number of samples to create the approximation function with input space dimension n is selected using the following sequence

$$s(n, i) = \begin{cases} i^d & \text{if } n \leq 9 \\ 10^i & \text{if } n \geq 10. \end{cases} \tag{13}$$

This equation creates a sequence of the form $\{2^d, 3^d, 4^d, \ldots\}$ for 9 input dimensions or less, for larger dimensions the sequence is $\{10^2, 10^3, 10^4, \ldots\}$. For given dimension and index, the sequence s(d,i) determines the number of samples following a good compromise on the number of samples to generate per iteration. Alternative formulations can be used as well.

The accuracy ε and confidence δ levels are selected, then sPACCis computed using (12) and s_{val} is selected from the sequences defined by (13) such that $s(n, s_{val} - 1) \leq s_{PACC} \leq s(n, s_{val})$. In other terms, the minimum number of samples used to create the training, testing and validation datasets is selected such that it satisfies the Chernoff bound for the given ε and δ.

The number of *iteration* is set to 0 and the function approximator $\hat{f}(\theta_0)$ is created using the initial model parameters θ0. In case that a neural network is used, then the model parameters are the network weights and biases.

Smart Sampling during Initialization

The sampling method LCVT or HA is selected based on the right plot of Fig. 5 and using the current input space dimension n and the number of samples s to generate. Once the sampling method is selected then the set of inputs $X = \{x_1, \ldots, x_s\}$ with $X \subset \mathbb{R}^n$ is created.

In case that a probability density function of f() is available then the input sequence is resampled using the important sampling technique as described in Section 2.7.

In the next step, the output $Y = \{f(x_i)\}$ for $i = 1$ with $Y \subset \mathbb{R}^m$ is created. This is usually the most time consuming part of the sampling generation process, the processing time needed for computing f(x) for only a few samples is usually larger than the time needed for creating the complete input sequence.

Finally the dataset $Train = \{x_i, y_i\}$ for $i = 1$ to s samples is created and the *Test* dataset is initialized with an empty set.

It is important to note that by using LCVT or HA the input space is uniformly covered independently of the number of samples generated.

Smart Sampling during Iteration

First the *Train* dataset from the previous iteration is used as the new *Test* dataset and then a new *Train* dataset is created using the generic procedure described in the previous subsection. The number of samples of the *Train* dataset is increased in each iteration according to the sequences defined by (13) and starting with $s(n,s_{val})$. In other words, the first training dataset contains the same amount of samples as the testing dataset (satisfying the Chernoff bound) and then new samples are created in each iteration.

It is worth noticing that the test and training datasets generated in each smart sampling iteration are complementary (no duplicated inputs) and they cover the full input/output space.

Incremental Function Learning

The goal of learning is to build a PACC approximator of the target function f()

$$\hat{f}(\theta) \approx f() \tag{14}$$

with θ the parameter vector describing the family of models. In case that artificial neural networks are used as approximators, then θ represents the weights and biases of the network.

In each iteration, the parameter vector θ is updated based on the information present in training dataset *Train.* Overfitting is avoided using the early-stopping technique based on the average approximation error over the testing dataset *Test*.

The training algorithm is selected according to the dimensions of the network input/output and the number of training samples. For small- and medium-sized dimensionality the Levenberg–Marquardt algorithm can be used, but with large- and high-size dimensions a backpropagation algorithm working in parallel over a multi-core architecture (Schuessler & Loyola, 2011) is preferred.

The smart sampling procedure described in the previous subsections ensures that the size of the training dataset is continuously increased in every iteration. That means that the target function is learned from the training data in an incremental way during each iteration.

The proposed incremental function learning optimizes the function approximation problem by learning first the ground structure of the target function with relative few sampling points that cover the complete input/output space and then fine-tuning the model parameters to better approximate the target function using more and more sampling points in every iteration. This process is visualized in Fig. 7 showing the incremental learning of the Rosenbrock function in 5-D, the learned function accurately approximates the target function after a few iterations.

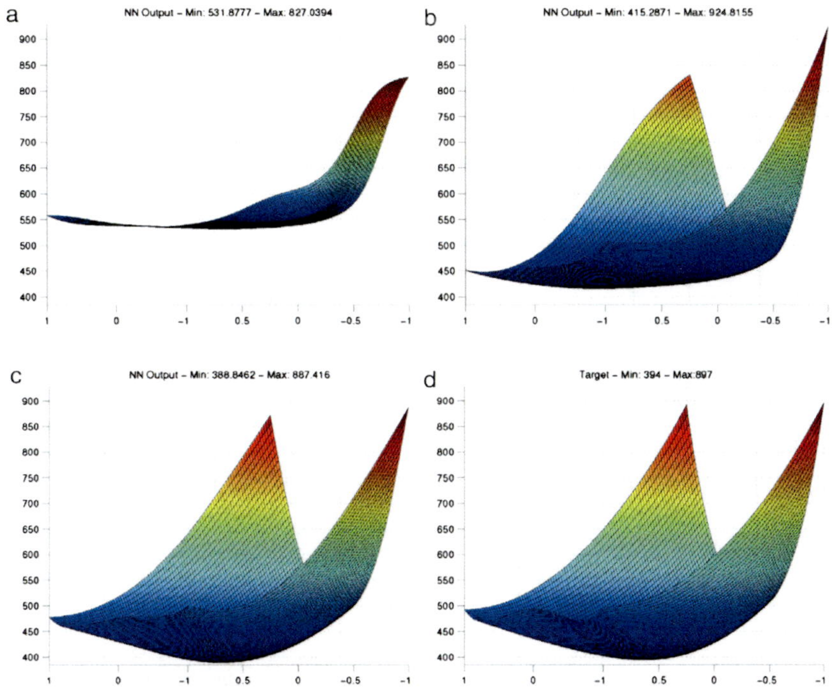

Figure. 7. Incremental learning of the Rosenbrock function in 5-D input space. The plots show the approximation results from (a) 4^5 samples, (b) 6^5 and (c) 8^5 samples. After a few iterations the learned function accurately approximates the tar get function (d).

Check Convergence

The iteration over smart sampling for creating training datasets and incremental function learning is repeated until pre-defined convergence criteria are reached.

The first criteria are the convergence of the statistical moments of the target function output from iteration to iteration as described in Section 3.3. The difference between the statistical moments of Yfrom the current iteration $Y_i = f(X_i)$ and the previous iterations $Y_{i-1} = f(X_{i-1})$ should be smaller than a given delta to ensure that the target function is well represented by the generated sampling points.

Interestingly, using the minimum number of samples sPACC for calculating the expectation E[f(X)] of the function f(X) by estimating the empirical mean $\hat{E}_{Sm}\left[f(X)\right] = 1/n \sum_{i}^{Sm} f(x_i)$ ensures that the following inequality

$$P\left(\left|\hat{E}_{Sm}\left[f(X)\right] - E\left[f(X)\right]\right| \leq \varepsilon\right) \geq 1 - \delta \tag{15}$$

holds for any accuracy level ε∈(0,1) and confidence δ∈(0,1)(Alippi, 2014). That means that the convergence check on the first statistical moment is actually a convergence check on the expectation of the target function as function of the number of samples.

The second criteria are the convergence of the histograms of the network parameters θ from iteration to iteration. Following the Bayesian formalism applied to neural networks (MacKay, 1995), learning means changing our belief about the network parameters θ from the prior P(θ) to the posterior P(θ|T) as a consequence of seeing the training data *Train*

$$P(\theta|Train) = \frac{P(Train|\theta)P(\theta)}{P(Train)}. \tag{16}$$

P(Train|θ) is the likelihood and P(Train) is called the evidence for the neural network. In other terms, the posterior distribution incorporates the prior knowledge and the information conveyed by the training dataset.

The Occam factor (MacKay, 1995) that penalizes a neural network for having the parameters θ defined as

$$Occam\, factor = \frac{\sigma_{\theta|Train}}{\sigma_{\theta}} \tag{17}$$

where $\sigma_{\theta|Train}$ is the width of the evidence and the prior is uniform on a large interval σ_{θ}. The second convergence criteria are reached when the ratio of the *Occam factors* from iteration to iteration is close to one. This ratio is difficult to compute directly but it can be approximated by the ratio of the FWHM from the histogram of the network parameters θ_i and θ_{i-1} from the previous and current iteration.

The third and last stop criterion is the check that a maximum allowed number of iterations are reached.

Determine Approximator Accuracy and Confidence Levels

The PACC formalism allows a probabilistic performance estimation of the accuracy τ and confidence η of $\hat{f}(x)$ from Eq. (11) by defining a p(γ) such that

$$p(\gamma) = \Pr\left(\left|f(x) - \hat{f}(x)\right| \leq \gamma\right). \tag{18}$$

Following the method from Alippi (2014), this probability can be approximated with $\hat{p}(\gamma)$ defined as

$$\hat{p}(\gamma) = \frac{1}{s}\sum_{i=1}^{s} L\left(\left|f(x_i) - \hat{f}(x_i)\right| \leq \gamma\right),$$

$$L(u(x_i) \leq \gamma) = \begin{cases} 1 & \text{if } u(x_i) \leq \gamma \\ 0 & \text{if } u(x_i) > \gamma \end{cases} \tag{19}$$

and evaluated over the finite dataset $x_i \in Val$. The *Val* dataset is created using the generic procedure described in Section 4.2 and using $s(n, s_{val})$ sampling points, that means that the *Val* dataset has at least sPACC element and therefore the Chernoff bound holds.

The value $\bar{\gamma}$ is defined as the smallest value from an arbitrary incremental set of points $\{\gamma_1, \ldots, \gamma_k\}$ with $\gamma_i < \gamma_j, \forall i < j$ $\gamma_i < \gamma_j, \forall i < j$ from which $\hat{p}(\gamma_i) \geq 1 - \varepsilon, \forall \gamma_i \geq \bar{\gamma}$. Selecting $k \geq s_{PACC}$ points, the Chernoff bound discrepancy $|p(\bar{\gamma}) - \hat{p}(\bar{\gamma})| \leq \varepsilon$ holds with probability $1 - \delta$ and therefore $p(\bar{\gamma}) \geq 1 - \varepsilon$.

The PACC accuracy of $\hat{f}$ is then $\tau = \bar{\gamma}$ with confidence $\eta = 1 - \varepsilon$.

RESULTS

This section shows the results of using the SSIFL algorithm for two benchmark functions and one function from a real-world problem. Note that the selected asymmetric benchmark functions with multiple minima and maxima are very complex not only from the sampling but also from the function approximation point of view.

Regression of 5-D Input Space Benchmark Function

The Shifted Ackley function FSA defined in (2) with an input space dimension of 5 is approximated using the SSIFL algorithm. This function is characterized by a nearly flat outer region and a deep hole at the shifted center, see left plot in Fig. 3. Therefore we use a Gaussian probability distribution important sampling function center at the hole.

The expected accuracy and confidence levels are set to ε=0.03and δ=0.01; then sPACC is calculated using (12) given 2944 as result and the corresponding sval is set to 55=3125.

The initial feedforward neural network with three hidden layers of 25–10–5 neurons is created and the Levenberg–Marquardt algorithm is used to train the NN. Moreover, the activation functions used are tangent sigmoid for the hidden layers and linear for the output layer.

In the first iteration the input sequence for the *Test* and *Train*datasets are created using LCVT with 5^5 and 6^5 samples respectively, the input sequences are resampled using the Gaussian importance sampling function and the Shifted Ackley function is evaluated in the resampled input sequences to generate the target output. The NN is then trained with the *Test* and *Train*datasets.

In the second iteration the *Test* dataset is replaced with the *Train*dataset from the previous iteration containing a total of 6^5 samples. The new *Train* dataset is created by first generating a LCVT input sequence with 7^5 new samples, the input sequence is resampled with the Gaussian importance sampling function and the corresponding target outputs are computed using the FSA function. The NN from the previous iteration is re-trained with the new *Train*and *Test* datasets.

A total of three iterations are needed before the convergence criteria are reached. Table 1 summarizes the smart sampling (target function mean and standard deviation) and incremental function learning (epochs needed, RMS on the training dataset, and Occam factor) results for each iteration.

Table 1. SSIFL results for the Shifted Ackley function FSA in 5 dimensions

Iteration	Train samples	Test samples	FSAmean	FSAS-tdDev	NN epochs	NN MSE	*Occam ratio*
1	65(7776)	5^5	−118.757	1.0441	34	0.000688	0.9381
2	75(16807)	6^5	−118.759	1.0446	17	0.000626	1.0659
3	85(32768)	7^5	−118.454	1.0454	17	0.000598	0.9378

Finally a validation dataset *Val* with sval=45 samples is created and it is used to compute the accuracy and confidence levels of the final neural network. The results are 0.9959 and 0.99 respectively, as mean value of the target function is −118.454, a mean accuracy at the 0.84% level is reached.

Regression of 100-D Input Space Benchmark Function

The Rosenbrock function defined as FRO=∑i=1n/2[100(x2i−12−x2i−12)2+(x2i−1−1)2] with an input space dimension of 100 is approximated using the SSIFL algorithm. First the expected accuracy ε=0.02 and confidence δ=0.03 are defined; then sPACC=5250 is calculated and finally the corresponding sval is set to 10^4.

The initial feedforward neural network with one hidden layer of 100 neurons is created; the NN will be trained using the conjugate gradient backpropagation algorithm. Furthermore, the activation functions used are tangent sigmoid for the hidden layers and linear for the output layer.

The NN is trained with *Test* and *Train* datasets created by evaluating the target function FRO on HA input sequences with $10^4$and 10^5 samples respectively. In the second iteration the *Test*dataset is replaced with the *Train* dataset from the previous iteration containing a total of 10^4 samples and a new *Train* dataset is created evaluating the Rosenbrock function over a new input HA sequence with 10^6 samples. The NN from the previous iteration is re-trained using backpropagation and taking as input the new *Train*and *Test* datasets.

The third and last iteration contains 10^7 new *Train* samples. Finally the accuracy and confidence levels of the final neural network are computed using a validation dataset *Val* with sval=45 samples. The results for each single iteration are summarized in Table 2.

Table 2. SSIFL results for the Rosenbrock function FRO in 100 dimensions.

Iteration	**Train samples**	**Test samples**	FRO**mean**	FROS**td-Dev**	**NN epochs**	**NN MSE**	***Occam ratio***
1	10^5	10^4	5830.62	789.85	42	459 310	0.9964
2	10^6	10^5	5809.56	768.17	272	308 080	0.9575

Regression of 62-D Output Space Function from a Real-World Problem

The operational processing of atmospheric remote sensing data usually requires the usage of very complex and extremely time consuming radiative transfer (Gimeno García et al., 2012) models. In this subsection we apply the SSIFL algorithm to accurately approximate a radiative transfer model problem characterized by a 5 dimension input space representing the satellite viewing geometry and surface conditions. In contrast to the benchmark functions, the function from this real-world problem has not a single output value but the output space covers 62 dimensions.

The expected accuracy and confidence levels are set to ε=0.05and δ=0.02; the calculated s_{PACC} is 921 and the corresponding s_{val} is set to 5^5=1024.

The initial feedforward neural network is created using one hidden layer with 70 neurons. The hidden layers are activated by a tangent sigmoid function and the output layer by a linear function. The weights are initialized using the Nguyen–Widrow algorithm and the training is performed using the conjugate gradient backpropagation.

In the first iteration the input sequence for the *Test* and *Train*datasets are created using HA with 5^5 and 6^5 samples respectively. The forward model is evaluated using both 5-dimensional input sequences in order to generate the corresponding outputs in the 62-dimension space. The NN is then trained with the *Test* and *Train* datasets.

In the second iteration the *Test* dataset is replaced with the *Train*dataset from the previous iteration containing a total of 6^5 samples and a new *Train* dataset is created using 7^5 new samples. The NN from the previous iteration is re-trained with the new *Train* and *Test*datasets.

This iteration process is repeated seven times until the convergence criteria are reached. In the final iteration the *Test* and *Train* datasets contain a total of 11^5 (161 051) and 12^5 (248 832) new samples, Table 3 summarizes the smart sampling and incremental function learning results for each iteration.

Table 3. SSIFL results for the real-world remote sensing (FRTM) function with 5 input and 62 output dimensions

Iteration	**Train samples**	**Test samples**	FRTM**Av. mean**	FRTM**Av. StdDev**	**NN epochs**	**NN MSE**	***Occam ratio***
1	5^5	4^5	0.082274	0.069293	9700	2.0e–06	1.0100
2	6^5	5^5	0.082301	0.069333	4315	1.7e–06	0.9564
3	7^5	6^5	0.082296	0.069340	50 662	8.2e–07	1.0268
4	8^5	7^5	0.082333	0.069367	12 918	6.9e–07	0.9851
5	9^5	8^5	0.082319	0.06935	142	6.9e–07	0.9823
6	10^5	9^5	0.082291	0.069338	70 625	5.7e–07	1.0141
7	11^5	10^5	0.082289	0.069343	200 000	2.8e–07	1.0081
8	12^5	11^5	0.082286	0.069342	61 291	2.5e–07	0.9797

It is important to note that the generation of the almost half a million sample points is a very CPU demanding task; the time needed for training the NNs is minor compared with the time needed for computing the radiative transfer modeling outputs.

In the final step the validation dataset *Val* with s_{val}=45 samples is created and it is used to compute the accuracy 0.001606684 and confidence levels 0.98 of the final neural network.

Regression of 280-D Input Space Function from a Real-World Problem

In the previous subsections we presented the results of applying SSIFL to regression problems where the samples are created as part of the algorithm with the constraint of minimizing the number of calls to the target function. In this subsection we show how to apply SSIFL to problems with pre-existing datasets that follow an unknown probability distribution function. The SSIFL algorithm can be easily adapted for such problems: the smart sampling is not used for creating new patterns, but it is used for resampling patterns from the pre-existing dataset.

The smart sampling technique described in Sections 4.2 and 4.3is adapted as follows. After creating the set of s inputs X={x1,…,xs} with X⊂Rn and n the input space dimension, the patterns {ei,yi} with e⊂Rn and y⊂Rm from the pre-existing dataset that are closest to xi are selected as the *Train* dataset. Depending on the type of the input attributes different distance measurements can be applied to find the ei closest to xi.

We apply the adapted SSIFL to the "Blog feedback" dataset (Buza, 2014) with 280 input dimensions downloaded from UCI and consisting of 52 397 and 7624 pre-existing training and validation data.

In the first SSIFL iteration the input sequences xi using HA are created with 10^2 and 10^3 samples and 280 dimensions. *Test* and *Train* datasets are then created by resampling from the pre-existing $\{e_i,y_i\}$ patterns that are nearest to the HA sequence points, the closest patterns are selected using the Euclidean distance between the points ei and xi. Note that the distance in all dimensions are equally weighted by normalizing the input attributes of the pre-existing data to [0, 1]. A feedforward neural network with one hidden layer of 100 neurons is trained using the resampled datasets.

Table 4. SSIFL results for the real-world blog problem with 280 input dimensions.

Iteration	**Train samples**	**Test samples**	**NN epochs**	**NN MSE**	***Occam ratio***
1	1000	100	159	0.0003592	4.1621
2	10 000	1000	106	8.161e–05	0.8903
3	41 397	10 000	124	0.000478	1.2640

In the second iteration the *Test* dataset is replaced with the *Train*dataset from the previous iteration and a new *Train* dataset is created resampling 10^4 patterns from the remaining pre-existing dataset. The third and final iteration uses the remaining 41 397 samples from the pre-existing dataset as the *Train* dataset. The NNs from the previous iterations are re-trained with the new *Train*and *Test* datasets. See Tables 3 and 4 for a summary of the results from applying SSIFL to this problem.

Finally the accuracy of the trained NNs is assessed computing the mean square error (MSE) over the pre-existing validation dataset containing 7624 samples. The NN trained with 10 000 samples is already quite accurate with only a 8.4% worst MSE than the one from the NN trained with the complete dataset. As comparison the MSE of NNs trained with 10 000 randomly subsamples are on average 66.1% worst. In this way we demonstrate that SSIFL optimally sub-samples and accurately approximates the underlying function of pre-existing datasets.

CONCLUSIONS

In this paper we presented a novel approach (SSIFL) for computing a regression model that accurately approximates functions defined over high dimensional input spaces. This has been achieved under the constraint of minimizing the number of calls to the target function for generating the training data. We smartly combine previously published deep theoretical methods and tools into a practical sampling and function learning algorithm and we showed the applicability of the proposed technique to high-dimensional regression problems.

The sampling method used for generating the training dataset plays a key-role in order to reach high performance of the regression model. Different sampling methods have been evaluated and systematically analyzed in this study. Following a comprehensive discrimination analysis, in this paper we showed that the LCVT and HA sampling sequences are superior in handling data in high dimensionality and therefore they have been selected among all the other techniques for the proposed smart sampling technique. However, due to high computational effort needed for calculating the sequence, it might be more convenient to use the HA method in very high dimensional spaces, which has been already pointed out in this study to reach good achievements in reasonable computational time.

Smart sampling allows the easy incorporation of *a priori* information about the probability distribution of the target function (importance sampling). The number of samples to be created (i.e., the calls to the target function) is minimized by incrementally creating the new training dataset until it contains an optimal representation of the target function.

Incremental function learning takes place in each iteration: the training datasets from previous iterations are reused as the test datasets for the current iteration. The novel SSIFL technique proposed in this paper confirmed to be very efficient in order to minimize the mean square error in each function learning iteration. Learning takes place in an incremental way, the regression model is improved in each iteration based on the model from the previous iteration and the new train and test datasets.

Finally, the accuracy and confidence levels of the regression function are determined using a probabilistic performance estimation method.

The SSIFL algorithm has been proven to scale well for very high input dimensions dataset as shown with two benchmark functions with 5-D and 100-D. Furthermore the practical applicability and goodness of SSIFL method has been confirmed in two real-world applications with 62-D and 280-D. It is worth noting that SSIFL is (a) a generic approach that may be used for problems where the training patterns are either created or resampled in an optimal way and (b) it can be combined with any regression tool such as neural networks, support vector machines and many others.

ACKNOWLEDGMENTS

We are grateful to the reviewers for their constructive input to the application of SSIFL to real-world problems. This research work was supported by the German Aerospace Center (DLR).

REFERENCES

1. Alippi, C. (2014). Intelligence for embedded systems—a methodological approach. Switzerland: Springer.
2. Aziza, M., & Tayarani-N., M.-H. (2014). An adaptive memetic Particle Swarm Optimization algorithm for finding large-scale Latin hypercube designs. Engineering Applications of Artificial Intelligence, 36, 222–237.
3. Beachkofski, B. K., & Grandhi, R. V. (2002). Improved distributed hypercube sampling. In 43rd structures, structural dynamics, and materials conference (pp. 22–25).
4. Braaten, E., & Weller, G. (1979). An improved low-discrepancy sequence for multidimensional quasi-Monte Carlo integration. Journal of Computational Physics, 33(2), 249–258.
5. Buza, K. (2014). Feedback prediction for blogs. In Data analysis,

machine learning and knowledge discovery (pp. 145–152). Springer.
6. Cervellera, C., & Macciò, D. (2013). Learning with Kernel smoothing models and low-discrepancy sampling. IEEE Transactions on Neural Networks and Learning Systems, 24(3), 504–509.
7. Chernoff, H. (1952). A measure of asymptotic efficiency for tests of a hypothesis based on the sum of observations. The Annals of Mathematical Statistics, 23(4), 493–507.
8. Cioppa, T. M. (2002). Efficient nearly orthogonal and space filling experimental designs for high-dimensional complex models. (Ph.D. dissertation), California: Naval Postgraduate School.
9. Du, Q., Gunzburger, M., & Ju, L. (2010). Advances in studies and applications of centroidal Voronoi tessellations. Numerical Mathematics: Theory, Methods and Applications, 3(2), 119–142.
10. Enăchescu, C. (2013). Supervised learning using an active strategy. In The 7thinternational conference interdisciplinarity in engineering (pp. 220–228).
11. Fang, K.-T., Ma, C.-X., & Winker, P. (2002). Centered L2 discrepancy of random sampling and latin hypercube design, and construction of uniform designs. Mathematics of Computation, 71(237), 275–296.
12. Flato, G., et al. (2013). Evaluation of climate models in Climate Change 2013: The Physical Science Basis. Contribution of Working Group I to the Fifth Assessment Report of the Intergovernmental Panel on Climate Change. Cambridge University Press, (pp. 741–866).
13. Gimeno García, S., Trautmann, T., & Venema, V. (2012). Reduction of radiation biases by incorporating the missing cloud variability by means of downscaling techniques: a study using the 3-d mocart model. Atmospheric Measurement Techniques, 5, 2261–2276.
14. Gnecco, G., Kůrková, V., & Sanguineti, M. (2011). Some comparisons of complexity in dictionary-based and linear computational models. Neural Networks, 24(2), 171–182.
15. Gordon, C., Cooper, C., Senior, C. A., Banks, H., Gregory, J. M., Johns, T. C., et al. (2000). The simulation of SST, sea ice extents and ocean heat transports in a version of the Hadley Centre coupled model without flux adjustments. Climate Dynamics,16(2–3), 147–168.

16. Grell, G. A., et al. (2005). Fully coupled "online" chemistry within the WRF model. Atmospheric Environment, 39(37), 6957–6975.
17. Halton, J. H. (1960). On the efficiency of certain quasi-random sequences of points in evaluating multi-dimensional integrals. Numerische Mathematik, 2(1), 84–90.
18. Han, J., & Pan, H.-L. (2011). Revision of convection and vertical diffusion schemes in the NCEP global forecast system. Weather and Forecasting, 26, 520–533.
19. Haupt, S. E., Pasini, A., & Marzban, C. (2009). Artificial intelligence methods in the environmental sciences. Netherlands: Springer.
20. Hilbert, M., & López, P. (2011). The world's technological capacity to store, communicate, and compute information. Science, 332(6025), 60–65.
21. Hsieh, W. W., & Tang, B. (1998). Applying neural network models to prediction and data analysis in meteorology and oceanography. Bulletin of the American Meteorological Society, 79(9), 1855–1870.
22. Jin, Y., & Hammer, B. (2014). Computational intelligence in big data. IEEE Computational Intelligence Magazine, (3), 12–13.
23. Kainen, P. C., Kůrková, V., & Sanguineti, M. (2012). Dependence of computational models on input dimension: Tractability of approximation and optimization tasks. IEEE Transactions on Information Theory, 58(2), 1203–1214.
24. Kazimipour, H., Li, X., & Qin, A. K. (2013). Initialization methods for large scale global optimization. In IEEE congress on evolutionary computation (pp. 2750–2757).
25. Kolmogorov, A. (1957). On the representation of continuous functions of several variables by superposition of continuous functions of one variable and addition. Doklady Akademii Nauk SSSR, 114, 953–956.
26. Krasnopolsky, V. M., & Chevallier, F. (2003). Some neural network applications in environmental sciences. Part II: advancing computational efficiency of environmental numerical models. Neural Networks, 16, 335–348.
27. Krasnopolsky, V. M., & Schiller, H. (2007). Neural network emulations for complex multidimensional geophysical mappings:

Applications of neural network techniques to atmospheric and oceanic satellite retrievals and numerical modeling. Reviews of Geophysics, 45, RG3009.

28. Leung, Y.-W., & Wang, Y.(2001). An orthogonal genetic algorithm with quantization for global numerical optimization. IEEE Transactions on Evolutionary Computation, 5(1), 41–53.
29. Liang, J. J., Suganthan, P. N., & Deb, K. (2005). Novel composition test functions for numerical global optimization. In Proceedings of the IEEE swarm intelligence symposium (pp. 68–75).
30. Loyola, D. (2006). Applications of neural network methods to the processing of earth observation satellite data. Neural Networks, 19(2), 168–177.
31. Lynch, P. (2006). The ENIAC integrations. In The emergence of numerical weather prediction (pp. 206–208). Cambridge: University Press.
32. Lynch, C. (2008). Big data: How do your data grow? Nature, 455(7209), 28–29.
33. MacKay, D. J. C. (1995). Probable networks and plausible predictions—a review of practical Bayesian methods for supervised neural networks. Network: Computation in Neural Systems, 6(3), 469–505.
34. Marsaglia, G., & Tsang, W. W. (2000). The Ziggurat method for generating random variables. Journal of Statistical Software, 5(8).
35. Matsumoto, M., & Nishimura, T. (1998). Mersenne Twister: A 623-dimensionally equidistributed uniform pseudorandom number generator. ACM Transactions on Modeling and Computer Simulation, 8(1), 3–30.
36. McKay, M. D., Beckman, R. J., & Conover,W. J.(1979). A comparison of three methods for selecting values of input variables in the analysis of output from a computer code. Technometrics, 21(2), 239–245.
37. Menut, L., Bessagnet, B., Khvorostyanov, D., Beekmann, M., Blond, N., Colette, A., et al. (2013). CHIMERE 2013: a model for regional atmospheric composition modelling. Geoscientific Model Development, 6, 981–1028.

38. Morcrette, J.-J. (1991). Radiation and cloud radiative properties in the ECMWF operational weather forecast model. J. Geophys. Res., 96D, 9121–9132.

39. Moskowitz, B., & Caflisch, R. (1996). Smoothness and dimension reduction in quasiMonte Carlo methods. Mathematical and Computer Modelling, 23(89), 37–54.

40. Niederreiter, H. (1992). Random number generation and quasi-Monte Carlo methods. Society for Industrial and Applied Mathematics.

41. O'Leary, D. E. (2013). Artificial intelligence and big data. IEEE Intelligent Systems, 28(2), 96–99.

42. Patelli, E., Pradlwarter, H. J., & Schuëller, G. I. (2011). On multinormal integrals by Importance Sampling for parallel system reliability. Structural Safety, 33(1), 1–7.

43. Peng, L., Wang, Y., Dai, G., & Cao, Z. (2012). A novel differential evolution with uniform design for continuous global optimization. Journal of Computers, 7(1), 3–10.

44. Perkins, T., Adler-Golden, S., Matthew, M. W., Berk, A., Bernstein, L. S., Lee, J., et al. (2012). Speed and accuracy improvements in FLAASH atmospheric correction of hyperspectral imagery. Optical Engineering, 51(11).

45. Qazi, M., & Linshu, H. (2006). Nearly-orthogonal sampling and neural network etamodel driven conceptual design of multistage space launch vehicle. Computer-Aided Design, 38, 595–607.

46. Rahnamayan, S., Tizhoosh, H. R., & Salama, M. M. (2007). Quasi-oppositional differential evolution. In IEEE congress on evolutionary computation (pp. 2229–2236).

47. Rahnamayan, S., Tizhoosh, H. R., & Salama, M. M. (2008). Opposition-based differential evolution. IEEE Transactions on Evolutionary Computation, 12(1), 64–79.

48. Richter, R., Heege, T., Kiselev, V., & Schläpfer, D. (2014). Correction of ozone influence on TOA radiance. International Journal of Remote Sensing, 35(23), 8044–8056.

49. Robinson, D., & Atcitty, D. (1999). Comparison of quasi- and pseudo- Monte Carlo sampling for reliability and uncertainty

analysis. In 40th structures, structural dynamics, and materials conference.

50. Romero, V. J., Burkardt, J. V., Gunzburger, M. D., & Peterson, J. S. (2006). Comparison of pure and Latinized centroidal Voronoi tessellation against various other statistical sampling methods. Reliability Engineering & System Safety, 91(10–11), 1266–1280.
51. Sacks, J., Welch, W. J., Mitchell, T. J., & Wynn, H. P. (1989). Design and analysis of computer experiments. Statistical Science, 4(4), 409–423.
52. Sagaut, P. (2006). Large eddy simulation for incompressible flows (3rd ed.). Springer.
53. Schuessler, O., & Loyola, D. (2011). Parallel training of artificial neural networks using multithreaded and multicore CPUs. In Lecture notes in computer science: Vol. 6593. Adaptive and natural computing algorithms (pp. 70–79).
54. Swiler, L. P., Slepoy, R., & Giunta, A. A. (2006). Evaluation of sampling methods in constructing response surface approximations. In 47th AIAA/ASME/ ASCE/AHS/ASC structures, structural dynamics, and materials conference (pp. 1–24).
55. Tang, K., Yao, X., Suganthan, P. N., MacNish, C., Chen, Y.-P., & Chen, C.-M. et al. (2008).
56. Benchmark functions for the CEC'2008 special session and competition on large scale global optimization. Nature Inspired Computation and Applications Laboratory, USTC, China.
57. Thiémard, E. (2001). An algorithm to compute bounds for the star discrepancy. Journal of Complexity, 17(4), 850–880.
58. Tripathi, S., Srinivas, V. V., & Nanjundiah, R. S. (2006). Downscaling of precipitation for climate change scenarios: A support vector machine approach. Journal of Hydrology, 330(3–4), 621–640.
59. Vapnik, V. N. (2006). Estimation of dependences based on empirical data. New York: Springer.
60. Yiteng, Z., Yew-Soon, O., & Tsang, I. W. (2014). The emerging "Big Dimensionality". IEEE Computational Intelligence Magazine, (3), 14–26.

CHAPTER 11

HADOOP NEURAL NETWORK FOR PARALLEL AND DISTRIBUTED FEATURE SELECTION

Victoria J. Hodge, Simon O'Keefe, Jim Austin

Advanced Computer Architecture Group, Department of Computer Science, University of York, York, YO10 5GH, UK

ABSTRACT

In this paper, we introduce a theoretical basis for a Hadoop-based neural network for parallel and distributed feature selection in Big Data sets. It is underpinned by an associative memory (binary) neural network which is highly amenable to parallel and distributed processing and fits with the Hadoop paradigm. There are many feature selectors described in the literature which all have various strengths and weaknesses. We present the implementation details of five feature selection algorithms constructed using our artificial neural network framework embedded in

Hadoop YARN. Hadoop allows parallel and distributed processing. Each feature selector can be divided into subtasks and the subtasks can then be processed in parallel. Multiple feature selectors can also be processed simultaneously (in parallel) allowing multiple feature selectors to be compared. We identify commonalities among the five features selectors. All can be processed in the framework using a single representation and the overall processing can also be greatly reduced by only processing the common aspects of the feature selectors once and propagating these aspects across all five feature selectors as necessary. This allows the best feature selector and the actual features to select to be identified for large and high dimensional data sets through exploiting the efficiency and flexibility of embedding the binary associative-memory neural network in Hadoop.

Keywords: Hadoop MapReduce Distributed Parallel Feature selection Binary neural network

INTRODUCTION

The meaning of "big" with respect to data is specific to each application domain and dependent on the computational resources available. Here we define "Big Data" as large, dynamic collections of data that cannot be processed using traditional techniques, a definition adapted from (Franks, 2012; Zikopoulos & Eaton, 2011). Today, data is generated continually by an increasing range of processes and in ever increasing quantities driven by Big Data mechanisms such as cloud computing and on-line services. Business and scientific data from many fields, such as finance, astronomy, bioinformatics and physics, are often measured in terabytes (10^{12}bytes). Big Data is characterised by its complexity, variety, speed of processing and volume (Laney, 2001). It is increasingly clear that exploiting the power of these data is essential for information mining. These data often contain too much noise (Liu, Motoda, Setiono, & Zhao, 2010) for accurate classification (Dash & Liu, 1997; Han & Kamber, 2006), prediction (Dash & Liu, 1997; Guyon & Elisseeff, 2003) or outlier detection (Hodge, 2011). Thus, only some of the features (dimensions) are related to the target concept (classification label or predicted value). Also, if there are too many data features then the data points become sparse. If data is too sparse then distance measures such as the popular Euclidean distance and the concept of nearest neighbours become less applicable (Ertöz, Steinbach, & Kumar, 2003). Many machine learning algorithms are adversely affected by this noise and these superfluous features in terms of both their accuracy and their ability to generalise. Consequently, the data must be pre-processed by the classification or prediction algorithm itself or by a separate feature selection algorithm to prune these superfluous features (Kohavi & John, 1997; Witten & Frank, 2000).

The benefits of feature selection include: reducing the data size when superfluous features are discarded, improving the classification/ prediction accuracy of the underlying algorithm where the algorithm is adversely affected by noise, producing a more compact and easily understood data representation and reducing the execution time of the underlying algorithm due to the smaller data size. Reducing the execution time is extremely important for Big Data, which has a high computational resource demand on memory and CPU time.

In this paper, we focus on feature selection in vast data sets for parallel and distributed classification systems. We aim to remove noise and reduce redundancy to improve classification accuracy. There is a wide variety of techniques proposed in the machine learning literature for feature selection including Correlation-based Feature Selection (Hall, 1998), Principal Component Analysis (PCA) (Jolliffe, 2002), Information Gain (Quinlan, 1986), Gain Ratio (Quinlan, 1992), Mutual Information Selection (Wettscherek, 1994), Chi-square Selection (Liu & Setiono, 1995), Probabilistic Las Vegas Selection (Liu & Setiono, 1996) and Support Vector Machine Feature Elimination (Guyon, Weston, Barnhill, & Vapnik, 2002). Feature selectors produce feature scores. Some feature selectors also select the best set of features to use while others just rank the features with the scores. For these feature rankers, the best set of features must then be chosen by the user, for example, using greedy search (Witten & Frank, 2000).

It is often not clear to the user which feature selector to use for their data and application. In their analysis of feature selection, Guyon and Elisseeff (2003) recommend evaluating a variety of feature selectors before deciding the best for their problem. Therefore, we propose that users exploit our framework to run a variety of feature selectors in parallel and then evaluate the feature sets chosen by each selector using their own specific criteria. Having multiple feature selectors available also provides the opportunity for ensemble feature selection where the results from a range of feature selectors are merged to generate the best set of features to use. Feature selection is a combinatorial problem so needs to be implemented as efficiently as possible particularly on big data sets. We have previously developed a k-NN classification (Hodge & Austin, 2005; Weeks, Hodge, O'Keefe, Austin, & Lees, 2003) and prediction algorithm (Hodge, Krishnan, Austin, & Polak, 2011) using an associative memory (binary) neural network called the Advanced Uncertain Reasoning Architecture (AURA) (Austin, 1995). This multi-faceted k-NN motivated a unified feature selection framework exploiting the speed and storage efficiency of the associative memory neural network. The framework lends itself to parallel and distributed processing across multiple nodes allowing vast data sets to be processed. This could be done by processing the data at the same geographical location using a single machine with multiple processing cores (Weeks, Hodge, & Austin, 2002) or at the same geographical location using multiple compute nodes (Weeks et al., 2002) or even distributed processing of the data at multiple geographical locations.

Data mining tools such as Weka (Witten & Frank, 2000), Matlab, R and SPSS provide feature selection algorithms for data mining and analytics. However, these products are designed for small scale data analysis. Researchers have parallelised individual feature selection algorithms using MapReduce/Hadoop (Chu et al., 2007; Reggiani, 2013; Singh, Kubica, Larsen, & Sorokina, 2009; Sun, 2014). Data mining libraries such as Mahout (https://mahout.apache.org) and MLib (https://spark.apache.org/mllib/) and data mining frameworks such as Radoop (https://rapidminer.com/products/radoop/) include a large number of data mining algorithms including feature selectors. However, they do not explicitly tackle processing reuse with a view to multi-user and multi-task resource allocation. Zhang, Kumar, and Ré (2014) developed a database systems framework for optimised feature selection providing a range of algorithms. They observed that there are reuse opportunities that could yield orders of magnitude performance improvements on feature selection workloads as we will also demonstrate here using AURA in an

Apache Hadoop (https://hadoop.apache.org/) framework.

The main contributions of this paper are:

- To extend the AURA framework to parallel and distributed processing of vast data sets in Apache Hadoop,
- To describe five feature selectors in terms of the AURA framework. Two of the feature selectors have been implemented in AURA but not using Hadoop (Hodge, Jackson, & Austin, 2012; Hodge, O'Keefe, & Austin, 2006) and the other three have not been implemented in AURA before,
- To theoretically analyse the resulting framework to show how the five feature selectors have common requirements to enable reuse.
- To theoretically analyse the resulting framework to show how we reduce the number of computations. The larger the data set then the more important this reduction becomes.
- To demonstrate parallel and distributed processing in the framework allowing Big Data to be analysed.

In our AURA framework, the feature selectors all use one common data representation. We only need to process any common elements once and can propagate the common elements to all feature selectors that require them. Thus, we can rapidly and efficiently determine the best feature selector and the best set of features to use for each data set under investigation. In Section 2, we discuss AURA and related neural networks and how to store and retrieve data from AURA, Section 3 demonstrates how to implement five feature selection algorithms in the AURA unified framework and Section 4 describes parallel and distributed feature selection using AURA. We than analyse the unified framework in Section 5 to identify common aspects of the five feature selectors and how they can be implemented in the unified framework in the most efficient way. Section 6 details the overall conclusions from our implementations and analyses.

BINARY NEURAL NETWORKS

AURA (Austin, 1995) is a hetero-associative memory neural network (Palm, 2013). An associative memory is addressable through its contents and a hetero-associative memory stores associations between input and output vectors where the vectors are different (Palm, 2013). AURA uses binary Correlation Matrix Memories (CMMs): binary hetero-associative matrices that store and retrieve patterns using matrix calculus. They are non-recursive and fully connected. Input vectors (stimuli) address the CMM rows and output vectors address the CMM columns. Binary neural networks have a number of advantages compared to standard neural networks including rapid one-pass training, high levels of data compression, computational simplicity, network transparency, a partial match capability and a scalable architecture that can be easily mapped onto high performance computing platforms including parallel and distributed platforms (Weeks et al., 2002). AURA is implemented as a C++ software library.

Previous parallel and distributed applications of AURA have included distributed text retrieval (Weeks et al., 2002), distributed time-series signal searching (Fletcher, Jackson, Jessop, Liang, & Austin, 2006) and condition monitoring (Austin, Brewer, Jackson, & Hodge, 2010). This new development will augment these existing techniques and is aimed at these same domains. It will couple feature selection, classification and prediction with the speed and storage efficiency of a binary neural network allowing parallel and distributed data mining. This makes AURA ideal to use as the basis of an efficient distributed machine learning framework. A more formal definition of AURA, its components and methods now follows.

AURA

The AURA methods use binary input I and output O vectors to efficiently store records in a CMM M as in Eq. (1) using the binary rule (Palm, 2013).

$$\boldsymbol{M} = \bigvee \boldsymbol{I_j} \boldsymbol{O_j^T} \quad \text{where } \vee \text{ is logical OR.} \tag{1}$$

Training (construction of a CMM) is a single epoch process with one training step for each input–output association (each $I_j O_j^T$ in Eq. (1)) which equates to one step for each record j in the data set. Thus, the trained CMM M represents $\{(\boldsymbol{I_1} \times \boldsymbol{O_1^T}), (\boldsymbol{I_2} \times \boldsymbol{O_2^T}), \ldots (\boldsymbol{I_n} \times \boldsymbol{O_n^T})\}$ superimposed using bitwise ***or***. $\boldsymbol{I_j O_j^T}$ is an estimate of the weight matrix

W(j) of the synaptic connections of the neural network as a linear associator with binary weights. W(j) forms a mapping representing the association described by the jth input/output pair of vectors. As a consequence of using unipolar elements {0,1} throughout, the value at each matrix component w_{ij} means the existence of an association between elements i and j. The trained CMM M is then effectively an encoding (correlation) of the N weight matrices W for all N records in the data set. Individual weights within the weight matrix update using a generalisation of Hebbian learning (Hebb, 1949) where the state for each synapse (matrix element) is binary valued. Every synapse can update its weight independently using a *local learning rule* (Palm, 2013). Local learning is biologically plausible and computationally simple allowing parallel and rapid execution. The learning process is illustrated in Fig. 1.

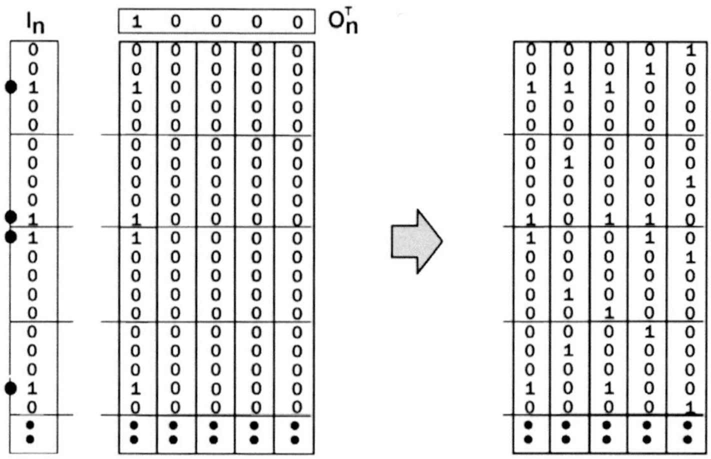

Figure. 1: Showing a CMM learning input vector I_n associated with output vector O_n on the left.

The CMM on the right shows the CMM after five associations $I_j O_j^T$. Each column of the CMM represents a record. Each row represents a feature value for qualitative features or a quantisation of feature values for quantitative features and each set of rows (shown by the horizontal lines) represents the set of values or set of quantisations for a particular feature.

For feature selection, the data are stored in the CMM which forms an index of all features in all records. During training, the input vectors

I_j represent the feature and class values and are associated with a unique output vector O_j representing a record. Fig. 1 shows a trained CMM. In this paper, we set only one bit in the vector O_j indicating the location of the record in the data set, the first record has the first bit set, the second record has the second bit set etc. Using a single set bit makes the length of O_j potentially large. However, exploiting a compact list representation (Hodge & Austin, 2001) (more detail is provided in Section 4.3.1) means we can compress the storage representation.

Data

The AURA feature selector, classifier and predictor framework can handle qualitative features (symbolic and discrete numeric) and quantitative features (continuous numeric).

The raw data sets need pre-processing to allow them to be used in the binary AURA framework. Qualitative features are enumerated and each separate token maps onto an integer (Token↦Integer) which identifies the bit to set within the vector. For example, a SEX_TYPE feature would map as (F↦0) and (M↦1). Any quantitative features are quantised (mapped to discrete bins) (Hodge & Austin, 2012). Each individual bin maps onto an integer which identifies the bit to set in the input vector. Next, we describe the simple equi-width quantisation. We note that the Correlation-Based Feature Selector described in Section 3.2uses a different quantisation technique to determine the bin boundaries. However, once the boundaries are determined, the mapping to CMM rows is the same as described here.

To quantise quantitative features, a range of input values for feature F_f map onto each bin. Each bin maps to a unique integer as in Eq. (2) to index the correct location for the feature in I_j. In this paper, the range of feature values mapping to each bin is equal to subdivide the feature range into b equi-width bins across each feature.

$$R_{f_i} \rightarrow \boldsymbol{bins}_{f_k} \mapsto \boldsymbol{Integer}_{f_k} + \boldsymbol{offset}\left(\boldsymbol{F_f}\right)$$
$$\text{where } \boldsymbol{F_f} \in \boldsymbol{F}, \boldsymbol{f_i} \text{ is a value of } \boldsymbol{F_f} \text{ and } \boldsymbol{cardinality}\left(\boldsymbol{Integer}_{f_k}\right) \equiv \boldsymbol{cardinality}\left(\boldsymbol{bins}_{f_k}\right). \quad (2)$$

In Eq. (2), offset(F_f) is a cumulative integer offset within the binary vector for each feature F_f, → is a many-to-one mapping and ↦ is a one-to-one mapping. The offset for the next feature F_{f+1} is given by offset(F_{f+1})=offset(F_f)+nBins(F_f)where nBins(F_f) is the number of bins for feature Ff.

For each record in the data set.

For each feature.

Calculate bin for feature value.

Set bit in vector as in Eq. (2).

AURA Recall

To recall the matches for a query (input) record, we firstly produce a recall input vector R_k by quantising the target values for each feature to identify the bins (CMM rows) to activate as in Eq. (3). During recall, the presentation of recall input vector R_k elicits the recall of output vector O_k as vector R_k contains all of the addressing information necessary to access and retrieve vector O_k. Recall is effectively the dot product of the recall input vector R_k and CMM M, as in Eq. (3) and Fig. 2.

$$\boldsymbol{S}^T = \boldsymbol{R}_k^T \cdot \boldsymbol{M}. \tag{3}$$

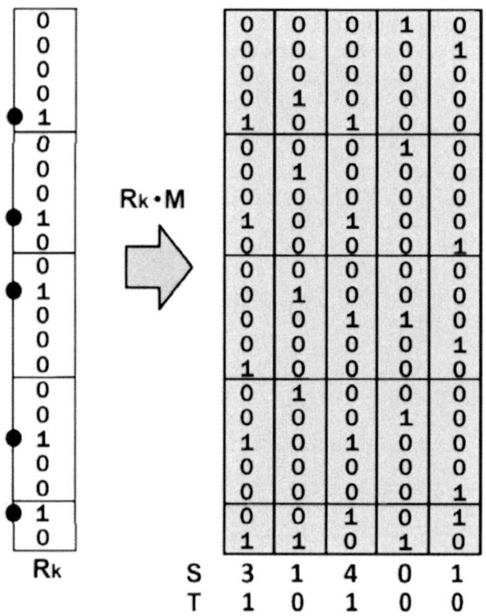

Figure. 2: Showing a CMM recall. Applying the recall input vector R_k to the CMM M retrieves a summed integer vector S with the match score for each CMM column. S is then thresholded to retrieve the matches. The threshold here

is either Willshaw with value 3 retrieving all columns that sum to 3 or more or L-Max with value 2 to retrieve the 2 highest scoring columns.

If Rk appeared in the training set, we get an integer-valued vector S (the summed output vector), composed of the required output vector multiplied by a weight based on the dot product of the input vector with itself. If the recall input Rk is not from the original training set, then the system will recall the output Ok associated with the closest stored input to Rk, based on the dot product between the test and training inputs.

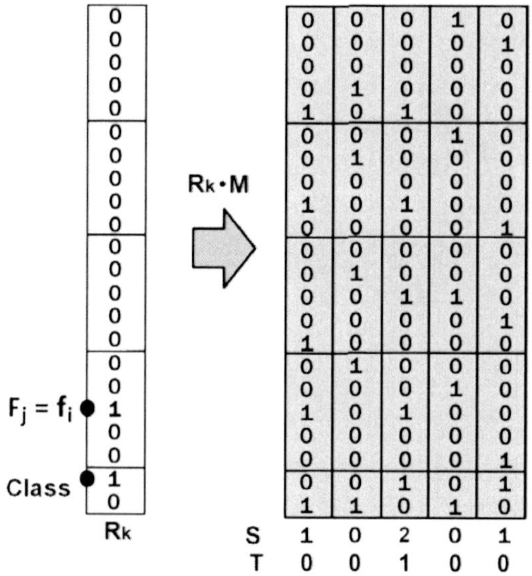

Figure. 3: Diagram showing the feature value row and the class values row excited to determine co-occurrences ($C=c\wedge F_j=f_i$).

Matching is a combinatorial problem but can be achieved in a single pass in AURA. AURA can also exploit the advantages of sparse vectors (Palm, 2013) during recall by only activating regions of interest. If the input vector R_k has 1000 bits indexing 1000 CMM rows then only the rows addressed by a set bit in the input vector need be examined (as shown in Figs. 2 and 3). For a 10 bit set vector then only 10 of the 1000 rows are activated. The input pattern R_k would be said to have a saturation of (10/1000=0.01). The total amount of data that needs to be examined is reduced by a factor that is dependent on this saturation providing that the data is spread reasonably evenly between the rows and the CMM is implemented effectively. Using smart encoding schemes can bring the performance improvement resulting from very low saturation input patterns to over 100-fold (Weeks et al., 2002).

The AURA technique thresholds the summed output S to produce a binary output vector T as given in Eq. (4).

$$\boldsymbol{T_j} = \begin{cases} 1 & \text{if } \boldsymbol{S_j} \geq \theta \\ 0 & \text{otherwise.} \end{cases} \tag{4}$$

For exact match, we use the Willshaw threshold (Willshaw, Buneman, & Longuet-Higgins, 1969) to set θ. This sets a bit in the thresholded output vector for every location in the summed output vector that has a value higher than or equal to θ. The value of θvaries according to the task. If there are ten features in the data and we want to find all stored records that match the ten feature values of the input vector then we set θ to 10. Thus, for full match θ=b1, where b1 is set to the number of set bits in the input vector. For partial matching, we use the L-Max threshold (Casasent & Telfer, 1992). L-Max thresholding essentially retrieves at least L top matches. Our AURA software library automatically sets θ to the highest integer value that will retrieve at least L matches.

Feature selection described in Section 3 requires both exact matching using Willshaw thresholding and partial matching using L-Max thresholding.

FEATURE SELECTION

There are two fundamental approaches to feature selection (Kohavi & John, 1997; Witten & Frank, 2000): (1) filters select the optimal set of features independently of the classifier/predictor algorithm while (2) wrappers select features which optimise classification/prediction using the algorithm. We examine the mapping of five filter approaches to the binary AURA architecture. Filter approaches are more flexible than wrapper approaches as they are not directly coupled to the algorithm and are thus applicable to a wide variety of classification and prediction algorithms. Our method exploits the high speed and efficiency of the AURA techniques as feature selection is a combinatorial problem.

We examine a mutual information approach Mutual Information Feature Selection (MI) detailed in Section 3.1 that analyses features on an individual basis, a correlation-based multivariate filter approach Correlation-based Feature Subset Selection (CFS) detailed in Section 3.2 that examines greedily selected subsets of features, a revised Information Gain approach Gain Ratio (GR) detailed in Section 3.3, a feature dependence approach Chi-Square Feature selection(CS) detailed in Section 3.4 which is univariate, and a univariate feature relevance approach Odds Ratio (OR) detailed in Section 3.5.

Univariate filter approaches such as MI, CS or OR are quicker than multivariate filters as they do not need to evaluate all combinations of subsets of features. The advantage of a multivariate filter compared to a univariate filter lies in the fact that a univariate approach does not account for interactions between features. Multivariate techniques evaluate the worth of feature subsets by considering both the individual predictive ability of each feature and the degree of redundancy between the features in the set.

All five feature selection algorithms have their relative strengths. We refer the reader to Forman (2003) and Varela, Martins, Aguiar, and Figueiredo (2013) for accuracy evaluations of these feature selectors. These papers show that the best feature selector varies with data and application. Using the CFS attribute selector, Hall and Smith (1998) found significant improvement in classification accuracy of k-NN on five of the 12 data sets they evaluated but a significant degradation in accuracy on two data sets. Hence, different feature selectors are required for different data sets and applications.

We note that the CFS as implemented by Hall (1998) uses an entropy-based quantisation whereas we have used equi-width quantisation for the other feature selectors (MI, GR, CS and OR). We plan to investigate unifying the quantisation as a next step. For the purpose of our analysis in Section 5, we assume that all feature selectors are using identical quantisation. We assume that all records are to be used during feature selection.

Mutual Information Feature Selection

Wettscherek (1994) described a mutual information feature selection algorithm. The mutual information between two features is "*the reduction in uncertainty concerning the possible values of one feature that is obtained when the value of the other feature is determined*" (Wettscherek, 1994). MI is defined by Eq. (5):

$$MI\left(F_j, C\right) = \sum_{i=1}^{b(F_j)} \sum_{c=1}^{nClass} p(C = c \bigwedge F_j = f_i) \cdot \log_2 \left(\frac{p(C = c \bigwedge F_j = f_i)}{p(C = c) \cdot p(F_j = f_i)} \right). \quad (5)$$

To calculate $p(C=c \wedge F_j=f_i)$, we use AURA to calculate $\frac{n(BVf_i \wedge BVc)}{N}$.

AURA excites the row in the CMM corresponding to feature value f_i of feature F_j and the row in the CMM corresponding to class value c as shown in Fig. 3. By thresholding the output vector S at Willshaw threshold value = 2, we obtain a thresholded output vector with a bit set for every co-occurrence. We can count these set bits to determine the co-occurrence count. Furthermore, p(C=c) is the count of the number of set bits n(BVc) in the binary vector (CMM row) for c and $p(F_j=f_i)$ is the count of the number of set bits $n(BVf_i)$ in the binary vector (CMM row) for fias used by GR.

The MI calculated using AURA for qualitative features is given by Eq. (6) where N is the number of records in the data set, $rows(F_j)$ is the number of CMM rows for feature F_j and ***nClass***is the number of classes:

$$MI\left(F_j, C\right) = \sum_{i=1}^{rows(F_j)} \sum_{c=1}^{nClass} \frac{n(BVf_i \wedge BVc)}{N} \cdot \log_2 \left(\frac{\frac{n(BVf_i \wedge BVc)}{N}}{\frac{n(BVf_i)}{N} \cdot \frac{n(BVc)}{N}} \right). \quad (6)$$

We can follow the same process for real/discrete ordered numeric features in AURA. In this case, the mutual information is given by Eq. (7):

$$MI\left(F_j, C\right) = \sum_{i=1}^{bins(F_j)} \sum_{c=1}^{nClass} \frac{n(BVb_i \wedge BVc)}{N} \cdot \log_2 \left(\frac{\frac{n(BVb_i \wedge BVc)}{N}}{\frac{n(BVb_i)}{N} \cdot \frac{n(BVc)}{N}} \right) \quad (7)$$

where bins(F_j) is the number of bins (effectively the number of rows) in the CMM for feature F_j and BVb_i is the CMM row for the bin mapped to by feature value f_i.

The MI feature selector assumes independence of features and scores each feature separately so it is the user's prerogative to determine the number of features to select. The major drawback of the MI feature selector along with similar information theoretic approaches, for example Information Gain, is that they are biased towards features with the largest number of distinct values as this splits the training records into nearly pure classes. Thus, a feature with a distinct value for each record has a maximal information score. The CFS and GR feature selectors make adaptations of information theoretic approaches to prevent this biasing.

Correlation-Based Feature Subset Selection

Hall (1998) proposed the Correlation-based Feature Subset Selection (CFS). It measures the strength of the correlation between pairs of features. CFS favours feature subsets that contain features that are highly correlated to the class but uncorrelated to each other to minimise feature redundancy. CFS is thus based on information theory measured using Information Gain. Hall and Smith (1997) used a modified Information Gain measure, Symmetrical Uncertainty, (***SU***) given in Eq. (8) to prevent bias towards features with many distinct values (Section 3.1). ***SU***estimates the correlation between features by normalising the value in the range [0, 1]. Two features are completely independent if SU=0 and completely dependent if SU=1.

$$\boldsymbol{SU\left(F_j, G_l\right) = 2.0 \cdot \left[\frac{Ent\left(F_j\right) - Ent\left(F_j \mid G_l\right)}{Ent\left(F_j\right) + Ent(G_l)}\right]} \qquad (8)$$

where the entropy of a feature F_j for all feature values f_i is given as Eq. (9):

$$\boldsymbol{Ent}\left(F_j\right) = -\sum_{i=1}^{n(F_j)} p(f_i)\log_2(p(f_i)) \tag{9}$$

and the entropy of feature Fj after observing values of feature Glis given as Eq. (10):

$$\boldsymbol{Ent}\left(F_j \mid G_l\right) = -\sum_{k=1}^{n(G_l)} p(g_k) \sum_{i=1}^{n(F_j)} p(f_i \mid g_k)\log_2(p(f_i \mid g_k)). \tag{10}$$

Any quantitative features are discretised using Fayyad and Irani's entropy quantisation (Fayyad & Irani, 1993). The bin boundaries are determined using Information Gain and these quantisation bins map the data into the AURA CMM as previously.

CFS has many similarities to MI when calculating the values in Eqs. (8)–(10) and through using the same CMM (Fig. 3) as noted below.

In the AURA CFS, for each pair of features (F_j,G_l) to be examined, the CMM is used to calculate Ent(F_j),Ent(G_l) and Ent(F_j|G_l) from Eqs. (8)–(10). There are three parts to the calculation.

- Ent(F_j) requires the count of data records for the particular value f_i of feature F_j which is n(BVf_i) in Eq. (6) for qualitative and class features and n(BVb_i) in Eq. (7) for quantitative features. AURA excites the row in the CMM corresponding to feature value f_i of feature F_j. This row is a binary vector (BV) and is represented by BVf_i. A count of bits set on the row gives n(BVf_i) from Eq. (6)and is achieved by thresholding the output vector S_k from Eq. (4) at Willshaw value 1.
- Similarly, Ent(G_l) counts the number of records where feature G_l has value g_k.
- Ent(F_j|G_l) requires the number of co-occurrences of a particular value f_i of feature F_j with a particular value g_k of feature G_ln(BVf_i∧BVg_k) for qualitative features and n(BVb_i∧BVb_k) for quantitative features and between a feature and the class n(BVf_i∧BVc) and n(BVb_i∧BVc) for qualitative and quantitative features respectively. If both the feature value row and the class values row are excited then the summed output

vector will have a two in the column of every record with a co-occurrence of f_i with c_j as shown in Fig. 3. By thresholding the summed output vector at a threshold of two, we can find all co-occurrences. We represent this number of bits set in the vector by $n(BVf_i \wedge BVc)$ which is a count of the set bits when BVc is logically *and*ed with BVf_i.

CFS determines the feature subsets to evaluate using forward search. Forward search works by greedily adding features to a subset of selected features until some termination condition is met whereby adding new features to the subset does not increase the discriminatory power of the subset above a pre-specified threshold value. The major drawback of CFS is that it cannot handle strongly interacting features (Hall & Holmes, 2003).

Gain Ratio Feature Selection

Gain Ratio (GR) (Quinlan, 1992) is a new feature selector for the AURA framework. GR is a modified Information Gain technique and is used in the popular machine learning decision tree classifier C4.5 (Quinlan, 1992). Information Gain is given in Eq. (11) for feature F_j and the class C. CFS (Section 3.2) modifies Information Gain to prevent biasing towards features with the most values. GR is an alternative adaptation which considers the number of splits (number of values) of each feature when calculating the score for each feature using normalisation.

$$\boldsymbol{Gain}\left(\boldsymbol{F_j}, \boldsymbol{C}\right) = \boldsymbol{Ent}\left(\boldsymbol{F_j}\right) - \boldsymbol{Ent}(\boldsymbol{F_j} \mid \boldsymbol{C}) \quad (11)$$

where $Ent(F_j)$ is defined in Eq. (9) and $Ent(F_j|C)$ is defined by Eq. (10). Then Gain Ratio is defined as Eq. (12):

$$\boldsymbol{GainRatio}\left(\boldsymbol{F_j}, \boldsymbol{C}\right) = \frac{\boldsymbol{Gain}(\boldsymbol{F_j}, \boldsymbol{C})}{\boldsymbol{IntrinsicValue}(\boldsymbol{F_j})} \quad (12)$$

where ***IntrinsicValue*** is given by Eq. (13):

$$\boldsymbol{IntrinsicValue}\left(\boldsymbol{F_j}\right) = \sum_{p=1}^{V} \frac{\boldsymbol{S_p}}{\boldsymbol{N}} \boldsymbol{log_2}\left(\frac{\boldsymbol{S_p}}{\boldsymbol{N}}\right) \quad (13)$$

and V is the number of feature values ($n(F_j)$) for qualitative features and number of quantisation bins $n(b_i)$ for quantitative features and S_p is a subset of the records that have $F_j=f_i$ for qualitative features or map to the quantisation bin $b_{in}(f_i)$ for quantitative features.

To implement GR using AURA, we train the CMM as described in Section 2.1 We can then calculate Ent(F_j) and Ent(F_j|C)as per the CFS feature selector described in Section 3.2 to allow us to calculate Gain(F_j,C). To calculate IntrinsicValue(F_j) we need to calculate the number of records that have particular feature values. This is achieved by counting the number of set bits n(BVf_i) in the binary vector (CMM row) for fi for qualitative features or n(BVb_i) in the binary vector for the quantisation bin b_i for quantitative features. We can store counts for the various feature values and classes as we proceed so there is no need to calculate any count more than once.

The main disadvantage of GR is that it tends to favour features with low Intrinsic Value rather than high gain by overcompensating towards a feature just because its intrinsic information is very low.

Chi-Square Algorithm

We now demonstrate how to implement a second new feature selector in the AURA framework. The Chi-Square (CS) (Liu & Setiono, 1995) algorithm is a feature ranker like MI, OR and GR rather than a feature selector; it scores the features but it is the user's prerogative to select which features to use. CS assesses the independence between a feature (F_j) and a class (C) and is sensitive to feature interactions with the class. Features are independent if CS is close to zero. Forman (2003) and Yang and Pedersen (1997) conducted evaluations of filter feature selectors and found that CS is among the most effective methods of feature selection for classification.

Chi-Square is defined as Eq. (14):

$$\chi^2\left(F_j, C\right) = \sum_{i=1}^{b(F_j)} \sum_{c=1}^{nClass} \frac{N * (wz - yx)^2}{(w+y) * (x+z) * (w+x) * (y+z)} \qquad (14)$$

where b(F_j) is the number of bins (CMM rows) representing feature F_j, nClass is the number of classes, w is the number of times f_i and c co-occur, x is the number of times f_i occurs without c,y is the number of times c occurs without f_i,z is the number of times neither c nor f_i occur. Thus, CS is predicated on counting occurrences and co-occurrences and, hence, has many commonalities with MI, CFS and GR.

- Fig. 3 shows how to produce a binary output vector ($BVf_i \wedge BVc$) for qualitative features or ($BVb_i \wedge BVc$) for quantitative features listing the co-occurrences of a feature value and a class value. It is then simply a case of counting the number of set bits (1s) in the thresholded binary vector T in Fig. 3 to count w.
- To count x for qualitative features, we logically subtract ($BVf_i \wedge BVc$) from the binary vector (BVf_i) to produce a binary vector and count the set bits in the resulting vector. For quantitative features, we subtract ($BVb_i \wedge BVc$) from (BVb_i) and count the set bits in the resulting binary vector.
- To count y for qualitative features, we can logically subtract ($BVf_i \wedge BVc$) from (BVc) and count the set bits and likewise for quantitative features we can subtract ($BVb_i \wedge BVc$) from BVc and count the set bits.
- If we logically *or* (BVf_i) with (BVc), we get a binary vector representing $(F_j=f_i) \vee (C=c)$ for qualitative features. For quantitative features, we can logically *or* (BVb_i) with (BVc) to produce $(F_j=bin(f_i)) \vee (C=c)$. If we then logically invert this new binary vector, we retrieve a binary vector representing z and it is simply a case of counting the set bits to get the count for z.

As with MI and OR, CS is univariate and assesses features on an individual basis selecting the features with the highest scores, namely the features that interact most with the class.

Odds Ratio

The third new feature selector is Odds Ratio (OR) (see Forman, 2003). OR is another feature ranker. Standard OR is a two-class feature ranker although it can be extended to multiple classes. It is often used in text classification tasks as these are often two-class problems. It performs well particularly when used with Naïve Bayes Classifiers. OR reflects relevance as the likelihood (odds) of a feature occurring in the positive class normalised by that of the negative class. OR has many commonalities with MI, CFS and GR but particularly with CS where it requires the same four calculations w, x, y and z (defined above in Section 3.4). Odds Ratio is defined by Eq. (15):

$$OR\left(F_j, C\right) = \sum_{i=1}^{b(F_j)} \frac{wz}{yx} \tag{15}$$

where $b(F_j)$ is the number of bins (CMM rows) representing feature F_j, w is the number of times f_i and c co-occur, x is the number of times f_i occurs without c, y is the number of times coccurs without f_i, z is the number of times neither c nor f_i occur. Thus, OR is predicated on counting occurrences and co-occurrences. To avoid division by zero the denominator is set to 1 if $\frac{yx}{}$ evaluates to 0.

PARALLEL AND DISTRIBUTED AURA

Feature selection is a combinatorial problem so a fast, efficient and scalable platform will allow rapid analysis of large and high dimensional data sets. AURA has demonstrated superior training and recall speed compared to conventional indexing approaches (Hodge & Austin, 2001) such as hashing or inverted file lists which may be used for data indexing. AURA trains 20 times faster than an inverted file list and 16 times faster than a hashing algorithm. It is up to 24 times faster than the inverted file list for recall and up to 14 times faster than the hashing algorithm. AURA k-NN has demonstrated superior speed compared to conventional k-NN (Hodge & Austin, 2005) and does not suffer the limitations of other k-NN optimisations such as the KD-tree which only scales to low dimensionality data sets (McCallum, Nigam, & Ungar, 2000). We showed in Hodge et al. (2006) that using AURA speeds up the MI feature selector by over 100 times compared to a standard implementation of MI.

For very large data sets, the data may be processed in parallel on one compute node (such as a multi-core CPU) or across a number of distributed compute nodes. Each compute node in a distributed system can itself perform parallel processing.

Parallel AURA

In Weeks et al. (2002), we demonstrated a parallel search implementation of AURA. AURA can be subdivided across multiple processor cores within a single machine or spread across multiple connected compute nodes. This parallel processing entails "striping" the CMM across several parallel subsections. The CMM is effectively subdivided vertically across the output vector as shown in Fig. 4. In the data, the number of features m is usually much less than the number of records N, m≪N. Therefore, we subdivide the data along the number of records N (column stripes) as shown in the leftmost example in Fig. 4.

Splitting the data across multiple CMM stripes using columns means that the CMM can store data as separate rows within a single stripe. Each record is contained within a single stripe. Each separate CMM stripe outputs a thresholded vector from that CMM stripe.

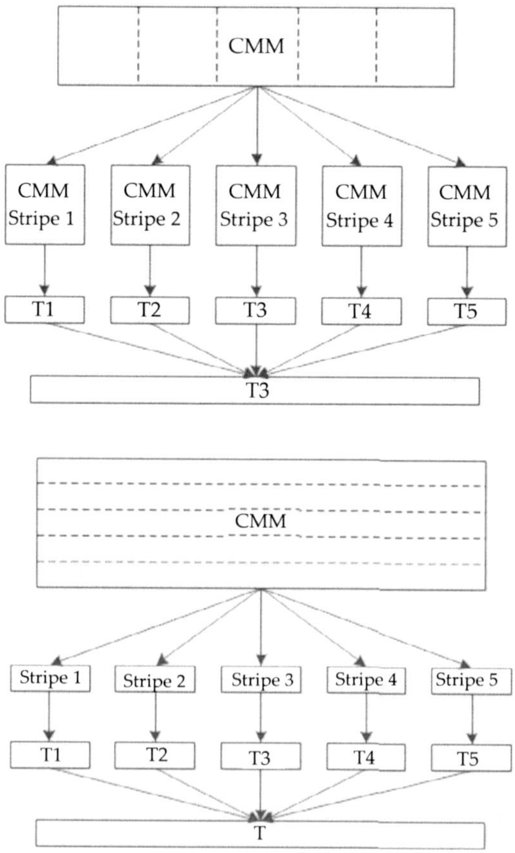

Figure. 4: If a CMM contains large data it can be subdivided (striped) across a number of CMM stripes. In the left hand figure, the CMM is striped vertically (by time) and in the right hand figure the CMM is striped horizontally (be feature subsets). On the left, each CMM stripe produces a thresholded output vector T_n containing the top k matches (and their respective scores) for that stripe. All

$\{T_n\}$ are aggregated to form a single output vector T which is thresholded to list the top matches overall. On the right, each stripe outputs a summed output vector S_n. All S_n are summed to produce an overall summed output vector which is thresholded to list the top matches overall.

If the number of features is large then it is possible to subdivide the CMMs further. The CMM is divided vertically by the records (column stripes) as before and then the column stripes are subdivided by the input features (row stripes). Subdivision by input features (row stripes) is shown in the rightmost diagram in Fig. 4. Dividing the CMM using the features (row stripes) makes assimilating the results more complex than assimilating the results for column stripes. Each row stripe produces a summed output vector containing column subtotals for those features within the stripe. The column subtotals need to be assimilated from all row stripes that hold data for that column. Thus, we sum these column subtotals to produce a column stripe vector C holding the overall sum for each column in that stripe. Row striping involves assimilating integer vectors of length c where c is the number of columns for the column subdivision (column stripe).

Distributed AURA

There are two central challenges for distributed feature selection: firstly, maintaining a distributed data archive so that data does not have to be moved to a central repository and secondly, orchestrating the search process across the distributed data. Different data and applications will have different criteria that they wish to optimise. These could be optimising communication overhead, processing speed, memory usage or combinations of these criteria. Hence, there is unlikely to be a single best technique for distribution.

To distribute AURA, we use the striping mechanisms detailed in the previous section. However, rather than spreading the stripes within the cores of a multicore processor, we distribute the stripes across computers within a distributed network. The stripes need to be distributed for maximum efficiency. This can be to maximise processing speed, to minimise memory usage, to minimise communication overhead or a combination of criteria. Distributing the stripes requires an efficient distribution mechanism to underpin the procedure.

Orchestrated search with minimal data movement is provided by the open source software project: Apache Hadoop (Shvachko, Hairong, Radia, & Chansler, 2010). Hadoop operates on the premise that "*moving computation is cheaper than moving data*" (Borthakur, 2008). Hadoop allows the distributed processing of large data sets across clusters of commodity servers. It provides load balancing, is highly scalable and has a very high degree of fault tolerance. It is able to run on commodity hardware due to its ability to detect and handle failures at the application layer. There are multiple copies of the stored data so, if one server or node is unavailable, its data can be automatically replicated from a known good copy. If a compute node fails then Hadoop automatically re-balances the work load on the remaining nodes. Hadoop has demonstrated high performance for a wide variety of tasks (Borthakur et al., 2011). It was initially aimed at batch processing tasks so is ideally suited to the task of feature selection where the feature selector is trained with the training data and feature selection is run once on a large batch of test data. Hadoop is currently developing real-time processing capabilities. In this paper, we focus on batch processing and the implementation details of the five feature selectors using AURA with Hadoop.

Hadoop is highly configurable and can be optimised to the user's specific requirements, for example, optimising to minimise memory overhead, optimising for fastest processing or optimising to reduce communication overhead. Hence, we do not attempt to evaluate Hadoop here. Instead, we focus on describing how to map AURA CMMs to Hadoop to create a feature evaluation framework.

There are two parts of Hadoop that we consider here: YARN which assigns work to the nodes in a cluster and the Hadoop Distributed File System (HDFS) which is a distributed file system spanning all the nodes in the Hadoop cluster with a single namespace.

YARN (Kumar et al., 2013) supersedes MapReduce in Hadoop. YARN is able to run existing MapReduce applications. YARN decouples resource management and scheduling from the data processing. This means that data can continue to be streamed into the system simultaneously with MapReduce batch jobs. YARN has a central resource manager that reconciles Hadoop system resources according to constraints such as queue capacities or user-limits. Node manager agents monitor the processing operations of individual nodes in the cluster. The processing is controlled by an ApplicationMaster which negotiates resources from the central resource manager and works with the node manager agents to execute and monitor the tasks. The actual MapReduce procedure, divides (maps) the processing into separate chunks which are processed in parallel. The outputs of the processing tasks are combined (reduced) to generate a single result. The input and output data for MapReduce can be stored in HDFS on the same compute nodes used for processing the MapReduce jobs. This produces a very high aggregate bandwidth across the cluster. The user's applications specify the input/output locations and supply *map* and *reduce* functions via implementations of appropriate interfaces and/or abstract-classes. The framework takes care of distributing the software/configuration, scheduling tasks, monitoring the tasks and re-executing any failed tasks.

HDFS links together the file systems on many local nodes to make them into one big file system. HDFS assumes nodes will fail, so it achieves reliability by replicating data across multiple nodes. Processing data in situ on local nodes is efficient compared to moving the data over the network to a single processing node. This local processing architecture of Hadoop has resulted in very good performance (Rutman, 2011) on cheap computer clusters even with relatively slow network connections (such as 1 Gig Ethernet) (Rutman, 2011). Hence, Hadoop is ideal to underpin our distributed processing architecture.

Hadoop Feature Selection

Feature selection is a two part procedure. A training phase described in Section 2.1 trains the data into the CMMs. A test phase then applies test data to the trained CMMs and correlates the results to produce feature selections. Each compute node holds a CMM, CMM stripe or set of CMM stripes that stores all local data. During training, CMMs are not immutable as each association in Eq. (1) changes the underlying CMM so Hadoop MapReduce is not a suitable paradigm for CMM training. Hence, the CMMs are trained in a conventional fashion and uploaded to HDFS once trained. If the data stored in a node's CMM exceed the memory capacity of that node then the CMM is subdivided into stripes as described in Section 4.1 and shown in Figs. 4 and 5. The set of all CMM stripes at a node stores all data for that node. Every CMM stripe across the distributed system has to be coordinated so that record identifiers (such as timestamps) are matched to allow the CMM sum and threshold. Sum and threshold is column-based and relies on columns representing the same datum. When the results from different CMMs are unified then the columns from the various CMMs need to be aligned. The system is very flexible; we only need to access relevant CMM stripes so we can access subsets of data. The approach is a combination of the striping described above in Section 4.1 and the CMM distribution described in Section 4.2 with Hadoop orchestrating the search.

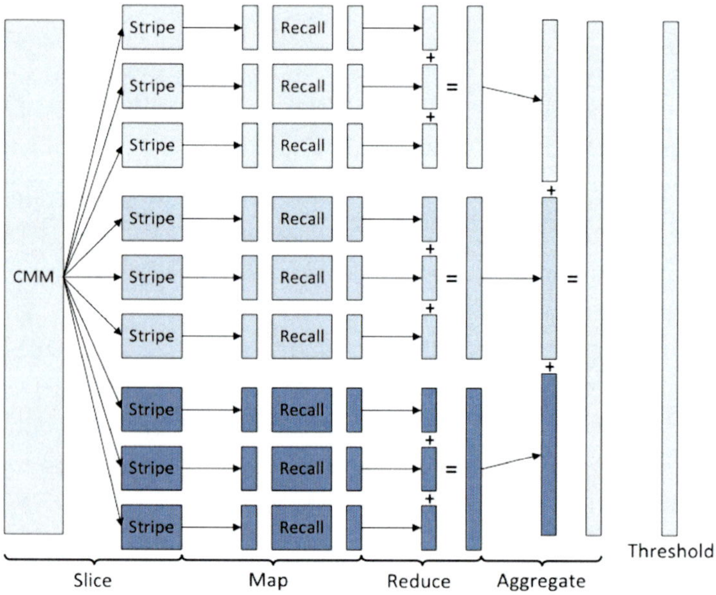

Figure. 5: Figure showing distributed AURA recall in Hadoop. In the figure, there are three distributed compute nodes as shown by the shading with three CMM stripes per node (3 CPU cores per node and one stripe per core). Thus, the top three stripes are on one compute node spread across three cores. In the map phase, the required input vectors are applied to the CMM stripes and the summed output vector is recalled for each stripe. The summed output vector can be thresholded now or later following aggregation as described in Section 4.3.1. During the reduce phase, these output vectors are aggregated at each compute node giving three aggregated vectors. Finally, the three vectors are combined.

While the CMMs are being trained it is expedient to generate a MapReduce input file of input vectors to be used to produce the feature selections. These files will be split into batches by the MapReduce software and the results will be correlated to produce the feature selection scores. There is one input file per CMM stripe and the input vectors in each file represent the set of input vectors for recall to produce the feature selections.

Each CMM stripe that receives a search request, executes the recall process described in Section 2.3. The candidate matches are the set of stored patterns that are close to the query in the feature space. In Hadoop the processing is coordinated by MapReduce (Shvachko et al., 2010). Hadoop YARN schedules the MapReduce tasks independently of the problem being solved. There is one Map job for each input file. Therefore, we model feature selection as a series of MapReduce jobs with each job representing one CMM stripe and the tasks are batches of file iterations (batch processing subsets of records) from the test data. The tasks are processed in parallel on distributed nodes. Each CMM stripe is read into a job. The recall function for CMM stripes is written as a Map task. Each MapReduce job invokes multiple Map tasks, each task represents a batch of recalls for a subset of input records, the batches execute in parallel. The Hadoop Mapper keeps track of the output vector versus record ID pairs so we know which output vector is associated with which record. The Reduce tasks perform the integer output vector thresholding as described in Section 2.3 and write the data back into the file associated with the CMM stripe. Multiple feature selectors can be run in parallel, each executing as a series of MapReduce jobs. The CMMs for feature selection are immutable so subsequent iterations do not depend on the results (or changes) of the CMMs.

This whole MapReduce process has to be coordinated. If the MapReduce process is running at a single location then it can be coordinated as a Java class that initiates the individual jobs and then coordinates the results from all jobs to produce the feature selection scores. If the processing is geographically distributed then it needs a more complete coordinator. This can be achieved using for example the UNIX curl command and a monitor process that determines when curl has collected new data. Alternatively, it can be achieved using a distributed stream processor such as Apache Flume (https://flume.apache.org/) or Storm (https://storm.incubator.apache.org/). Essentially, whichever tool is used this is a three part process: initiate the feature selection process at each of the distributed nodes; retrieve the results data from the distributed nodes; and, monitor when the results have been returned from all nodes and combine them into a single unified result.

Stripe Vectors

For Big data, the CMMs are too big to store in a one computer's memory. Hence, they need to be striped across multiple computers as in Figs. 4 and 5. Each CMM stripe returns a vector representing the matching results for the input vector with respect to that CMM stripe. Palm (2013) has extensively analysed representations in associative memories and found that sparse representations are optimal because the number of matrix operations is proportional to the number of set bits in the vectors. A sparse pattern will have fewest set bits and require fewest operations. For our feature selector, each CMM stripe can return its results as

- an integer vector S_k (un-thresholded),
- a thresholded vector T_k or
- a list of the set bits in the thresholded vector.

Option 1 is the least efficient as, potentially, every column could have an integer score so the vector would be an integer vector of length N where N is the number of data records stored. This integer vector can be thresholded for option 2 which produces a binary vector. A binary vector requires less storage capacity than an integer vector (1 bit per element for the binary vector compared to 16 or 32 bits per element for the integer vector). For option 3, we would return a list of the set bits. For this we can exploit a compact list representation for representing binary vectors (Hodge & Austin, 2001). This compact list representation is similar to the pointer representation used in associative memories (Bentz, Hagstroem, & Palm, 1997). It ensures that retrieval is proportional to the number of set bits in the thresholded output vector so is fast and scalable. The feature selection process produces a large set of output vectors from the CMM stripes; namely, all vectors necessary for all feature selectors. Option 3 allows AURA to be used for distributed processing with data sets of millions of records while using a relatively small amount of memory and with a massively reduced communication overhead. For example, if there were 10,000,000 records in the data set then a vector would have 10,000,000 elements. If only three records match (records; 8, 10 and 11) then processing {8,10,11} as indices requires much less time, memory and communication bandwidth compared to processing 10,000,000 binary digits. Hence, wherever possible we use option 3.

The results need to be amalgamated for each feature selector to produce the feature scores for that feature selector. The system maintains an index of what data are stored where and what each datum represents so the coordinating node can coordinate the matching, receive all matching data and determine the set of best matches across all searchable data. Each feature selector will have a separate amalgamate program running at the coordinating node. This program uses the required vectors and set bit counts returned from AURA to produce the feature score as described in Sections 3 and 5.

ANALYSIS OF AURA FEATURE SELECTION

We demonstrate theoretically using a worked example that our framework vastly reduces the number of required computations compared to processing the feature selectors separately. The worked example provides an easy and simple illustration of the method on a small data size. We envisage using the feature selector on Big Data sets where Big Data refers to data sets that require at a minimum multiple CPUs but more likely multiple compute nodes to process in tractable time for the application. The larger the data set and the more time critical the data processing then the more important our computation reduction will become. MI, CFS, CS, OR and GR can all use a single CMM representation for the data such as the CMM in Fig. 6. This overall CMM is amenable to striping across the processing nodes to allow Hadoop processing in a similar fashion to Figs. 4 and 5. The framework is underpinned by Hadoop which has been thoroughly evaluated in the literature (Kumar et al., 2013). Hadoop is highly configurable large data set framework that can be optimised to the user's specific requirements, for example, optimising to minimise memory overhead, optimising for fastest processing or optimising to reduce communication overhead. Hence, we do not attempt to evaluate Hadoop itself here but just focus on how we minimise the number of feature selection computations to minimise processing. Users will use our framework to select the best feature selector for their data and application using their own specific criteria.

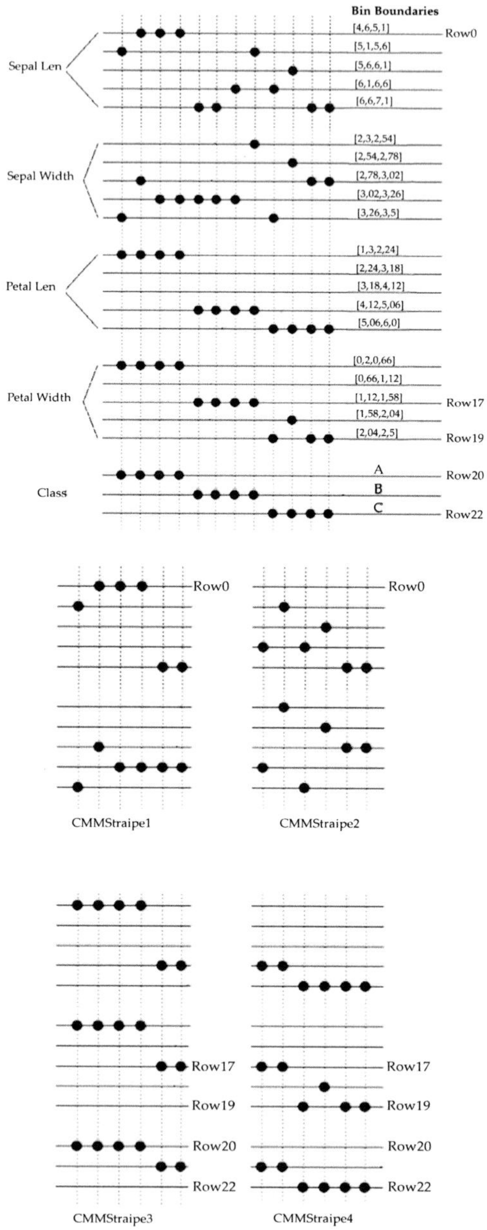

Figure. 6: The 12 records from the iris data set, quantised and trained into a single AURA CMM (left) and subdivided across 4 stripes of the CMM (right). The letters in rows 20–22 indicate the class of the record: A = Iris-setosa, B = Iris-versicolour, C = Iris-virginica.

The feature selectors in Section 3 have many commonalities when implemented in the unified AURA framework. We can demonstrate the commonalities by analysing 12 records from the Iris data set (Fisher, 1936). The Iris data are illustrated in Fig. 6 (left) when trained into the CMM. The 12 records have been trained into a CMM using the four features and the class. Each feature is quantitative and has been subdivided into five quantisation bins of equal width. Fig. 6 (right) shows the same data divided into four CMM stripes (***CMMStripe1***, ***CMMStripe2***, ***CMMStripe3*** and ***CMMStripe4***). The horizontal (row-based) striping means that the features "*sepal len*" and "*sepal width*" are in the top stripes and "*petal len*", "*petal width*" and the *class* are in the bottom two stripes. The vertical (column-based) striping means that the first 6 data records are stored in the left two stripes and the other 6 records in the right two stripes. If the data were time-series or sequential, the column-based striping would form two time frames with the oldest data in the left two stripes and the newest data in the right two stripes. The input vectors are stored in a file for each CMM or CMM stripe. These files can then be batch processed in the Hadoop framework described. Within the evaluation, we consider how the data and CMMs would be accommodated in our Hadoop framework.

MI, CFS, CS, OR and GR all use BVf_i (the binary vector where $(F_j=f_i)$), BVb_i (the binary vector representing the quantisation bin $bin(f_i)$) and ***BVc*** (the binary vector representing all records that have class label c). These only need to be extracted once and used in each feature selector as appropriate. For example in Fig. 6, if we want all records where $1.12 \leq petalwidth < 1.58$then we activate row 17 of the CMM. We can then Willshaw threshold the resultant integer output vector S (0000**1111**0000) at level 1 and retrieve the binary thresholded vector T with a bit set for every matching record (bits 4, 5, 6, 7). For the Hadoop distributed version, only the relevant CMM stripes are queried in Fig. 6 (right). In this case, activating row 17 of ***CMMStripe3*** and ***CMMStripe4*** queries the relevant data. ***CMMStripe3*** will output thresholded vector T3 with bits 5 and 6 set and ***CMMStripe4*** will output T_4 with bits 7 and 8 set. T_3 and T_4 can be concatenated to form a single vector thresholded vector T (as in Fig. 4) with bits 4, 5, 6 and 7 set. For the Hadoop distributed version, each CMM stripe ***CMMStripeX*** outputs a list of the indices of the set bits in TX which are collected by the coordinator.

CFS, GR and MI all require $nBVf_i$ a count of the number of data records where a particular feature has a particular value F_j=fiand BVc a count of the number of records where the class has a particular label C=c. To count the number of records where 1.12≤petalwidth<1.58, we retrieve the binary thresholded vector as above and count the number of set bits (bits 4, 5, 6 and 7 are set giving 4 matching records). For the Hadoop approach, we coordinate the retrieval as above, concatenate the lists to produce a single overall list of set bits and count the list length. T_3 has bits 4 and 5 set and T_4 has bits 6 and 7 set giving 4 matching records in total.

CFS, CS, OR, GR and MI all use (BVf_i∧BVc) and (BVb_i∧BVv) for qualitative and quantitative features respectively. For example, we can find all records where 4.6≤sepallen<5.1*and* the class is A by activating rows 0and 20 of the CMM, thresholding S (1**222**000000) at Willshaw level 2 to give T with three bits set: column 1, 2 and 3 in Fig. 6(left). This takes more coordinating in the Hadoop framework as the data for the feature value may not be stored with the data for the class; they may be in different CMM stripes. In Fig. 6 (right), activate row 0 in ***CMMStripe1*** and ***CMMStripe2*** and then activate row 20 in ***CMMStripe3*** and ***CMMStripe4***. The coordinating program needs to correlate the sections of the vector for the feature value and correlate the sections of the vector for the class to form a single vector. ***CMMStripe1*** needs to be added (summed) with the output integer vector of ***CMMStripe3*** to give S_{1+3} and ***CMMStripe2*** needs to be added (summed) with the output integer vector of ***CMMStripe4*** to give S_{2+4}. The summed vectors can then be thresholded at 2 to give T_{1+3} with bits 1, 2 and 3 set (three matching records) and T_{2+4} with no bits set (no matches). The two thresholded output vectors are concatenated to produce T with bits 1, 2 and 3 set. If the thresholded vectors are stored as lists of indices (see Section 4.3.1) then this is simply a task of finding the common indices between the two vectors.

MI, CFS, CS, OR and GR all also need a count of the conjunction, that is n(BVf_i∧BVc) and n(BVb_i∧BVc) for qualitative and quantitative features respectively. Hence, we retrieve the binary thresholded vector T as above and count the set bits.

Rather than calculating these elements multiple times, we can take advantage of the commonalities by calculating each common value, binary vector or count only once and propagating the result to each feature selector that requires it. Following these common calculations, all necessary calculations will have been made for MI and GR. CFS just requires the pairwise feature versus feature analyses ($BVb_i \wedge BVb_k$). These are performed in the same way as the feature versus class analyses above. CS and OR require the manipulation of some of the binary vectors to produce the logical *or* vectors. This requires the coordination of the vectors. To find (BVb_i)∨(BVc), we combine the list of set bits for (BVb_i) with the list of set bits for (BVc) and count the resulting list length. By calculating the common elements first, the remainder of the calculations can be performed for each feature selector using either this CMM and processing the algorithms in series or by generating multiple copies of the CMM and processing them in parallel if sufficient processing capacity is available.

Once all of the binary vectors have been retrieved by the distributed Hadoop system, they need to be processed to calculate the feature scores as per Section 3 using the various feature selectors. A coordinator program organises this in parallel. There is one feature score calculation process per feature selector (currently five feature selectors are described here).

For the Iris data set, there are 20 feature row activations $20*BVb_i$ and three class activations $3*BVc$. To calculate ($BVb_i \wedge BVc$) requires 20×3=60 calculations. Hence, there are 83 common calculations (20+3+60) across all five feature selectors. CFS then needs to calculate ($BVb_i \wedge BVb_k$) which would require 19! calculations if every feature value was compared to every other. However, CFS uses greedy forward search so that the number of comparisons is minimised (Hall, 1998) to a worst case of (202−20)/2=190. We have already extracted all $20*BVb_i$ binary vectors so CFS needs 190 logical *and*s but no CMM accesses. We have saved a minimum of 20 CMM accesses for BVb_i and a maximum of 190 CMM accesses for worst case forward search. Manipulating the binary vectors can be performed at the coordinating node and in parallel as a Hadoop batch process. CS requires the logical *or* vectors ($BVb_i \vee BV_c$). Again, we already have all $20*BVb_i$ binary vectors and all $3*BVc$binary vectors so there are 20×3=60 logical *or*s to perform. Thus, we have saved a minimum of $20*BV_{bi}+3*BV_c=23$CMM accesses and potentially 60 CMM accesses if all 60 *or*operations were performed in the CMM. Thus

MI requires 83 calculations, GR also requires 83, CFS requires 83 plus 190 and CS requires 83 plus 60. Without our reductions there would be 83+83+83+190+83+60 calculations. We have reduced this to 83+190+60. Additionally, 190+60 of these can use vectors already extracted so there is no need to access the CMM. We have saved 3∗83=249 recalls from the CMM by finding common aspects, have removed a minimum of 20+23 further CMM recalls and have reduced the other calculations to logical operations on stored binary vectors. The minimum saving on CMM recalls is given by Eq. (16).

$$Saving = (3 \times (n(BVb_i) + n(BVc) + (n(BVb_i)\, n(BVc)))) + ((2 \times n(BVb_i)) + n(BVc)) . \quad (16)$$

CONCLUSION

Massive and complex data sources pose challenges for data mining but they also hold many opportunities. New information can be uncovered, vast timelines of data are available for analysis and the data models learned will be increasingly rich as the training data expands. How the data is represented needs to be carefully considered including careful preparation such as cleaning and selecting feature subsets. In this paper we have introduced a distributed processing framework for feature selection using the AURA neural network and Apache Hadoop. There are currently five feature selectors available which may be used independently or coupled with the AURA k-NN for classification or prediction.

All five feature selectors can use a single trained CMM. We have identified common aspects of the five feature selectors when they are implemented in the AURA framework and indicated how these common aspects may be processed as a common block. All remaining aspects of the feature selectors can then be implemented in parallel using duplicate copies of the trained CMM as compute resources allow. CMMs lend themselves to distributed processing as they can be striped (split) using both row-based and column-based striping. The CMM created for feature selection can be used directly for the AURA k-NN for classification or prediction and any unwanted features (those not selected by the feature selection) can simply be ignored (masked off). Alternatively, the CMM can be retrained with only the required data if processing speed and memory usage at recall time are the primary concern.

The AURA neural architecture has demonstrated superior training and recall speed compared to conventional indexing approaches such as hashing or inverted file lists (Hodge & Austin, 2001) and an AURA-based implementation of the MI feature selector was over 100 times faster than a standard implementation (Hodge et al., 2006). This is further augmented by using the scalability of Hadoop. This combined platform allows rapid processing of feature selectors on large and high dimensional data sets that cannot be processed on standard computers. We envisage using the method on data sets that require at a minimum multiple CPUs but more likely multiple compute nodes to process. The method is also best suited to data mining and analytics that processes a Big Data file in a longer term processing run such as overnight rather than on-line transaction processing which requires near real-time updating. The user can then evaluate the feature sets chosen by the feature selectors against their own data to determine the best feature selector and the best set of features. Additionally, each feature selector (MI, CFS, GR, CS and OR) generates scores for the features which can be used to weight the features during machine learning.

The technique is flexible and easily extended to other feature selection algorithms. By implementing a range of feature selectors in a single framework, we can also investigate ensemble feature selection where the results from a range of feature selectors are merged to generate a consensus overview of the best set of features to use.

We will investigate whether we can use Apache Spark, the in-memory data analytics and cluster computing framework (https://spark.apache.org/) to underpin the AURA feature selection framework. Apache Spark is closely coupled with Hadoop and allows YARN and MapReduce jobs to be run. Spark enables in-memory computing and is reputed to be up to 100 times faster than MapReduce (see https://spark.apache.org/). CMMs are optimised for in-memory processing so fit well with the Spark paradigm. A related development, Optimised Row Columnar (ORC) file format is currently being adopted by Spark. ORC is a file storage format that is tightly integrated with HDFS and provides optimisations for both read performance and data compression. An ORC file divides the data into groups of row data called stripes. This fits with the stripes used in AURA CMMs and would allow a direct mapping from ORC data file stripes to CMM stripes for optimised performance.

We plan to use the feature selection framework that we have developed in this paper in conjunction with the AURA k-NN for traffic analysis (Hodge et al., 2012; Hodge, Krishnan, Austin, & Polak, 2010; Hodge et al., 2011), condition monitoring (Austin et al., 2010) and railway infrastructure monitoring in the NEWTON Project (Hodge, O'Keefe, Weeks, & Moulds, 2015).

ACKNOWLEDGEMENT

This work was supported by UK Engineering and Physical Sciences Research Council (Grant EP/J012343/1).

REFERENCES

1. Austin, J. (1995). Distributed associative memories for high speed symbolic reasoning. In R. Sun & F. Alexandre (Eds), IJCAI'95 working notes of workshop on connectionist-symbolic integration: from unified to hybrid approaches. Montreal, Quebec (pp. 87–93).
2. Austin, J., Brewer, G., Jackson, T., & Hodge, V. (2010). AURA-Alert: The use of binary associative memories for condition monitoring applications. In Procs 7th int'l conf. on condition monitoring and machinery failure prevention technologies (pp. 699–711). Red Hook, USA: Curran Associates.
3. Bentz, H., Hagstroem, M., & Palm, G. (1997). Selection of relevant features and examples in machine learning. Neural Networks, 2(4), 289–293.
4. Borthakur, D. (2008). HDFS architecture guide. HADOOP APACHE PROJECT http://pristinespringsangus.com/hadoop/docs/hdfs_design.pdf.
5. Borthakur, D., Gray, J., Sarma, J. S., Muthukkaruppan, K., Spiegelberg, N., Kuang, H., . . . Aiyer, A. (2011). Apache Hadoop goes realtime at facebook. In Procs 2011
6. ACM SIGMOD international conference on management of data (pp. 1071–1080). New York: ACM.
7. Casasent, D., & Telfer, B. (1992). High capacity pattern recognition associative processors. Neural Networks, 5(4), 251–261.

8. Chu, C., Kim, S., Lin, Y., Yu, Y., Bradski, G., Ng, A., & Olukotun, K. (2007). Map-reduce for machine learning on multicore. In Advances in neural information processing systems. Vol. 19 (pp. 281–288). MIT Press.
9. Dash, M., & Liu, H. (1997). Feature selection for classification. Intelligent Data Analysis, 1(3), 131–156.
10. Ertöz, L., Steinbach, M., & Kumar, V. (2003). Finding clusters of different sizes, shapes, and densities in noisy, high dimensional data. In Proc. 3rd SIAM int'l conf. on data mining (pp. 47–58).
11. Fayyad, U., & Irani, K. (1993). Multi-interval discretization of continuous-valued attributes for classification learning. In Procs int'l joint conf. on artificial intelligence (pp. 1022–1027). San Mateo, USA: Morgan Kaufmann.
12. Fisher, R. (1936). The use of multiple measurements in taxonomic problems. Annual Eugenics, 7(2), 179–188.
13. Fletcher, M., Jackson, T., Jessop, M., Liang, B., & Austin, J. (2006). The signal data explorer: A high performance grid based signal search tool for use in distributed diagnostic applications. In CCGrid 2006—6th IEEE int'l symp. on cluster computing and the grid. Singapore (pp. 217–224).
14. Forman, G.(2003). An extensive empirical study of feature selection metrics for text classification. Journal of Machine Learning Research, 3, 1289–1305.
15. Franks, B. (2012). Taming the big data tidal wave: finding opportunities in huge data streams with advanced analytics. Hoboken, NJ: John Wiley & Sons.
16. Guyon, I., & Elisseeff, A. (2003). An introduction to variable and feature selection. Journal of Machine Learning Research, 3, 1157–1182.
17. Guyon, I., Weston, J., Barnhill, S., & Vapnik, V. (2002). Gene selection for cancer classification using support vector machines. Machine Learning, 46(1), 389–422.
18. Hall, M. (1998). Correlation-based feature subset selection for machine learning. (Unpublished doctoral dissertation), Hamilton,

New Zealand: University of Waikato.

19. Hall, M., & Holmes, G. (2003). Benchmarking attribute selection techniques for discrete class data mining. IEEE Transaction on Knowledge & Data Engineering, 15(6), 1437–1447.
20. Hall, M., & Smith, L. (1997). Feature subset selection: a correlation based filter approach. In Int'l conf. on neural information processing and intelligent information systems (pp. 855–858). Berlin: Springer.
21. Hall, M., & Smith, L. (1998). Practical feature subset selection for machine learning. In Procs of the 21st Australian computer science conf. (pp. 181–191). Berlin: Springer.
22. Han, J., & Kamber, M. (2006). The Morgan Kaufmann series in data management systems, Data mining: concepts and techniques. Elsevier.
23. Hebb, D.(1949). The organization of behavior: a neuropsychological theory. New York: Wiley.
24. Hodge, V. (2011). Outlier and anomaly detection: A survey of outlier and anomaly detection methods. LAP LAMBERT Academic Publishing.
25. Hodge, V., & Austin, J. (2001). An evaluation of standard retrieval algorithms and a binary neural approach. Neural Networks, 14(3).
26. Hodge, V., & Austin, J. (2005). A binary neural k-nearest neighbour technique. Knowledge and Information Systems, 8(3), 276–292.
27. Hodge, V., & Austin, J. (2012). Discretisation of data in a binary neural k-nearest neighbour algorithm. Tech. Report YCS-2012-473. UK: University of York, Department of Computer Science.
28. Hodge, V., Jackson, T., & Austin, J. (2012). A binary neural network framework for attribute selection and prediction. In 4th int'l conf. on neural computation theory and applications. Barcelona, Spain.
29. Hodge, V., Krishnan, R., Austin, J., & Polak, J. (2010). A computationally efficient method for on-line identification of traffic control intervention measures. In 42nd annual UTSG conf., University of Plymouth, UK.
30. Hodge, V., Krishnan, R., Austin, J., & Polak, J. (2011). Short-term traffic prediction using a binary neural network. In 43rd Annual UTSG Conf., Milton Keynes, UK.

31. Hodge, V., O'Keefe, S., & Austin, J. (2006). A binary neural decision table classifier. NeuroComputing, 69(16–18), 1850–1859.
32. Hodge, V., O'Keefe, S., Weeks, M., & Moulds, A. (2015). Wireless sensor networks for condition monitoring in the railway industry: A survey. IEEE Transactions on Intelligent Transportation Systems, 16(3), 1088–1106.
33. Jolliffe, I. (2002). Principal component analysis (2nd ed.). New York, USA: SpringerVerlag. Kohavi, R., & John, G. (1997). Wrappers for feature subset selection. Artificial Intelligence Journal, 97(1–2), 273–324. Special Issue on Relevance.
34. Kumar, V., et al. (2013). Apache Hadoop YARN: yet another resource negotiator. In Proceedings of the 4th annual symposium on cloud computing, (SOCC'13). New York, NY, USA: ACM.
35. Laney, D. (2001). 3D Data management: Controlling data volume, velocity, and variety. Application Delivery Strategies by META Group.
36. Liu, H., Motoda, H., Setiono, R., & Zhao, Z. (2010). Feature selection: An ever evolving frontier in data mining. Journal of Machine Learning Research, 10, 4–13.
37. Liu, H., & Setiono, R. (1995). Chi2: Feature selection and discretization of numeric attributes. In Procs IEEE 7th int'l conf. on tools with artificial intelligence (pp. 338–391). IEEE.
38. Liu, H., & Setiono, R. (1996). A probabilistic approach to feature selection—a filter solution. In Procs of 13th int'l conf. on machine learning (pp. 319–327). San Mateo, USA: Morgan Kaufmann.
39. McCallum, A., Nigam, K., & Ungar, L. (2000). Efficient clustering of high-dimensional data sets with application to reference matching. In Procs 6th ACM SIGKDD int'l conf. on knowledge discovery & data mining (pp. 169–178). New York: ACM.
40. Palm, G. (2013). Neural associative memories and sparse coding. Neural Networks, 37, 165–171.
41. Quinlan, J. (1986). Induction of decision trees. Machine Learning, 1, 81–106.
42. Quinlan, J. (1992). C4.5 programs for machine learning. San Mateo, USA: Morgan Kaufmann.

43. Reggiani, C. (2013). Scaling feature selection algorithms using MapReduce on Apache Hadoop. (Master's thesis), Italy: Politecnico di Milano, Retrieved from https:// www.politesi.polimi.it/bitstream/10589/81201/1/2013_07_Reggiani.pdf.
44. Rutman, N. (2011). Map/reduce on lustre, (white paper). Technical report. Havant, UK: Xyratex Technology Limited.
45. Shvachko, K., Hairong, K., Radia, S., & Chansler, R.(2010). The Hadoop distributed file store system. In Procs IEEE 26th symp. on mass storage systems and technologies (pp. 1–10). IEEE.
46. Singh, S., Kubica, J., Larsen, S., & Sorokina, D. (2009). Parallel large scale feature selection for logistic regression. In SIAM international conference on data mining. SDM (pp. 1172–1183).
47. Sun, Z. (2014). Parallel feature selection based on MapReduce. In Lecture notes in electrical engineering: Vol. 277. Computer engineering and networking (pp. 299–306). Springer.
48. Varela, P., Martins, A., Aguiar, P., & Figueiredo, M. (2013). An empirical study of feature selection for sentiment analysis. In 9th conference on telecommunications. Conftele 2013, Castelo Branco.
49. Weeks, M., Hodge, V., & Austin, J. (2002). A hardware accelerated novel IR system. In Procs 10th euromicro workshop (PDP-2002). IEEE Computer Society.
50. Weeks, M., Hodge, V., O'Keefe, S., Austin, J., & Lees, K. (2003). Improved AURA k-nearest neighbour approach. In Artificial neural nets problem solving methods (pp. 663–670). Berlin: Springer.
51. Wettscherek, D. (1994). A study of distance-based machine learning algorithms. (Unpublished doctoral dissertation), Corvallis, USA: Oregon State University. Willshaw, D., Buneman, O., & Longuet-Higgins, H. (1969). Non-holographic associative memory. Nature, 222, 960–962.
52. Witten, I., & Frank, E. (2000). Data mining: practical machine learning tools and techniques with Java implementations. San Mateo, USA: Morgan Kaufmann.
53. Yang, Y., & Pedersen, J. (1997). A comparative study on feature selection in text categorization. In Procs 14th int'l conf. on machine learning (pp. 412–420).

54. Zhang, C., Kumar, A., & Ré, C. (2014). Materialization optimizations for feature selection workloads. In Proceedings of the 2014 ACM SIGMOD international conference on management of data (pp. 265–276). ACM.
55. Zikopoulos, P., & Eaton, C. (2011). Understanding big data: analytics for enterprise class Hadoop and streaming data. New York: McGraw-Hill.

CHAPTER 12

IMPROVED SYSTEM IDENTIFICATION USING ARTIFICIAL NEURAL NETWORKS AND ANALYSIS OF INDIVIDUAL DIFFERENCES IN RESPONSES OF AN IDENTIFIED NEURON

Alicia Costalago Meruelo[a], David M. Simpson[a], Sandor M. Veres [b], Philip L. Newland[c]

[a] Institute of Sound and Vibration, University of Southampton, Southampton, UK

[b] Department of Autonomous Control and Systems Engineering, University of Sheffield, Sheffield, UK

[c] Centre for Biological Sciences, University of Southampton, Southampton, UK

ABSTRACT

Mathematical modelling is used routinely to understand the coding properties and dynamics of responses of neurons and neural networks. Here we analyse the effectiveness of Artificial Neural Networks (ANNs) as a modelling tool for motor neuron responses. We used ANNs to model the synaptic responses of an identified motor neuron, the fast extensor motor neuron, of the desert locust in response to displacement of a sensory organ, the femoral chordotonal organ, which monitors movements of the tibia relative to the femur of the leg. The aim of the study was threefold: first to determine the potential value of ANNs as tools to model and investigate neural networks, second to understand the generalisation properties of ANNs across individuals and to different input signals and third, to understand individual differences in responses of an identified neuron. A metaheuristic algorithm was developed to design the ANN architectures. The performance of the models generated by the ANNs was compared with those generated through previous mathematical models of the same neuron. The results suggest that ANNs are significantly better than LNL and Wiener models in predicting specific neural responses to Gaussian White Noise, but not significantly different when tested with sinusoidal inputs. They are also able to predict responses of the same neuron in different individuals irrespective of which animal was used to develop the model, although notable differences between some individuals were evident.

Keywords: Artificial neural network Metaheuristic algorithm Proprioception Grasshopper Motor neuron Individual differences

INTRODUCTION

Mathematical modelling has been used for many years to understand and describe biological systems, including their dynamics (Gamble & DiCaprio, 2003), the effects of experimental manipulation (Marder & Taylor, 2011), the impact of noise and variability (Sarkar, Christini, & Sobie, 2012) and how they change with ageing and disease (Horn, Ruppin, Usher, & Herrmann, 1993).

Linear and non-linear models, such as those derived by Wiener's method, have been widely used to quantify the behaviour of the nervous system (Kondoh, Okuma, & Newland, 1995; Marmarelis, 2004; Marmarelis & Naka, 1972). While these methods are powerful and provide quantitative descriptions of the linear dynamic transfer characteristics of a system (Marmarelis & Naka, 1973a) they can contain estimation errors due to background noise (Dewhirst, Angarita-Jaimes, Simpson, Allen, & Newland, 2012) and possibly overtraining of models

(Tötterman & Toivonen, 2009). These estimation errors in turn produce erroneous predictions of the system's responses (Gamble & DiCaprio, 2003). Moreover, Wiener methods are not always applicable (Angarita-Jaimes et al., 2012) as the system expansion of the Wiener representation does not necessarily converge for all input functions (Palm & Poggio, 1977). In addition, such mathematical models have, in general, been fitted to the response to a stimulus of an individual (Dewhirst et al., 2012; DiCaprio, 2003; Marmarelis & Naka, 1972; Newland & Kondoh, 1997), without verifying if a particular model is a good representation of the population as a whole (Marder & Taylor, 2011). Using an average response to represent the population responses may be misleading due to different characteristics inherent to each individual (Goldman, Golowasch, Marder, & Abbott, 2001).

Recently, there has been an interest in modelling dynamic systems using Artificial Neural Networks (ANNs) since they have been found to model accurately many continuous functions (Haykin, 1999). In particular, they have been applied to chemical processes, plant identification and controller structures (Xing & Pham, 1995), and have been shown to have good predictive performance in simulations of non-linear dynamic systems (Hunt, Sbarbaro, Zbikowski, & Gawthrop, 1992), financial markets (White, 1988), classification (Suraweera & Ranasinghe, 2008) and pattern recognition (Bishop, 1995). The choice of ANNs to model so many different systems is, in part, due to their flexibility, adaptability and generalisation capabilities (Benardos & Vosniakos, 2007) and their easy application in software and hardware devices (Hunt et al., 1992; Twickel, Büschges, & Pasemann, 2011). They have been applied successfully as robot locomotion controllers (Beer, Chiel, Quinn, Espenschied, & Larsson, 1992; Chiel, Beer, Quinn, & Espenschied, 1992; Cruse et al., 1995) by imitating the nervous systems that produce motion in insects' legs. Taken together these characteristics make them a powerful tool for non-linear model description and control system implementation.

The nervous system of insects, such as the desert locust, is relatively simple compared to mammals and a number of the constituent neurons with specific function are identifiable in different animals (Burrows, 1996) making them ideal to test the potential of ANNs to model neuronal responses, to analyse their generalisation abilities from one animal to the next and to understand individual differences in identified neurons.

Imposed movement of the tibia, relative to the femur of the locust results in a resistance reflex that opposes the applied movement. This reflex can aid during stance and walking, in situations such as tripping or under the influence of other external forces (Field & Matheson, 1998). The Fast Extensor Tibiae (FETi) motor neuron, which is activated during reflex movements, is identifiable in every animal and has been studied for many decades and various mathematical models have been developed to understand its dynamics (Dewhirst et al., 2012; Dewhirst, Simpson, Allen, & Newland, 2009; Newland & Kondoh, 1997). The linear and non-linear properties of the FETi responses are well known, however, computational limitations in parameter estimation and the noise and variability of individual recordings have yet to be understood in detail (Angarita-Jaimes et al., 2012; Dewhirst et al., 2012). These challenges provide the motivation to test and validate ANNs as an effective mathematical method to model and describe the neural response across individuals.

The aim of this work was threefold: first to develop a method to design ANNs to model FETi responses and to determine whether they provide an improved performance over previous mathematical models; second, to understand the generalisation properties of the ANNs across individuals and to different input signals, and third, to understand individual differences in an identified neuron between individuals. To address these issues the performance of the ANN models was measured by their ability to predict responses of an individual as well as different individuals and their responses to different input stimuli.

MATERIALS AND METHODS

Data Recording and Post-Processing

Adult male and female desert locusts (*Schistocerca gregaria*, Forskål) were mounted ventral surface uppermost in modelling clay and fixed firmly with one hind leg rotated through 90° and the femur–tibia angle set to 60° with the anterior face up. The apodeme of the femoral chordotonal organ (FeCO) was exposed by cutting a small window in the cuticle of the distal femur and grasped with a pair of forceps attached to a shaker (Ling Altec 101, LDS Test and Measurement) (Fig. 1). The apodeme was cut distally to avoid movement of the tibia thus opening the reflex-control loop. The thoracic ganglia were then exposed by removing the cuticle of the ventral thorax and also the air sacs and small trachea around the ganglia. A wax covered silver platform was then placed under the meta- and mesothoracic ganglia and the connectives cut posterior to the metathoracic ganglion. The ganglionic sheath was then softened by treating directly with protease (Sigma Type XIV) for 1 min (Newland & Kondoh, 1997). Intracellular recordings were made using glass microelectrodes, filled with 3M potassium acetate and with DC resistances of 50–80MΩ, by driving an electrode through the sheath and into the soma of FETi. FETi receives monosynaptic and polysynaptic inputs from the femoral chordotonal organ (FeCO) that monitors movements of the tibia about the femur and which together underlie a simple resistance reflex that resists imposed movements of the hind leg (Field & Matheson, 1998).

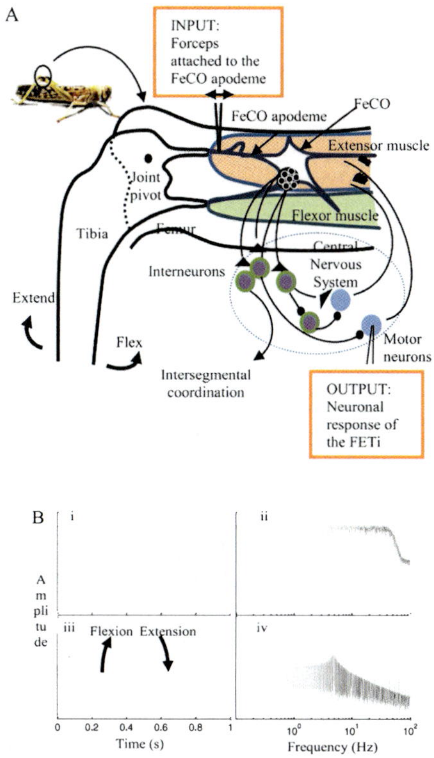

Figure. 1: (A) The components of the reflex control loop. The input was applied to the FeCO apodeme through a shaker and the output was the response of the

FETi motor neuron. (B) The stimuli applied to the system included a band-limited Gaussian White Noise input shown in (i) the time domain and (ii) the frequency domain. A 5 Hz sinusoidal input was also applied and is shown in (iii) in time domain and (iv) in frequency domain.

The synaptic signals recorded in FETi were amplified and digitised with a sampling frequency of 10,000 Hz using a data acquisition board (USB 2527 data acquisition card, Measurement Computing, Norton, MA, USA) and stored for later analysis. They were then re-sampled at 500 Hz in Matlab® following a 3rd order low-pass anti-aliasing filter with a cut-off frequency of 250 Hz to remove residual high frequency noise (the spectrum of the recording is very low above 150 Hz). This re-sampling reduced the size of the files to process and, therefore, the computational time, without removing frequency components of interest. A high pass Butterworth filter of 3rd order and cut-off frequency of 0.2 Hz was applied to eliminate any slow time varying drift. A relatively small sample size of five was selected so that individual differences and similarities could be readily described, in addition to comparing modelling methods. Since all the measurements from all animals were to be compared, and identical input signals were used in all animals, the responses were synchronised using cross-correlations between the input signals.

To move the chordotonal organ the shaker, on which the forceps grasping the apodeme of the FeCO was mounted, was driven with waveforms generated in MATLAB®, with a sampling frequency of 10,000 Hz. The displacements were applied to the apodeme via a digital-to-analogue (DA) converter (Fig. 1(C)) and included a Gaussian White Noise (GWN) (band limited to 0–50 Hz) and a sinusoidal (5 Hz) signal. The amplitude of the signals emulated the angular displacement of the tibia relative to the femur from 20°–100° approximately (Field & Burrows, 1982). We used GWN as an input to the system since it simultaneously excites all frequencies and amplitudes within a specified range and at thus reduces experimental and computational time. To test the ANN models with more realistic inputs, a sinusoid of 5 Hz was also applied to the FeCO (and later to the ANN) representing, approximately, the step frequency of gregarious locusts (Burns, 1973). The peak-to-peak displacement amplitude of the forceps was 1 mm, which represented the maximum linear displacement of the FeCO apodeme (Dewhirst et al., 2012; Field & Burrows, 1982). Previous studies have shown that the synaptic responses of FETi have a transient phase lasting approximately 3 s from the beginning of the FeCO stimulation, followed by a steady state period starting approximately after 10 s, with a transition period in-between (Dewhirst et al., 2012). Although ANNs are able to adapt during transient responses, to compare directly with previous methods, the signals used to train the networks were composed only of the steady state section of the responses (Dewhirst et al., 2012). All signals were visually inspected for the quality of the recording prior to analysis. The results are based on five successful recordings of 30 s duration of FETi responses to 50 Hz band-limited GWN inputs and 5 Hz sinusoidal inputs from five animals.

Artificial Neural Networks for System Identification

We adopted a dynamical artificial neural network, based on a Feed Forward Neural Network (FFNN), known as a Time Delay Neural Network (TDNN) (Waibel, Hanazawa, Hinton, Shikano, & Lang, 1989), to model FETi responses. This type of network uses delayed versions of the input to estimate the output, which turns a static FFNN into a dynamic network (Haykin, 1999), assuming that the response of FETi was a combination of the current and past input samples. The architecture of the ANN was formed by an input and an output layer and a series of hidden layers, each of which was formed by a determined number of nodes. A node is defined mathematically as follows:

$$y = f(\Sigma_i w_{ji} x_i + \theta_i) \quad (1)$$

where x and y are the i inputs and outputs respectively of the jth node, w_{ji} are the weights for each input, θ_i is the bias and f(x) is referred to as the activation function (Xing & Pham, 1995).

The activation function f(x) (Eq. (2)) used in the hidden layers of both networks is a sigmoid function, the most used activation function (Jain, Jianchang, & Mohiuddin, 1996; Xing & Pham, 1995).

$$f(x) = 1/(1 + e^{-x}). \quad (2)$$

This function was used to introduce a non-linearity in the estimated output. The output layer also contained a node, whose transfer function was a linear function instead of a sigmoid function, meaning that all the non-linear calculations occurred within the hidden layers.

The Matlab® timedelaynet built-in function was used to design the ANN network, specifying the number of layers and delays and the training function. A Levenberg–Marquardt back-propagation algorithm was used for training the network and to calculate the weights between nodes. This algorithm is a version of the back-propagation algorithm, chosen for its accuracy and fast convergence compared to classical back-propagation algorithms (Bishop, 1995; Webb, Lowe, & Bedworth, 1988). The number of samples in the input vector (i.e. the delayed input signal) was set to 100 samples (corresponding to an impulse response of 0.2 s in duration), based on preliminary optimisation studies. This value is similar to the length of the impulse response for the LNL model applied to the same data (120 samples, Dewhirst et al., 2012). Each time the network is trained, its performance depends on the initial value of the weights (which are usually initialised randomly). To remove this additional source of variation between individuals, the initial weight values were specified in the network description and the same set of initial values was used for each network.

Metaheuristic Algorithms for Ann Architecture Design

The performance of ANNs depends greatly on the number of nodes and layers of the ANN chosen. In this work we used a metaheuristic algorithm to indicate an optimal architecture. The algorithm was a combination of two metaheuristic algorithms: Evolutionary Programming (EP) (Eiben & Smith, 2003) and Particle Swarm Optimisation (PSO) (Kennedy & Eberhart, 1995). The algorithm combined the characteristics from metaheuristic optimisation methods proposed in the literature (Angeline, Saunders, & Pollack, 1994; Benardos & Vosniakos, 2007; Suraweera & Ranasinghe, 2008).

The algorithm is composed by a series of functions (Fig. 2). The first function of the algorithm creates a population of potential ANN's with individuals having specific architectures with up to 5 hidden layers and 32 nodes in each layer.

$$x_\eta = [n_1\ n_2\ n_3\ n_4\ n_5] \quad (3)$$

where x_η is a vector representing an individual's architecture, η is the specific individual within the population, and n_i are the number of nodes in each of the five layers of the ANN. The number of nodes in each layer was modified at every generation through variation operators, until an optimal solution was found. The optimal architecture was a trade-off between the performance of the network and its complexity (total number of nodes), with the performance being defined here as the prediction accuracy, while the complexity is defined by its size (i.e. the number of hidden layers and nodes). Once each individual from a population had been randomly initialised, the population 'evolved' over a number of generations.

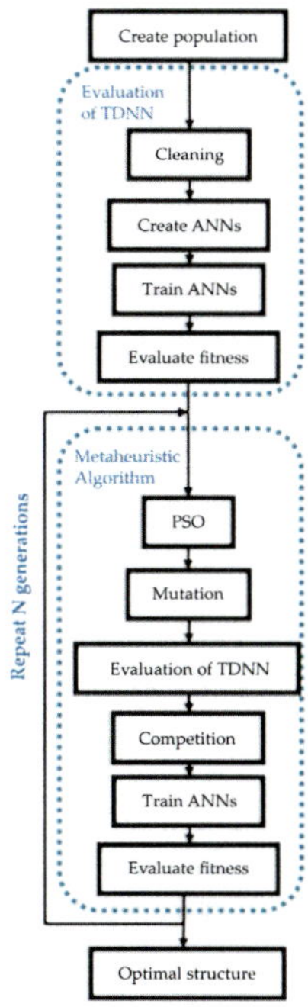

Figure. 2: Flow chart of the algorithm used to design the architecture of the feed-forward network combining the metaheuristic methods of Evolutionary Programming (EP) and Particle Swarm Optimisation (PSO). This algorithm was repeated either for a number of specified generations or until it converged, with every member of the population having the same unchanging architecture.

The algorithm contains a series of functions used in the generation of each ANN (Fig. 2). The first function was a cleaning function designed to avoid empty ANNs during the iteration process, i.e. networks that had no nodes in any layer, and were thus formed by just an input and an output. When the population was created and passed through the cleaning function, the networks were trained with the 50 Hz band-limited GWN response of FETi. The training signal was composed of two thirds of a recording of FETi from either one individual animal or the averaged response from all five animals. The training data (2/3 of the recording) was then divided into another three sections to use in the training process, one for training the weights (70%), one for validating them (15%) and one for testing its response (15%). The validation data is used during the calculation of the weights to check whether the network is improving and the testing data is used at the end of the training process to check whether the training has been successful.

To determine how close to the optimal solution the networks were, their fitness was calculated on the remaining third of the recording, not used with the training algorithm. The optimal performance was represented by the highest fitness in the search space, which was a combination of two traits: high performance and low complexity. To estimate the performance, the Normalised Mean Square Error (NMSE) was calculated (Eq. (4)), between the estimated output and the recorded output.

$$\mathrm{NMSE}(\%) = 100\left(\sum_{i=1}^{N}(y_i - \hat{y}_i)^2\right) / \left(\sum_{i=1}^{N}(y_i)^2\right) \tag{4}$$

where y_i is the measured output in the FETi at the ith sample and $\hat{y}_i$ is the estimated output of N samples, from the trained neural network.

The fitness function designed (Eq. (5)) seeks a compromise between the goodness-of-fit of the network and its complexity and is based on that used by Benardos and Vosniakos (2007).

$$fit = e^{(100/\mathrm{NMSE}(\eta) - a \cdot n(\eta))} \tag{5}$$

where NMSE is the Normalised MSE (%) as described in Eq. (4) of the individual η, and n(η) is the total number of hidden neurons of the individual network η. a is an adjustable parameter set in this case to a=0.002, which relates the importance of the size to the accuracy of the network. This value was chosen through trial and error methods, where high performances and small size networks were rewarded. Although the computational time is not directly taken into account in the fitness function, the networks are set to train for only 200 epochs (iterations of the back-propagation algorithm), and therefore those that had not converged, and needed a longer time to train, would result in a higher NMSE value and the faster ones would show higher fitness.

Once the networks had its fitness calculated, variation operators were applied to the population that were defined by the evolutionary algorithms (first PSO and then EP) and the fitness function. First, PSO was applied. The global and individual best ANNs were found to use with the PSO equations (Shi & Eberhart, 1998a), and the 'velocity' of each individual calculated (Eqs. (6a) and (6b)). The architecture of the individual represents its position (a vector of five parameters that represent the number of nodes in each layer), and the velocity of the individual is the change in the number of nodes in the architecture between the past iteration and the current iteration, is represented as another vector with five parameters. The global best was the architecture with the highest fitness obtained so far over the generations throughout the whole population. The individual best was the best performing architecture obtained by each individual over the generations. In each of the generations, the individuals were 'moved' with a certain 'velocity' towards the best performing architectures in the search space, namely those with highest fitness, according to the equation below.

$$v_\eta(t+1) = iv_\eta(t) + 2R_1(p_\eta - x_\eta(t)) + 2R_2(p_g - x_\eta(t)) \quad (6a)$$

$$x_\eta(t+1) = x_\eta(t) + v_\eta(t+1) \quad (6b)$$

where η is the individual, v_η is the velocity of the individual at generation t, i is the inertia weight, p_η is the personal best position of the individual, x_η is the actual position (i.e. number of hidden layers and nodes in each layer of the individual, as in Eq. (3)), p_g is the global best position of the population and R_1 and R_2 are random numbers in the range [0,1]. Using these 'velocities' (Eq. (6a)) the population is 'moved' towards those architectures that progressively represent the measured system better. The inertia weight i was set to 1.05 for this case, since it has been suggested that values between 0.8 and 1.2 are those that give the algorithm the best chance of finding the global optimum (Angeline et al., 1994; Shi &

Eberhart, 1998a). The value 1.05 was selected for its ability of finding the optimal value without failures (Shi & Eberhart, 1998a, 1998b).

Once the population was moved towards the global and personal best through PSO, it was mutated using EP. The mutation rate of the algorithm is a dynamic rate based on the evolutionary algorithm of Angeline et al. (1994), dependent on the fitness of the individuals (Eq. (7)). The purpose of the dynamic mutation rate is to encourage big changes for those individuals with poor performance and a fine tune search when the networks are near the optimum. Instead of using the maximum fitness achievable which is unknown, the fitness of the global best individual is used, and results in the global best architecture not being mutated in that generation.

$$R = 1 - \mathrm{fit}(\eta)/\mathrm{fit}_{\max} \tag{7}$$

where fit(η) is the fitness of the individual and fit_{max} is the fitness of the currently global best architecture.

To ensure that only the architectures with higher fitness are transmitted to the next generation, a competition function is included. This function compares the fitness of each couple of parent and offspring, retaining in the next generation the one with the higher fitness, assuring the survival of the fittest.

The optimisation process was repeated over a number of iterations, or generations, and the output of the algorithm was considered to be the optimal architecture and used to predict the data recorded from the FETi motor neuron.

Linear Nonlinear Linear (LNL) Model

The Linear–Nonlinear–Linear model, also known as Wiener Hammerstein model (Hunter & Korenberg, 1986), is a cascade model consisting of two dynamical linear elements separated by a static nonlinear element, of the form:

$$y(t) = \sum_{\sigma=0}^{T-1} g(\sigma) \sum_{q=0}^{Q} C^{(q)} \left(\sum_{\tau=0}^{T-1} h(\tau) u(t - \sigma - \tau) \right)^{q} \tag{8}$$

where g(σ) and h(τ) are the dynamic linear elements, $C^{(q)}$ represents the coefficients, u is the input and y the output.

This model is a subset of the Volterra series. The coefficients of the model can be calculated using different optimisation methods, such as the Least Squares method or a method developed by Korenberg and Hunter (1986) based on the Levenberg–Marquardt algorithm.

To compare the responses produced by the ANNs, the results from the LNL models developed by Dewhirst et al. (2012) are used, since they have been designed using the same recordings.

RESULTS

FETi Responses to FeCO Apodeme Displacement

Displacement of the apodeme of the FeCO evoked a synaptic response in FETi. Stretches of the apodeme, equivalent to flexions of the tibia, evoked depolarisations, whereas relaxation of the apodeme, equivalent to extensions of the tibia resulted in hyperpolarisation, typical of a resistance reflex (Fig. 3(A)) (Field & Matheson, 1998). The responses of FETi to displacement of the FeCO apodeme with a 50 Hz GWN stimulus resulted in the synaptic responses following, approximately, the input signal (Fig. 3(B)). The responses of FETi to the same stimulus input in different animals (n=5) appeared similar although there was some clear variability, or individual differences, between animals (Fig. 3(C)).

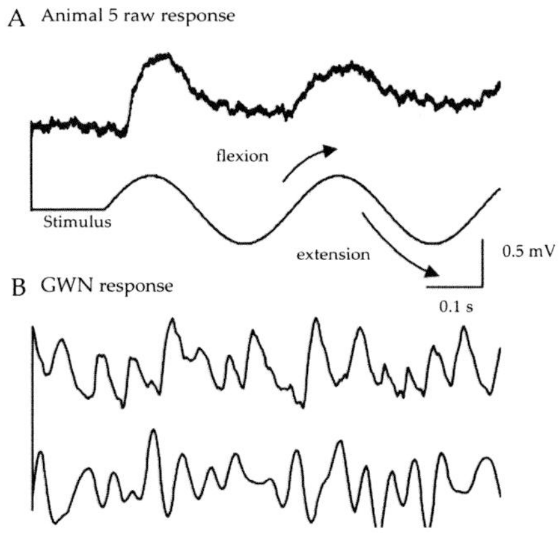

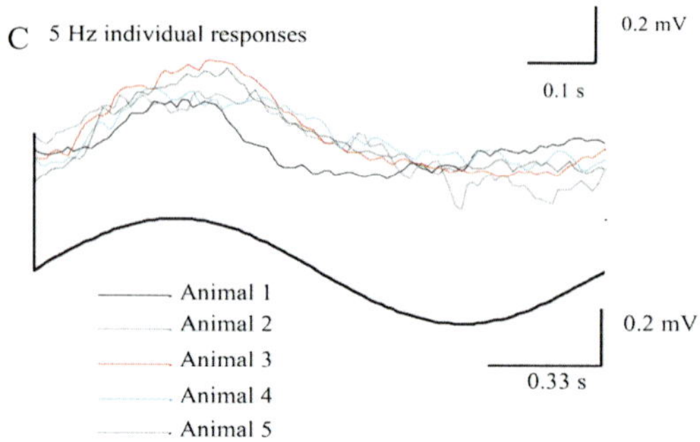

Figure. 3: Response of the FETi motor neuron to a 50 Hz band-limited GWN stimulus and a 5 Hz sinusoidal stimulus applied to the FeCO apodeme. (A) Raw response to 5 Hz before post-processing, including background noise and the stimulus onset. (B) Average GWN response after post processing. (C) Responses of FETi from 5 different animals to a 5 Hz sinusoidal stimulus.

To estimate a representative response of FETi across all individuals, an average response was calculated from the recorded FETi signals from 5 animals (shown in Fig. 3(B)). This average response was later used to determine whether it can be considered representative of the system. It was also used to study the inter-animal variability of the FETi motor neuron response and the ANN model generalisation.

Architectures of the ANN Models

The metaheuristic algorithm was run once for each FETi response and once for the average, a total of six times, over 50 generations. The output of the algorithm provided the architectures of the ANNs used to model each of the individual responses and were composed of up to two hidden layers, with each layer with between one and six nodes (Table 1).

Table 1. The architectures of the ANNs obtained for FETi in 5 different animals. Each network is composed of two hidden layers of up to 8 neurons in each

	Layer 1	Layer 2
Animal 1	6	1
Animal 2	4	1
Animal 3	6	2
Animal 4	8	7

Animal 5	8	3
Average response	**5**	**2**

The architectures for the individual responses and the average response were similar, although the algorithm allowed for networks with up to 32 neurons and 5 hidden layers. All networks were relatively small (Table 1) and were composed of two hidden layers with between 1 and 8 nodes in each layer.

Validation of the ANNs

The architectures obtained with the metaheuristic algorithm were trained for each of the FETi responses obtained from different animals (Fig. 3) using GWN as an input signal. An individual model, such as that for Animal 5 (Fig. 4(A)), was able to follow the response to the GWN signal quite well. The models for individual FETi responses from the different animals had a range of performance when tested with previously unseen data (the 1/3 of data not used in the network training) from the same individual. The model of FETi in Animal 3 had the best performance, with a NMSE of 15.5% (Table 2), indicating that the model was able to predict well, most of the FETi response. The models performed less well in other animals, with NMSEs of 29.4% in Animal 1, 18.1% in Animal 2 and 16.9% in Animal 5. The model for Animal 4 was the worst performing, with a NMSE of 33.7%. Together, these models produced an average NMSE of 21.6 ± 7.9% (NMSE ± Standard Deviation, SD), which represents the prediction error on unseen data from the each of the responses.

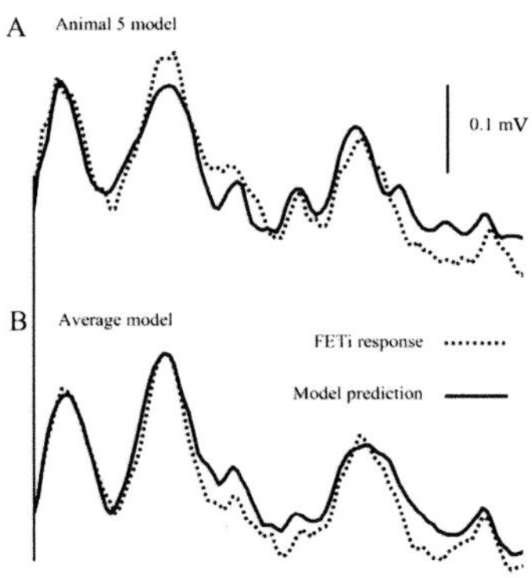

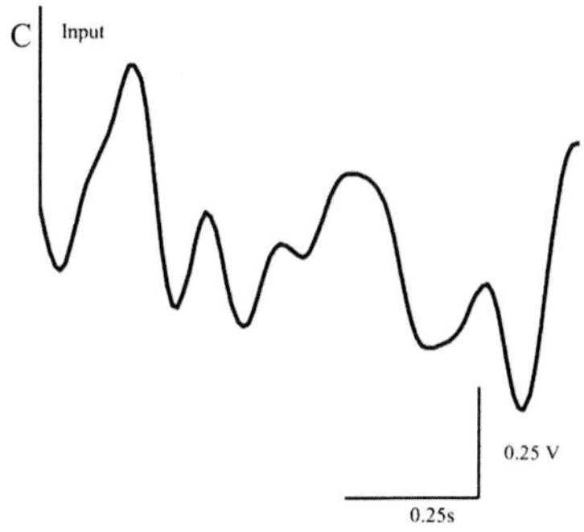

Figure. 4. ANN fit to Animal 5 and the average response. (A) The networks have been trained with two thirds of the FETi motor neuron response from Animal 5 to 50 Hz band-limited GWN. (B) Network trained with two thirds of the average response to 50 Hz band-limited GWN. (C) Input GWN signal applied to the system.

Table 2: NMSE (%) of the performances of the TDNN and LNL (from Dewhirst (2013)—using the same procedure on the same data) for GWN and the 5 Hz sine-wave

Model	**TDNN**		**LNL**	
Testing signal	**GWN**	**5 Hz**	**GWN**	**5 Hz**
Average response	15.8	8.4	36.9	13.4
Animal 1	29.4	39.4	32.1	13.4
Animal 2	18.1	40.0	27.5	16.3
Animal 3	15.5	9.0	30.7	44.9
Animal 4	33.7	26.5	41.2	82.9
Animal 5	16.9	14.2	27.8	33.3
Mean	*21.6*	*22.9*	*32.7*	*34.1*

An ANN optimised for the average response also produced a good performing model with 5 nodes in the first hidden layer and 2 in the second (Table 1). This model had a NMSE of 15.8% (Fig. 4(B)), slightly worse than that of the best performance seen using individual responses of FETi from different animals, but better than any other model. This improved performance may be expected since the averaging process reduces the additive noise in the recordings.

The performance of all the individual ANNs, expressed as NMSE, was compared to the Linear–Nonlinear–Linear (LNL) model designed by Dewhirst et al. (2012). Both models behaved similarly (Fig. 5), although the TDNN models consistently had lower prediction errors. The TDNN had a mean NMSE of 21.6% while the LNL models had a mean NMSE of 27.5%. A Mann–Whitney U test, chosen for the small sample size and the non-normality of the data, showed that the performances between the TDNN and the LNL were significantly different ($U=6.0, p=0.05$), which suggests that the ANNs were better at predicting FETi responses to GWN.

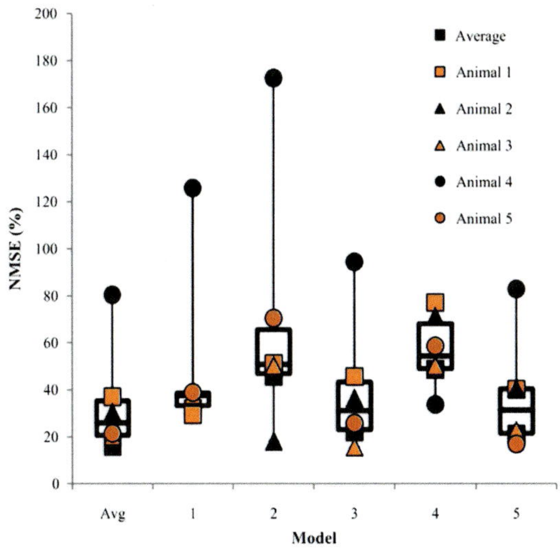

Figure. 5: Comparison of the NMSE of the TDNN and the LNL cascade models. The NMSE of the models was calculated when training and testing the models with 50 Hz band-limited GWN. Testing was carried out on data not used in training, but from the same individual. The NMSE was bigger for the LNL model in all individuals. The box-plots represent the median and quartiles and the whiskers extreme values of the same data.

Validating ANNs with Sinusoidal Inputs

To validate the ANN models estimated from a GWN input signal, their ability to predict a FETi response to 5 Hz sinusoidal stimulus applied to the FeCO was determined. An individual model, such as the model optimised for Animal 5 and trained with GWN predicted approximately the 5 Hz response from the same individual (Fig. 6(A)). The NMSE from individual responses ranged from 9.0% to 40.0% (Table 2). The best performing animal with GWN (NMSE of 15.5%) was also the best performing animal with a 5 Hz sinusoidal input (Animal 3 with NMSE = 9.0%). The worst performing animal with GWN (Animal 4), however, was not the worst performing with 5 Hz sinusoidal input (Animal 2 with NMSE = 40.0%). The performance of the TDNN trained with the average response and tested with a 5 Hz sinusoid had an NMSE of 8.4% (Fig. 6) showing that the GWN-trained ANN model was able to predict more than 91% of the responses to the sinusoidal input signal. The previous LNL models described by Dewhirst et al. (2012), when tested with 5 Hz sinusoidal inputs, showed a performance ranging from NMSE of 15% to 83%. A Mann–Whitney U test showed that while the means appear very different, the performances with 5 Hz input of the TDNN (with mean NMSE = 22.9%) and the LNL (with mean NMSE = 34.1%) were not significantly different ($U=13.0, p=0.49$), due in part to the considerable variability of both the ANN and the LNL models with different individuals (TDNN has considerably lower NMSE in only three of the five animals), and the small sample size. The ANNs can thus be considered at least as accurate, if not better, than LNL and Wiener methods (Dewhirst et al., 2012) at predicting unseen sinusoidal data from FETi recordings.

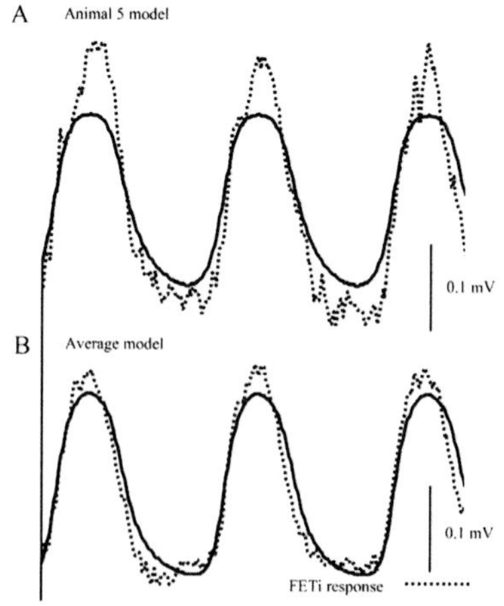

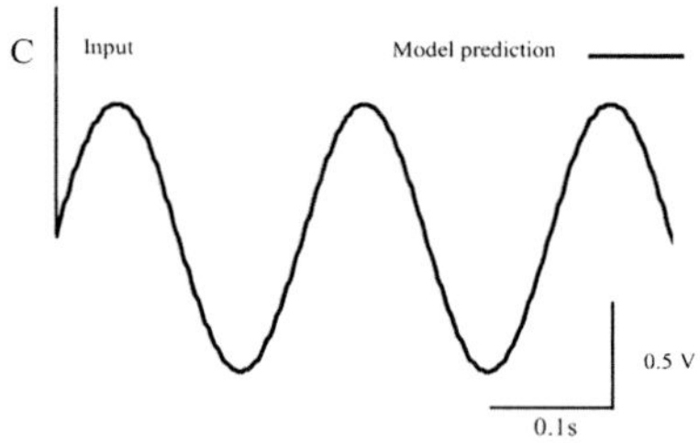

Figure. 6: Generalisation of the TDNN when trained with GWN and tested with a 5 Hz sinusoid. (A) Measured signal in Animal 5, and its prediction from the individual model optimised with GWN prediction. (B) Average signal and its prediction from the ANN model optimised on GWN stimuli and the averaged responses. (C) Input signal to the FeCO.

Generalisation of the ANNs to FETi Responses in different Animals

The FETi responses from different animals showed clear individual differences, or variability (Fig. 3). Understanding the underlying common responses of the same neuron between individuals and where the differences between individuals lie, could aid in a better understanding of how nervous systems generate behaviour. We therefore asked how accurately a model designed for one particular individual can predict the response from another individual. For this purpose, the ANNs trained with individual responses were tested with all the animal responses and their NMSE calculated (Fig. 7).

The performances of the ANNs showed an average NMSE of 47.7% (SD = 32.4%) for all tests. This indicates that any of the ANNs designed can predict on average 52.3% of the response in any individual FETi in different animals. The ANN trained with the average response had a lower NMSE when predicting the individual responses, with an average NMSE 34.2% (SD = 23.9%). A network trained with the average response could predict therefore an average of 65.8% of the responses in any individual.

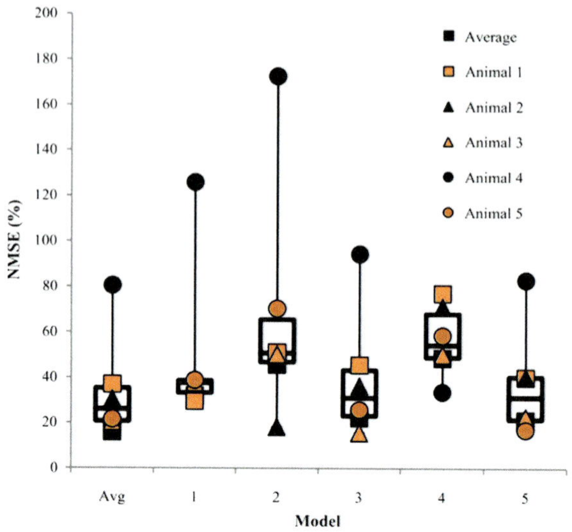

Figure. 7: NMSEs of all the ANN models to predict recordings from each of the other animal and the averaged response. The abscissa refers to the signal used in training and each symbol reflects the signal used in testing. The NMSE value was calculated after the networks were trained with 50 Hz band-limited GWN and tested with unseen 50 Hz band-limited GWN. The box-plots represent the median and the quartiles and the whiskers extreme values of the data. Each individual value is also shown. Results showed that no specific model was the best performing model and statistical tests indicate that there were no significant differences between model performances (Friedman's test, $p>0.05$).

To test whether there was a difference in the generalisation power of the models based on the training signal used, a Friedman test was applied. The test, run across the NMSEs from all models, was not significant ($\chi^2=8.93, p=0.063$). Therefore, no individual model was significantly better or worse than any other in generalising to other animals.

Although most of the individual responses were similar, all models fail to predict the synaptic responses from Animal 4, excluding the ANN trained with data from Animal 4 (black circles in Fig. 7). This may be considered an illustration of inter-subject variability of the FETi responses, where Animal 4 appears to behave differently from other individuals. Considering that not all the FETi responses can be predicted perfectly with any model, other factors might be present that influence the output and are not taken into account in the modelling process.

Variability between Individuals

In the previous section we focused on the six models estimated and asked how well each could predict the FETi responses. The counterpart to this question is addressed now, by focusing on each recording, and asking how well they are predicted by any of the models.

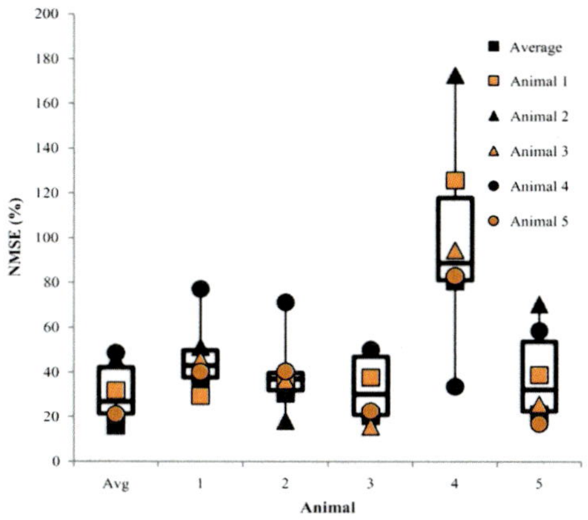

Figure. 8: NMSEs of all the animals when predicted by each of the ANN models. The abscissa refers to the animal used in testing and each symbol reflects the animal used in the model training. The box-plots represent the median and the quartiles and the whiskers extreme values of the data. The figure is based on same data as used in Fig. 7, but presented ordered by animal, rather than by model. The model from Animal 4 clearly provided the worst prediction, except in predicting the response of Animal 4. Statistical results showed that there were no significant differences between individuals (Friedman, $p>0.05$).

As seen in Fig. 8, Animal 3 and the average response to GWN were the best-predicted responses by all the individual models, with an average NMSE of 32.73% (SD = 15.4%) and 30.83% (SD = 13.7%) respectively. Animal 4 was the worst predicted individual response by any model (as already observed above), with an average NMSE of 98.3% (SD = 46.9%). A Friedman test was conducted to determine whether there was a difference for an individual response to be predicted by any model, i.e. the inter-subject variability. The test showed that the individuals were significantly different from each other when being predicted by any ANN model ($\chi^2=11.5, p=0.04$). A Wilcoxon signed-rank test with a Bonferroni correction (significance level of $p<0.003$) showed that the individuals, grouped by pairs, were not significantly different from each other ($p>0.046$).

DISCUSSION

The results of this study indicate that ANNs can be used to model responses of individual neurons in neural networks and can be used to explore and understand individual differences in their responses.

We have developed a method to design architectures for ANNs to model the synaptic responses of an identified motor neuron in a proprioceptive neural network. The metaheuristic algorithms optimised the ANN architectures to reduce computational time and improve their generalisation abilities. The ANNs were more accurate at predicting GWN responses in FETi than previously tested methods on the same neuron (Dewhirst et al., 2009). However, the ANNs were not significantly better at predicting the response to the sinusoidal input than the previously used methods (Dewhirst et al., 2009; Dewhirst et al., 2012; Newland & Kondoh, 1997).

We have shown, for the first time, that the models developed from a specific individual neuron can generalise well and predict the responses of the same neuron in other animals. This suggests that an ANN model could be used as a mathematical representation of the FETi neuron response to FeCO stimulation in any locust based on system estimates from one individual, with the caveat that predictions tend to be better when the same individual is used in training the model. This points to some individual differences and in our sample of five locusts, four were similar, but one (Animal 4) was clearly more distinct. There are a number of possible reasons for why the model fit is not perfect, even within each individual: there may be some adaptation (variation over time) in the response over the duration of stimulation; the low complexity of the models may not allow subtle features of the responses to be captured; additional inputs to the FETi not taken into account in the model; noise due to the electronic recording system. Perhaps the main contributor is spontaneous neuronal activity (noted even before stimulation commences) (Burrows, Laurent, & Field, 1988) and no input–output model would be able to account for that. In the current approach, all these factors would be captured in the mean-square error and contribute to the NMSE.

Comparison with other Modelling Methods

The results suggest that ANNs can be used as mathematical tools to study neural responses with a high degree of accuracy. Moreover, they have been shown to outperform previous LNL and Volterra methods ((Dewhirst, 2013; Dewhirst et al., 2009; Newland & Kondoh, 1997)) in predicting the responses of FETi to a GWN displacement of the FeCO apodeme, with an approximate improvement of 10% over LNL methods (Dewhirst, 2013) and a 25% improvement over Wiener methods (Newland & Kondoh, 1997). Both ANN and LNL methods have also been shown to be able to predict sinusoidal responses, where the model used was trained with 50 Hz band-limited GWN data from the same individual in which the synaptic responses of FETi to 5 Hz sinusoidal stimulation were recorded. The mean square error of the TDNN method was much lower than that obtained with Wiener methods (Dewhirst, 2013) but was not significantly different from LNL methods. This suggests that the ANNs and LNLs were similarly good at predicting responses to different inputs within the same animal. The small sample size together with considerable individual variability in the NMSE contributed to this result; clearly ANN is not better in every individual.

Other studies have also used Wiener methods to model neural responses to GWN stimuli. For example, Marmarelis and Naka (1973b) modelled the averaged intracellular responses of neurons in the catfish retina using Wiener kernels, with a NMSE of 25% or less, using modulated light as the white noise input stimulus. Proprioceptors in a chordotonal organ in the crab were also modelled using Wiener kernel methods (DiCaprio, 2003) and the NMSE of the second-order models was around 6%–26% when tested with data from the same individual. The NMSE obtained with ANNs in FETi in the current work is comparable to that obtained using Wiener models of the same neuron, but is not as accurate as other Wiener methods applied in other neuronal systems. The most likely explanation of which is that the studies by DiCaprio (2003) on the crab and Marmarelis and Naka (1973b) on the catfish focused on the input stages to the neural networks, whereas here we focused on the output, a motor neuron, FETi. The consequence of this is that FETi responses not only include those correlated with the stimulus input but also uncorrelated spontaneous synaptic responses typical of central neurons that clearly cannot be predicted from the input signal (see for example Burrows, 1987; Büschges, Kittmann, & Schmitz, 1994; Field & Burrows, 1982).

The advantage of using ANNs however, is that not only can they predict accurately the responses to an input, but they are also able to predict responses to other inputs and, furthermore, able to predict responses in different individuals using the same model. Such modelling technique provides the flexibility that any animal could be used to train the networks, namely an individual response to a stimulus or even the average response calculated across different individuals.

One should add that the comparison between models also depends on how the model order was chosen, and the conclusion that one model is 'better' than another has to be considered with some caution, especially when the performance does not differ by very much and there are quite large differences in performance in different individual animals.

Variability across Individuals

Goldman et al. (2001) and Marder and Taylor (2011) suggested that models based on individual or averaged responses may not be representative of the system across a population of individuals due to variability and individual differences even within identified neurons. A mathematical model able to capture the common response in different individuals would be able to describe better the responses within the population. However, as far as we know, no generic mathematical model of neural responses in identified neurons has been tested across different individual responses. Our results show that the individual responses of FETi can be predicted by any of the ANN models developed from FETi in any other animal, although usually with higher NMSE, independently of the data used to train the network. The model fitted to the averaged responses tended to give a better fit across all individuals (see Fig. 8), but this was not consistently observed.

Using the performances of the models, it has also been shown that the individual responses are not significantly different from one another. Previous studies have shown that mathematical models of FETi (Dewhirst et al., 2009; Newland & Kondoh, 1997) produced similar values of NMSE when the model was developed and tested with data from the same individual. Such models, however, were never used to test their applicability to the responses of other individuals. DiCaprio (2003) and Marmarelis and Naka (1973b) also used Wiener methods, but again did not test their applicability across individuals. It is notoriously difficult to interpret ANN models behaviour, and applying the models between individuals provides a means of assessing their similarity. The conclusion that models fit well between individuals suggests that results from one individual, or a small sample of individuals, provide a good approximation to the behaviour of the identified neurone *per se*.

The errors observed in the ANN models when tested with different individual responses can only partly be explained by differences across animals as a result of spontaneous activity found in central interneurons and motor neurons (Büschges, 1990; Field & Burrows, 1982). The observation that performance is consistently best when the model is trained on the same individual from which the (previously unseen) test data is taken, suggests that the models are distinct, not just that the noise levels differ. Variability has been shown in the synaptic properties of identified neurons, where responses across individuals differ, while the overall neural circuit behaviour remains the same, or similar, in all individuals (Marder & Taylor, 2011). Another explanation for these differences is small variations in the encoding of a stimulus in each individual. Schneidman, Brenner, Tishby, van Steveninck, and Bialek (2001)showed that approximately 30% of the information in an identified motor neuron in flies is specific to each individual, whereas the remaining 70% is a common response which is similar in the same motor neuron from the whole population of flies. Such a common response might also be present in the FETi, where the ability of the ANNs to predict part of the responses in all individuals represents a common response to all in the population, whereas the individuality of the responses, or uncorrelated spontaneous activity, remains as errors.

Nonetheless, the errors in the responses might not be entirely due to inherent differences between animals, but also may include differences within the same animal, such as variations over time, or across individuals, such as small changes in experimental methods (i.e. precise location of the electrodes). The nature of the experiments is such that repetition (with renewed insertion of electrodes) is not practicable.

Whether the average response is the most appropriate characteristic to represent a system is still debatable (Goldman et al., 2001; Marder

& Taylor, 2011). Our results suggest that an ANN model fitted to the responses averaged across individuals, provides a fairly good fit in all animals, and may thus be deemed as representative. The models developed were able to predict responses in different individuals, which show that FETi responses share common features across different individuals, but the responses contain some individuality that cannot be predicted by a generic model. Caution is recommended in assuming any model represents the behaviour of the identified neurone across a population.

In summary, the ANN models designed in this study have been shown to be significantly better than previously explored mathematical models of the FETi responses to GWN. ANNs thus provide a promising method in the study of the neural responses and similarly structured networks in locust and other species. The average response provides a robust basis for model fit, although further evidence of the presence of individual differences was found. The models designed with ANNs thus appear promising as a tool in analysing neuronal reflexes and may guide the future design of engineering control systems or medical assistive devices.

Conclusion

Here we show that ANNs can be used to model the synaptic responses of an identified motor neuron in a proprioceptive neural network to a GWN stimulus with more accuracy than previously used methods, such as Wiener or LNL models, and with undistinguishable accuracy following sinusoidal stimuli. We have developed a new method to optimise the ANN architecture, which reduces computational time and improves the generalisation abilities of the networks. ANNs thus appear promising tools for the analysis of neuronal networks. We show that ANN models are able to predict responses to other (previously unseen) inputs and, furthermore, the responses of the same motor neuron in other animals. Any ANN model was able to estimate the responses in any individual, although predictions tended to be better within the same animal, highlighting the common function of the motor neuron in different animals, as well as some individual differences.

ACKNOWLEDGEMENTS

Alicia Costalago Meruelo was supported by an EPRSC grant (EP/G03690X/1) from The Institute of Sound and Vibration Research and the Institute for Complex Systems Simulations at the University of Southampton. Data supporting this study are openly available from the University of Southampton repository at http://dx.doi.org/10.5258/SOTON/385038.

REFERENCES

1. Angarita-Jaimes, N., Dewhirst, O. P., Simpson, D. M., Kondoh, Y., Allen, R., & Newland, P. L. (2012). The dynamics of analogue signalling in local networks controlling limb movement. European Journal of Neuroscience, 36(9), 3269–3282.
2. Angeline, P. J., Saunders, G. M., & Pollack, J. B.(1994). An evolutionary algorithm that constructs recurrent neural networks. IEEE Transactions on Neural Networks, 5(1), 54–65.
3. Beer, R. D., Chiel, H. J., Quinn, R. D., Espenschied, K. S., & Larsson, P. (1992). A distributed neural network architecture for hexapod robot locomotion. Neural Computation, 4(3), 356–365.
4. Benardos, P. G., & Vosniakos, G.-C. (2007). Optimizing feedforward artificial neural network architecture. Engineering Applications of Artificial Intelligence, 20(3), 365–382.
5. Bishop, C. M. (1995). Neural networks for pattern recognition. New York, NY, USA: Oxford University Press.
6. Burns, M. D. (1973). The control of walking in orthoptera. Journal of Fish Biology, 58(1), 45–58.
7. Burrows, M. (1987). Parallel processing of proprioceptive signals by spiking local interneurons and motor neurons in the locust. The Journal of Neuroscience, 7(4), 1064–1080.
8. Burrows, M. (1996). The neurobiology of an insect brain. Oxford: Oxford University Press.
9. Burrows, M., Laurent, G. J., & Field, L. H. (1988). Proprioceptive inputs to nonspiking local interneurons contribute to local reflexes of a locust hindleg. The Journal of Neuroscience, 8(8), 3085–3093.

10. Büschges, A. (1990). Nonspiking pathways in a joint-control loop of the stick insect carausius morosus. Journal of Experimental Biology, 151, 133–160.
11. Büschges, A., Kittmann, R., & Schmitz, J. (1994). Identified nonspiking interneurons in leg reflexes and during walking in the stick insect. Journal of Comparative Physiology A, 174(6), 685–700.
12. Chiel, H. J., Beer, R. D., Quinn, R. D., & Espenschied, K. S. (1992). Robustness of a distributed neural network controller for locomotion in a hexapod robot. IEEE Transactions on Robotics and Automation, 8(3), 293–303.
13. Cruse, H., Bartling, C., Dreifert, M., Schmitz, J., Brunn, D. E., Dean, J., et al.
14. (1995). Walking: A complex behavior controlled by simple networks. Adaptive Behavior, 3(4), 385–418.
15. Dewhirst, O. P. (2013). Validation of nonlinear system identification models of the locust hind limb control system using natural stimulation. (Ph.D. thesis), UK: Southampton University.
16. Dewhirst, O. P., Angarita-Jaimes, N., Simpson, D. M., Allen, R., & Newland, P. L. (2012). A system identification analysis of neural adaptation dynamics and nonlinear responses in the local reflex control of locust hind limbs. Journal of Computational Neuroscience, 1–20.
17. Dewhirst, O. P., Simpson, D. M., Allen, R., & Newland, P. L. (2009). Neuromuscular reflex control of limb movement—validating models of the locusts hind leg control system using physiological input signals. In 4th international IEEE/EMBS conference on neural engineering, 2009 (pp. 689–692).
18. DiCaprio, R. A.(2003). Nonspiking and spiking proprioceptors in the crab: nonlinear analysis of nonspiking TCMRO afferents. Journal of Neurophysiology, 89(4), 1826–1836.
19. Eiben, A. E., & Smith, J. E. (2003). Introduction to evolutionary computing. SpringerVerlag.
20. Field, L. H., & Burrows, M. (1982). Reflex effects of the femoral chordotonal organ upon leg motor neurones of the locust. Journal of Fish Biology, 101(1), 265–285.

21. Field, L. H., & Matheson, T. (1998). Chordotonal organs of insects. In P. D. Evans (Ed.), Advances in insect physiology, Vol. 27 (pp. 1–228). Academic Press.
22. Gamble, E. R., & DiCaprio, R. A. (2003). Nonspiking and spiking proprioceptors in the crab: White noise analysis of spiking CB-chordotonal organ afferents. Journal of Neurophysiology, 89(4), 1815–1825.
23. Goldman, M. S., Golowasch, J., Marder, E., & Abbott, L. F. (2001). Global structure, robustness, and modulation of neuronal models. The Journal of Neuroscience, 21(14), 5229–5238.
24. Haykin, S. (1999). Neural networks: a comprehensive foundation (2nd ed.). Prentice Hall PTR.
25. Horn, D., Ruppin, E., Usher, M., & Herrmann, M. (1993). Neural network modeling of memory deterioration in alzheimer's disease. Neural Computation, 5, 736–749.
26. Hunt, K. J., Sbarbaro, D., Zbikowski, R., & Gawthrop, P. J. (1992). Neural networks for control systems—A survey. Automatica, 28(6), 1083–1112.
27. Hunter, I. W., & Korenberg, M. J. (1986). The identification of nonlinear biological systems: Wiener and Hammerstein cascade models. Biological Cybernetics, 55, 135–144.
28. Jain, A. K., Jianchang, M., & Mohiuddin, K. M. (1996). Artificial neural networks: a tutorial. Computer, 29(3), 31–44.
29. Kennedy, J., & Eberhart, R. (1995). Particle swarm optimization. In IEEE international conference on neural networks. proceedings. Vol. 4, 1944 (pp. 1942–1948).
30. Kondoh, Y., Okuma, J., & Newland, P. L. (1995). Dynamics of neurons controlling movements of a locust hind leg: Wiener kernel analysis of the responses of proprioceptive afferents. Journal of Neurophysiology, 73(5), 1829–1842.
31. Korenberg, M. J., & Hunter, I. W. (1986). The identification of nonlinear biologicalsystems—LNL cascade models. Biological Cybernetics, 55, 125–134.
32. Marder, E., & Taylor, A. L. (2011). Multiple models to capture the variability in biological neurons and networks. Nature Neuroscience,

14(2), 133–138.

33. Marmarelis, P. V. Z. (2004). Nonlinear dynamic modeling of physiological systems. Wiley.
34. Marmarelis, P. Z., & Naka, K. (1972). White-noise analysis of a neuron chain: an application of the Wiener theory. Science, 175(4027), 1276–1278.
35. Marmarelis, P. Z., & Naka, K. I. (1973a). Nonlinear analysis and synthesis of receptive-field responses in the catfish retina. I. Horizontal cell leads to ganglion cell chain. Journal of Neurophysiology, 36(4), 605–618.
36. Marmarelis, P. Z., & Naka, K. I. (1973b). Nonlinear analysis and synthesis of receptive-field responses in the catfish retina. II. One-input white-noise analysis. Journal of Neurophysiology, 36(4), 619–633.
37. Newland, P. L., & Kondoh, Y. (1997). Dynamics of neurons controlling movements of a locust hind leg III. Extensor tibiae motor neurons. Journal of Neurophysiology, 77(6), 3297–3310.
38. Palm, G., & Poggio, T. (1977). Wiener-like system identification in physiology. Journal of Mathematical Biology, 4(4), 375–381.
39. Sarkar, A. X., Christini, D. J., & Sobie, E. A. (2012). Exploiting mathematical models to illuminate electrophysiological variability between individuals. Journal of Physiology-London, 590(11), 2555–2567.
40. Schneidman, E., Brenner, N., Tishby, N., van Steveninck, R. R., & Bialek, W. (2001). Universality and individuality in a neural code. NIPS, 13, 159–165.
41. Shi, Y., & Eberhart, R. C. A (1998a). Modified particle swarm optimizer. In The 1998 IEEE international conference on evolutionary computation proceedings. IEEE world congress on computational intelligence (pp. 69–73).
42. Shi, Y., & Eberhart, R. C. (1998b). Parameter selection in particle swarm optimization. In Evolutionary programming VII (pp. 591–600). Springer.
43. Suraweera, N. P., & Ranasinghe, D. N. A (2008). Natural algorithmic

approach to the structural optimisation of neural networks. In 4th international conference on information and automation for sustainability (pp. 150–156).

44. Tötterman, S., & Toivonen, H. T. (2009). Support vector method for identification of Wiener models. Journal of Process Control, 19(7), 1174–1181.

45. Twickel, A., Büschges, A., & Pasemann, F. (2011). Deriving neural network controllers from neuro-biological data: implementation of a single-leg stick insect controller. Biological Cybernetics, 104(1–2), 95–119.

46. Waibel, A., Hanazawa, T., Hinton, G., Shikano, K., & Lang, K. J. (1989). Phoneme recognition using time-delay neural networks. IEEE Transactions on Acoustics, Speech and Signal Processing, 37(3), 328–339.

47. Webb, A. R., Lowe, D., & Bedworth, M. D. (1988). A comparison of nonlinear optimisation strategies for feed-forward adaptive layered networks. DTIC Document.

48. White, H. (1988). Economic prediction using neural networks: the case of IBM daily stock returns. In IEEE international conference on neural networks, Vol. 452 (pp. 451–458).

49. Xing, L., & Pham, D. T. (1995). Neural networks for identification, prediction, and control. Springer-Verlag New York, Inc.

CHAPTER 13

CHOICE REACHING WITH A LEGO ARM ROBOT (CORLEGO): THE MOTOR SYSTEM GUIDES VISUAL ATTENTION TO MOVEMENT-RELEVANT INFORMATION

Soeren Strauss, Philip J.W. Woodgate, Saber A. Sami, Dietmar Heinke

Centre for Computational Neuroscience and Cognitive Robotics, School of Psychology, University of Birmingham, Birmingham B15 2TT, United Kingdom

ABSTRACT

We present an extension of a neurobiologically inspired robotics model, termed CoRLEGO (Choice reaching with a LEGO arm robot). CoRLEGO models experimental evidence from choice reaching tasks (CRT). In a CRT participants are asked to rapidly reach and touch an item presented on the screen. These experiments show that non-target items can divert the reaching movement away from the ideal trajectory to the target item. This is seen as evidence attentional selection of reaching targets can leak into the motor system. Using competitive target selection and topological representations of motor parameters (dynamic neural fields) CoRLEGO is able to mimic this leakage effect. Furthermore if the reaching target is determined by its colour oddity (i.e. a green square among red squares or vice versa), the reaching trajectories become straighter with repetitions of the target colour (colour streaks). This colour priming effect can also be modelled with CoRLEGO. The paper also presents an extension of CoRLEGO. This extension mimics findings that transcranial direct current stimulation (tDCS) over the motor cortex modulates the colour priming effect (Woodgate et al., 2015). The results with the new CoRLEGO suggest that feedback connections from the motor system to the brain's attentional system (parietal cortex) guide visual attention to extract movement-relevant information (i.e. colour) from visual stimuli. This paper adds to growing evidence that there is a close interaction between the motor system and the attention system. This evidence contradicts the traditional conceptualization of the motor system as the endpoint of a serial chain of processing stages. At the end of the paper we discuss CoRLEGO's predictions and also lessons for neurobiologically inspired robotics emerging from this work.

Keywords: Target reaching Visual attention Transcranial direct current stimulation (tDCS), Motor cortex Neurobiologically inspired robotics LEGO Mindstorms

INTRODUCTION

Traditionally the motor system is seen as the endpoint of a serial chain of processing stages. This framework assumes that the motor system 'simply' reads out previous processing in order to execute a movement (e.g. Marr, 1982). However recent research has shown that the motor system and other processing stages are more closely linked. For instance, a series of studies showed that cognitive processes (e.g. language processing; Spivey, Grosjean, & Knoblich, 2005; numerical representation; Song & Nakayama, 2008a) can leak into reaching movements (see Song & Nakayama, 2009 for a review). This evidence for parallel feed forward processing is also complemented by evidence suggesting feedback influences. For instance, learning processes in the motor system can change the psychophysical judgment of perceptual stimuli (e.g. Ostry, Darainy, Mattar, Wong, & Gribble, 2010), and visual discrimination is better at locations that form movement targets than elsewhere (e.g. Deubel, Schneider, & Paprotta, 1998). The present paper focuses on the interplay of the motor system with visual selective attention.

Evidence for the interactions between the motor system and visual selective attention comes from studies by Song and Nakayama (Song & Nakayama, 2006, 2008b; see Song & Nakayama, 2009 for a review) and Woodgate, Strauss, Sami, and Heinke (2015) and will be summarized in the next section. In Section 3 we will present a neurobiologically inspired robotics model of this evidence. The model is called CoRLEGO (Choice reaching with a LEGO robot arm) and was first published by Strauss and Heinke (2012). Section 4 will present an extension of CoRLEGO to accommodate novel findings from Woodgate et al. (2015). Section 4 will discuss the implications of our work. Section 5 will summarize lessons for neurobiologically inspired robotics from our work.

Selective Attention and Reaching Movements

Visual scenes are highly complex. Selective attention is generally assumed to deal with this complexity by filtering out information irrelevant to the task at hand. The well-known biased competition theory (e.g. Desimone & Duncan, 1995) postulates that filtering of irrelevant items is implemented as a competition process in the brain. The various objects in scenes are assumed to compete for the observer's behavioural response. Competition can be biased by a wide range of factors, such as the physical properties of objects, the knowledge of the observer, the repetition of target features, etc. A commonly used experimental method to examine these factors is the visual search task (see Eckstein, 2011 and Müller & Krummenacher, 2006 for reviews). Typically in this task participants are asked to indicate the absence or presence of a pre-defined target object among non-target objects (distractors) by pressing a specific key on a keyboard. Within the biased competition framework the time it takes participants to press the key (reaction times) is assumed to be an index for the strength of the competition between items on the screen. A particularly well-known finding relevant to the present paper is that if the target items differ by one feature (e.g. a red square among green squares) there is relatively little competition resulting in short reaction times; also termed the pop-out effect (e.g. Treisman & Gelade, 1980). Nevertheless, studies by Maljkovic and Nakayama (1994) indicate that the repetition of target features (e.g. streaks of the same colour) can result in faster resolution of inter-item competition, i.e. the classic priming of pop-out (POP) effect. Note that in these tasks participants had to discriminate a feature of the target item rather than making a present/absent response. Target discrimination turned out to be a crucial pre-condition for the POP-effect which was not found in the present/absent task.

The choice reaching task (CRT) of Song and Nakayama combines the visual search task with reaching movements. In a CRT participants are asked to rapidly reach and touch an item presented on the screen. Song and Nakayama (2008b) employed two types of display. One type consisted of three squares and the movement target was indicated by its colour-oddity (e.g. a red square among green squares, or vice versa). The other display type contained just one item (either a green or red square). The odd-colour display can be considered to be a harder search task than the single target display. Song and Nakayama (2008b) found that search difficulty was reflected in two measures: the initiation latency of movement (IL) and the maximum deviation of the reach trajectory (MD) from the ideal path (straight line). When the task was easy IL was short and MD was small while when the task was hard IL was long and MD was larger. Moreover and consistent with the biased competition theory, Song and Nakayama interpreted the MD-effect as reflecting the competition between items on the screen during the selection process which 'leaks' into motor control (see also Song & Nakayama, 2009). Participants may begin their movements before selection is completed (especially in the mixed design) and, as a consequence, the non-target items (distractors) can divert the reaching movement away from the ideal trajectory to the target item. In another study, Song and Nakayama (2006) showed that when the target colour is repeated from trial-to-trial both latency and deviation are reduced. Hence both measurements index the POP-effect discussed earlier.

In Woodgate et al. (2015) we applied transcranial direct current stimulation (tDCS) to the motor cortex while participants completed a CRT. tDCS is an electrical neurostimulation technique that passes constant ultra-low currents to brain regions via a pair of electrodes placed on the scalp. Anodal stimulation refers to when the positively charged electrode (anode) is placed over the stimulated brain area whilst cathodal stimulation uses negatively charged current. tDCS is known to increase (anodal tDCS) or decrease (cathodal tDCS) excitability of the underlying cortex (e.g. Nitsche & Paulus, 2001). In both cases, there are measurable behavioural consequences (e.g. Reis et al., 2009). It is worth noting that on the face of it electrical stimulation on the scalp seems fairly non-specific. However a recent study combining tDCS over the motor cortex with electroencephalography (EEG) demonstrated effects predominantly on the motor cortex and functionally-related areas (Notturno, Marzetti, Pizzella, Uncini, & Zappasodi, 2014)

In fact, Woodgate et al. (2015) also stimulated participants over the motor cortex while they performed CRTs. They found that when the experimental design of the CRT was similar to Song and Nakayama (2008b) (i.e. randomly intermixed three item vs. single items trials with no target colour streaks) tDCS showed little influence on the participants' reaching behaviour. However, when the experiment contained streaks of target colour (similar to Song & Nakayama, 2006), tDCS affected the reaching trajectories. Woodgate et al. (2015) concluded from these findings that the motor system may be involved in the guidance of attention when target properties are predictable. Such an interpretation is consistent with findings referred to in the introduction that the motor system can influence perceptual processing (e.g. Deubel et al., 1998 and Ostry et al., 2010). However, there are many intriguing aspects of Woodgate's et al. (2015) findings. We will discuss them in more detail in Section 3 in the context of the extension of CoRLEGO.

CoRLEGO

CoRLEGO is based on Song and Nakayama's (2009) theory that the target selection process leaks into the motor stage (Strauss & Heinke, 2012). For the implementation of the target selection stage CoRLEGO draws on our laboratory's competition models of attention (e.g. Heinke & Backhaus, 2011Mavritsaki, Heinke, Allen, Deco, & Humphreys, 2011 and Zhao, Humphreys, & Heinke, 2012). The motor stage utilizes Erlhagen and Schoener's (2002) dynamic neural field (DNF) theory that postulates that movement parameters are encoded in a topological representation. In such a map (field), similar parameter values are encoded in a spatial neighbourhood whereas very different values are represented at locations that are far apart in the neural field. The output activation of the neural field indicates how likely it is that a particular parameter value influences the movement. In addition (Erlhagen & Schoener, 2002) assumes that the behaviour of dynamic neural fields is governed by local excitatory and global inhibitory connections (see Eq. (A.1)).

However, as the model is part of a real world setting inevitably not all mechanisms can be considered a true model of human processes and instead have to be considered as technical solutions. Despite this, CoRLEGO still operationalises central aspects of Song and Nakayama's (2009) theory. In the following we will first summarize the technical aspects of CoRLEGO and refer the reader to (Strauss & Heinke, 2012) for details. Subsequently we will give an overview of the model CoRLEGO is based on and summarize the results published by Strauss and Heinke (2012). Strauss and Heinke (2012) did not discuss how CoRLEGO relates to brain areas. We will remedy this shortcoming at the end of this section.

Setup and Technical Solutions

Fig. 1 shows the set-up. As the crucial effects of the CRT can be captured by two-dimensional movements (i.e. diversion from a straight line), we designed a planar LEGO robot arm with two joints. The robot arm was built with LEGO Mindstorms NXT kits. The total length of the arm is approximately 36 cm long (forearm 19 cm and upper arm 17 cm). The robot arm and its environment are filmed with a camera from a birds-eye view (see yellow box in top right corner of Fig. 1). The distance between camera and table is 90 cm. The birds-eye view constitutes a strong simplification from the real experimental set-up. However the exact viewing angle can be considered not to be crucial in the CRT-task. It also simplifies the necessary coordinate transformations from target coordinates to joint movements.

The photographs show that we used a normal desk light (grey object next to the camera in Fig. 1) to keep the lighting roughly constant. For an easier detection of the robot arm blue markers were attached to the arm base and to the end effector. For the search items red and green coloured markers were used. These markers were detected with a combination of simple standard computer vision methods such as HSV-colour space, thresholding and erosion filters (see Strauss & Heinke, 2012 for details). At the beginning of each experiment the parameters of the methods had to be calibrated in order to adapt them to the lighting conditions. The output formed three two-dimensional binary maps feeding into the green map, red map and hand map (blue marker) in CoRLEGO's model (see Fig. 2).

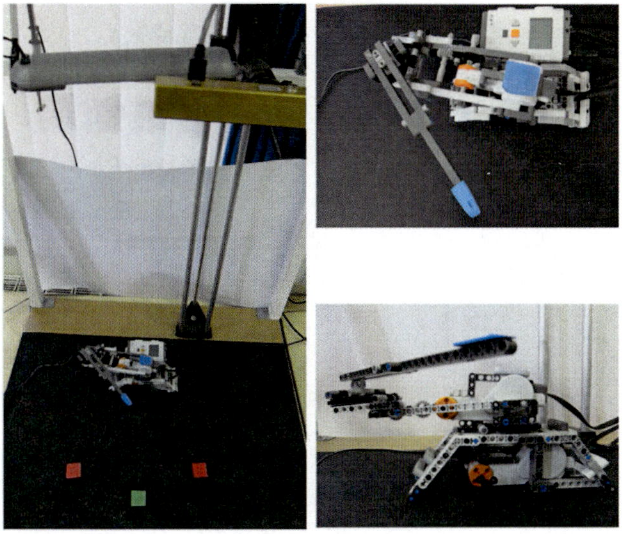

Figure. 1: Set-up of CoRLEGO with details of the arm. (For interpretation of the references to colour in this figure legend, the reader is referred to the web version of this article.)

The output of CoRLEGO's model is an activation blob which encodes a two-dimensional velocity vector in neural-field style (see next section for detailed description). To translate the neural-field encoding into speed commands for the LEGO motors, the distance of the centre of gravity of the activation blob to the centre of the V map was calculated. This Cartesian information was transformed into joint speeds following the standard approach of an approximation of the inverse of the Jacobian matrix.

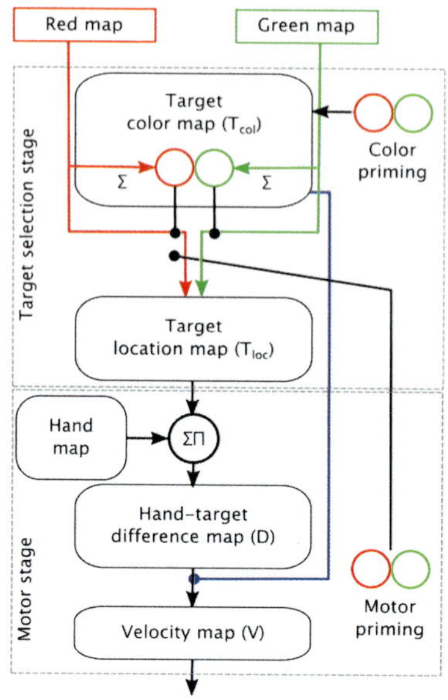

Figure. 2: Overview of CoRLEGO's architecture. The extensions present in the paper are the motor priming unit and the gating mechanism at the output of the D map controlled by a thresholded activation from target colour map (blue line). The dots in the graphics indicate modulatory connections. E.g. the target colour map weights the input of colour maps into the target location map (see main text for details). (For interpretation of the references to colour in this figure legend, the reader is referred to the web version of this article.)

Overview of CoRLEGO's Model

One of the inputs to the model are the red and green feature maps. The target selection stage consists of two maps, the target colour map and the target location map. The target colour map (T_{col} map, Fig. 2) detects the odd colour in the input image. The T_{col} map has two units representing the two colours, red and green. As input, each unit receives the total activation of the respective maps. A competition process then activates the unit with the larger input activation while the other unit's activation is suppressed. Hence, this losing unit indicates the target colour (odd colour). The input to the target location map (T_{loc} map) is the topologically summed activation from the green colour map and the red colour map. The summation is weighted by the inverted output of the T_{col} map (black dots in Fig. 2; Eq. (A.5)). Hence, once the odd colour is determined by the target colour map only the respective colour map is fed into the target location map. Consequently competitive processes in the target location map activate units at the position of the odd-colour target. This output activation feeds into the motor stage. However note that the competition in both maps operates in parallel. Hence even before the odd-colour is determined the target location map begins building up a representation of all items in the display including the distractors. This simultaneous representation of items will be crucial for simulating curved trajectories as we will explain below.

In the motor stage the output of the target location map is transformed into a topological representation of the distance between end effector and target (hand-centred target map; D map). This representation is generated through a spatial correlation between the end effector map and the T_{loc} map ($\sum\prod$-symbol in Fig. 2). To understand this it is important to remember that both maps encode locations through activation blobs as set out by the DNF framework. The target location map has a blob at the location of the target (once the selection processes are completed) and the end effector map has a blob at the position of the end effector. Consequently a spatial correlation between these maps determines the spatial "distance" between the blobs (in the same way a temporal correlation between two Dirac pulses determines the time delay between the two pulses). The spatial correlation is implemented in a way that the origin of the effector-centred coordinates is aligned with the centre of the D map. Note that a biologically plausible implementation of the spatial correlation can be achieved with sigma–pi units. Sigma–pi units were first proposed by McClelland, Rumelhart, and Hinton (1986) (see Heinke & Humphreys, 2003 for another example of an application).

The output of the D map feeds into the velocity map (V map), the core of the motor stage. The velocity map encodes the arm's velocity by using an activation blob. To be more specific, the blob's x- and y-coordinates encode the arm's velocity with respect to x- and y-directions. For instance a blob at the corner of a map corresponds to a maximal velocity in both directions while a blob in the centre of the map encodes zero velocity. Note that this velocity coding schema naturally aligns with the coding schema of the D map. A activation blob in centre of the D map encodes zero distance between target and endeffector. Furthermore, activation at the centre of the D map leads to an activation in the centre of the velocity map which in turn brings the arm to a halt. In other words, the encoding schema in the velocity map and the D map ensures that once the arm has reached the target the arm stops. It is also important to note that the activation in the velocity map changes in a moving blob style (Amari, 1977). The movement of the moving blob is directed by the target location via the Dmap. (This directing of the moving blob can be pictured as a ball, the moving blob, rolling on a planar surface whose slope can be modified.) As a consequence of the moving blob behaviour the arm changes direction and speed in a non-jerky human-like fashion.

During the course of a reach the moving blob behaviour plays out in the following way. At the beginning the blob is positioned at the centre of the velocity map (zero speed). Once the T_{loc} map begins to detect the target location the D map establishes the distance of the arm from the target and encodes it as an activation blob at the corresponding location. Consequently this blob in the D map directs the blob in the velocity map towards this position which encodes speed and movement directions in the velocity map resulting in a reach towards the target. However and importantly, competing distractors can also filter through to the velocity map which results in the distractors 'pulling' the moving blob in their direction allowing the model to simulate curved reaching trajectories found in the choice reaching tasks.

Finally, to model the POP-effect the model possesses "colour priming" units (see Fig. 2). These units influence the competition process in the target colour map by pre-activating the units within the map. If the model is set up to replicate the effect of target colour repetition the priming units pre-activate the colour unit corresponding to the odd-colour in the display. Thus, target selection on the whole is expedited and the influence of distractors is reduced, leading to shorter latencies and straighter trajectories. On the other hand, to simulate the switch of target colour at the end of a streak of repetitions the unit encoding the distractor colour is pre-activated and the influence of distractor is increased resulting in longer latencies and more curved trajectories. It is important to note that this architecture inherently links the two dependent measures, maximum deviation and initiation latency. This link is consistent with Song and Nakayama's and Woodgate et al.'s (2015) findings in the pre-tDCS conditions. However (Woodgate et al., 2015) also found that this link is disrupted in the tDCS conditions. We will return to this important aspect when we develop the extension of CoRLEGO in the second part of this paper.

Results

Strauss and Heinke's (2012) simulation results showed that CoRLEGO operates as expected. In particular, CoRLEGO is able to mimic Song and Nakayama's (2006)POP-effect with the help of the colour priming neurons. In addition, the moving blob mechanism is able to generate the typical bell-shaped velocity profile of human reaches.

However it is important to note that CoRLEGO's environment (see Fig. 1) is fairly noisy. The noise has several sources. For instance, the lighting conditions can change during reaches as well as from reach to reach. These variations results in the representations of the coloured squares in the colour maps to change in size thereby affecting the movement control. Note that the camera adds further to this noise. On the output side the mechanics of the LEGO arm and the motor servos add to the noise CoRLEGO has to deal with. In other words, this environment is reminiscent to the noise the brain has to deal with.

Interestingly, it turns out that for CoRLEGO to successfully execute the reaches in such a noisy environment the velocity map needs to be organized in a nonlinear fashion, meaning that the spatial resolution of the map changes with the speed the location encodes. To be more specific, at low velocities (the centre of the map) the map needs to have a fine-grained resolution, while at a high velocity, (the periphery of the map) the velocity resolution is coarser. This way the arm makes more precise movements when the end effector is close to the target (low velocity). Interestingly the structure of the map would match the structure of the retina and the visual cortex (e.g., Rovamo & Virsu, 1979), and is therefore a fairly plausible prediction. However, to date this prediction has not been tested.

Related Brain Areas

Since the aim of this paper is to provide a computational explanation for the results from our brain stimulation experiment, it is important to discuss how CoRLEGO can be related to the functionality of certain brain regions. The functions of the different maps in CoRLEGO can be related to assumed functions of brain areas. CoRLEGO's colour maps are assumed to be related to V4 in the visual cortex (e.g. Zeki, 1984). The velocity map with its encoding of the movement speed can be seen as part of the motor cortex. The maps in the target selection stage with their attentional function are traditionally linked to the posterior parietal cortex (PPC). Similarly, the priming neurons could be located in the parietal cortex as activation in this region is correlated with the POP effect (see Kristjansson & Campana, 2010 for a review). However, it has recently become clear that the PPC is also involved in motor control (see Baldauf & Deubel, 2010 for a good review). In particular the parietal reach region (PRR) has been implicated in reach-related processing, e.g. coordinate transformations, similar to the one implemented in the D map (see Yttri, Wang, Liu, & Snyder, 2014 for most recent evidence).

EXTENSION OF CORLEGO

In the earlier summary of Woodgate et al.'s (2015) findings we simply noted that the motor system may be involved in the guidance of attention when target properties are predictable. Even though these findings alone are interesting and novel, there are several intriguing details. In Section 3.1 we will present these details. In Section 3.2, to highlight the added value of CoRLEGO, we also discuss intuitively plausible explanations of the findings that can be ruled out in the framework of CoRLEGO. In Section 3.3, we present the extension of CoRLEGO we have chosen together with the simulation results.

Woodgate et al.'s (2015) Findings

Woodgate et al. (2015) found that tDCS influences reaching movements only when the experimental design contain streaks of target colour repetitions (predictable target colour). To be more specific, anodal tDCS reduced the maximum deviation (MD) from the ideal movement path while cathodal tDCS increased the deviation. Note that this qualitative difference between anodal and cathodal effect is consistent with findings in movement experiments (e.g. Reis et al., 2009). However and importantly this effect was not found in the initiation latency (IL) (compare Fig. 3(c) and (b)). This effect also contrasts with Song and

Nakayama's and our findings that MD and IL are normally linked, i.e. an experimental manipulation that reduces IL also reduces MD. To the best of our knowledge, this is the first time such dissociation between IL and MD has been found. In this paper we will use CoRLEGO to explain this intriguing dissociation.

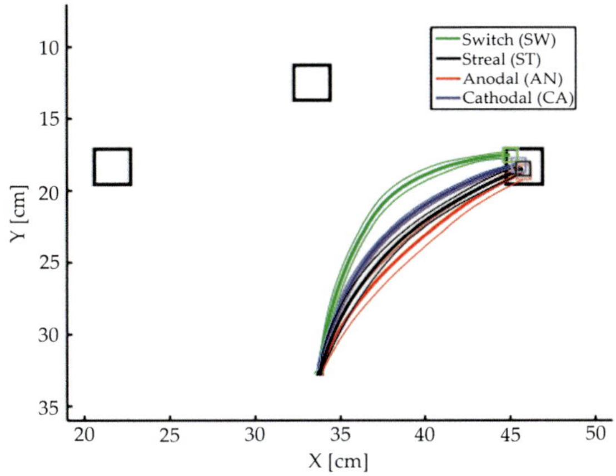

Figure. 3(a): (a) Averaged trajectories. The thin lines indicate the standard error from five trials in each condition.

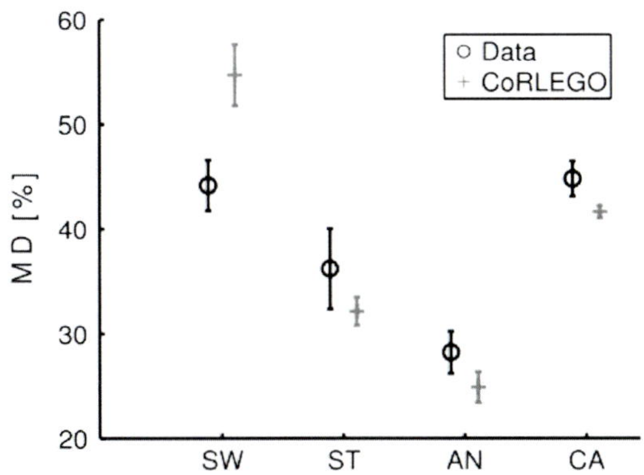

Figure. 3(b). (b) Maximum deviations (MD).

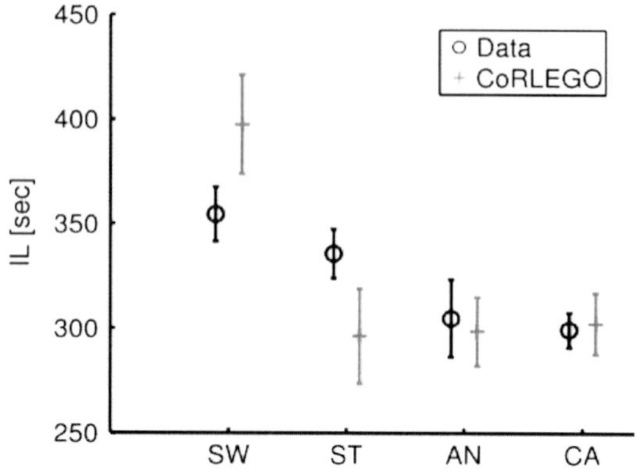

(c) Initiation latencies (IL).

Figure. 3(c). (c) Initiation latencies (IL).

Fig. 3. Results of CoRLEGO's extension. Figure (a) shows CoRLEGO reaching trajectories. Figures (b) and (c) document the two important characteristics of the trajectories, maximum deviation and initiation latency, and compares them with the empirical evidence by Woodgate et al. (2015). The error bars indicate the standard error.

The data also contains other effects commonly found in similar behavioural studies such as individual differences in participants' behaviour, practise effects, statistically non-significant experimental conditions, etc. As we were not interested in modelling these effects or conditions we reduced the data so that it constitutes a compact but veridical representation of the dissociation between IL and MD. The original study by Woodgate et al. (2015) first tested participants using Song and Nakayama's (2006)CRT with up to five target colour repetitions without applying tDCS (pre-tDCS condition). Subsequently participants performed in the same task but this time either anodal or cathodal tDCS was applied. Finally they completed yet another block of this task without tDCS (post-tDCS condition). There was little difference between different lengths of streaks and the effect in the post-tDCS condition was the same across the streaks. Therefore we model only the experimental outcome from the fifth colour repetition in the three conditions, the pre-tDCS condition, the anodal post-tDCS condition and the cathodal post-tDCS condition. Moreover, Woodgate et al.'s (2015) findings with respect to a tDCS-effect in the switch condition were weaker than for the streak condition due to a lack of statistical power. Therefore the switch condition from the post-tDCS condition was not included. Note that this reduction is not only justifiable from a scientific standpoint but also from a technical one as simulations with CoRLEGO are time-consuming. And the LEGO arm does not provide enough space to generate statistically significant effects in maximum deviation for all these conditions. However we still wanted to demonstrate the ability of the extension of CoRLEGO to reproduce the difference between the switch and the streak condition. Therefore we included the switch condition from the pre-tDCS condition.

Finally it is important to recap that the anodal and cathodal tDCS conditions were conducted with different participants groups to avoid an interference effect between these two types of stimulation. Due to individual differences the two groups performed differently in their pre-tDCS condition. Therefore we averaged across these two experimental groups. Also these individual differences do not allow for a direct comparison of the post-tDCS conditions. We therefore determined the change in performance caused by tDCS for each group. The change in performance was then added to the averaged pre-tDCS performance. The resulting values were used to evaluate the model's simulation results (see Fig. 3(b) and (c) for an illustration). Note that in order to simplify the terminology in the remaining part of the paper we simply refer to the averaged pre-tDCS conditions as the streak condition and the switch condition. To the tDCS effects corrected for individual differences we simply refer as the anodal tDCS condition and the cathodal tDCS condition.

Implausible Solutions with Corlego

The first obvious idea would be to assume that tDCS somehow influences the velocity map (e.g. its parameters) as it is assumed to be part of the motor cortex and the tDCS was applied to the motor cortex. This would be a parsimonious solution as it would not require the addition of more structures to CoRLEGO. However, any parameter change in the velocity map will affect the properties of the moving blob dynamic therefore affecting both dependent variables. In addition there is the problem that the tDCS effect depends on the predictability of the target colour. This perceptually-related information is not stored in the velocity map.

Interestingly an animal study by Zach, Inbar, Grinvald, Bergman, and Yaadia (2008)found that the motor cortex can form representations of colour information if the information is strongly related to movements (see Eisenberg, Shmuelof, Vaadia, & Zohary, 2011 for an fMRI-study with humans pointing in the same direction). Hence one can speculate that such representations are formed in the human brain when colours are highly movement-relevant (i.e. predictable of the position of movement target) and it is conceivable that our tDSC stimulation modulated these representations. These units, in turn, may feed into the colour priming units in the target selection stage (parietal cortex), i.e. supporting the priming effect. Such a pathway would be supported by well-known connections from the motor cortex to the parietal cortex. However, such a realization would not lead to a dissociation between IL and MD. Nevertheless, the possibility of perceptually related units in the motor cortex formed the basis for CoRLEGO's extension as discussed in the following section.

Plausible Implementation

The starting point of extending CoRLEGO was indeed priming neurons in the motor stage here termed motor priming units (see Fig. 2), as suggested in the previous section. However, rather than feeding them directly into the colour priming units they weight the influence of the colour target units at the input of the Tloc map (black line in Fig. 2; Eq. (A.6)). This weighting allows the modulation of the contrast between the two colour units. If the motor priming units are highly activated (anodal tDCS) the higher contrast between the two colour units lead to a short IL and straighter MD. In contrast, smaller activation of the motor priming units (cathodal tDCS) leads to less contrast consequently slower IL and more curvature. In other words, the motor priming units influence IL and MD similarly to the colour priming. Nevertheless the separate pathways allow us to implement an dissociation between MD and IL. We introduced a gating mechanism at the output of the T_{loc} map that only passes on output activation if the level of the T_{col} map surpasses a set threshold (see Eq. (A.7)). Thus the moving blob in the velocity map is only pushed off the centre once the threshold is passed. This mechanism ensures that IL only depends on perceptual priming and is little affected by the motor priming units (tDCS).

However, extensive tests of the gating mechanism at the output of the T_{loc} map under the noisy conditions of CoRLEGO did not lead to stable results (see Section 3.5 for an explanation of how the parameters were fitted). This was mainly due to the characteristics of how the priming (pre-activation) influences the build-up of activation in the dynamic neural fields. The longer the build-up the larger the activation differences between different levels of pre-activation. Hence to achieve a good priming effect in noisy conditions it is important to place the threshold (of the gating mechanism) relatively high. However such a delay in passing the threshold also leads to the completion of the target selection therefore resulting in little influence of distractors on the reaching trajectory. Therefore, we decided to apply the gating mechanism at the output of the D map (see blue line in Fig. 2), so that the D map acts as short-term storage for the distractor activation. This way the threshold can be fairly high, but because the D map (CoRLEGO's PRR) is slow to follow changes at the target location map the distractor can still can produce trajectory deviation. Interestingly, the design decision is also supported by very recent experimental evidence on the PRR (Yttri et al., 2014), where lesioning the PRR affects the onset of reaching but not the selection of movement targets (attention) or the control of movements as such.

Simulation Method, Data Analysis and Evaluation of Model

The size of the target objects was 3.5 cm × 3.5 cm. Targets were located on a virtual circle with the radius of 22 cm at 45°, 90°, and 135° (from left to right). The centre of this circle was the starting position of the robot arm's hand. The starting position was located 9 cm in front of the arm's base (shoulder). The odd-colour target object was placed on the right side. In each of the four conditions (switch, streak, anodal and cathodal tDCS) we conducted five trials.

For the purpose of data analysis the raw data of each trajectory was pre-processed with the following steps. First, a B-Spline (3rd order) of the raw data points was calculated which reduced noise and normalized the trajectories to 100 data points. Second, we applied a Butterworth filter (2nd order) with a cut-off frequency of 1/20 of the sampling rate of the camera in order to reduce the noise even more.

The maximum deviation (MD) was calculated by dividing the maximum deviation of the data points from a straight line by the length of this line. The straight line was determined by the start and end points of the trajectory. The initiation latency (IL) was defined as the point in time when the velocity was higher than 0.5 cm/s for the first time following the start of the simulation. After starting a trial the position of the end-effector was recorded until the target was reached. The arm was considered to have reached the target when the encoded velocity in the V map fell under a threshold value of approximately 0.5 cm/s.

The most obvious way of evaluating models is by comparing the quantitative differences between model and data, e.g. using sum of squared difference. For obvious reasons such a simple quantitative comparison is not adequate in context of CoRLEGO. Instead we focused on a qualitative assessment of CoRLEGO and examined whether the response pattern was qualitatively the same as human. For instance, CoRLEGO was expected to show no significant effect for IL from tDCS while should be a significant effect MD. These effects were analysed by using t-tests. It is worth noting that such a qualitative evaluation is seen as legitimate approach in the literature (see Heinke, 2009; Pitt, Kim, Navarro, & Myung, 2006; Wills & Pothos, 2012for more discussions). Furthermore, even though such a qualitative evaluation seems sufficient to us we also included an evaluation that assessed the relative magnitudes between conditions. Therefore we re-scaled CoRLEGO's responses with the ratio between the overall means of human responses and CoRLEGO's responses. This minimizes overall differences between human data and CoRLEGO while at the same time keeps relative differences between conditions.

Results

Before a test began the parameters of the image processing were adapted to the present lighting conditions. The parameters of the DNFs used very similar parameters as in Strauss and Heinke (2012) (see Appendix for details). These parameters are crucial for securing a selection of the target and a successful control of the reaching movement. For achieving the target behaviour only three parameters are crucial: the strength of the perceptual priming, the threshold of the gating mechanism and the weighting of the colour maps (motor priming) (see Appendix for details). The parameters were modified manually until we found a good approximation of the target behaviour particularly in terms of its qualitative pattern. The parameter fitting is not very complex as it turned out that the new architecture enables the user to establish good parameter values relatively independently. The simplicity of this process is important as the simulations with CoRLEGO are fairly time-consuming.

The parameter fitting procedure followed three steps. First a relatively high level of pre-activation of a colour unit (perceptual priming) had to ensure a fairly large difference between streak and switch condition in both MD and IL. But the level of pre-activation cannot be too high as otherwise the wrong colour would be selected in the switch condition. Also the streak condition needs to leave enough room for improvement in the MD-effect of the anodal tDCS condition. Second the threshold of the gating mechanism needs to be high enough so that there is a significant priming effect (see Section 3.3 for an explanation). But it cannot be set too high as an overly delayed IL would lead to a complete suppression of the distractors therefore exhibiting no curved trajectories. During these first two steps the weighting of the colour maps (motor priming) was set to a middle range value. This way the third step allowed an increase and decrease of this factor to modify MD for both the anodal and cathodal conditions. But the decrease needs to be limited so that the target colour map still has an influence on the target location map.

Fig. 3 and Table 1 show the results of the parameter fitting. The pairwise comparisons in Table 1 demonstrate that our model successfully mimics the pattern of the experimental outcome. A quantitative comparison between data and simulation in Fig. 3(b) and (c) indicates some minor differences. In particular the priming parameter for the switch condition led to higher switch effect on maximum deviation than in Woodgate et al.'s (2015) data. However, Woodgate et al.'s (2015) data showed an unusually small switching effect compared to other experiments we conducted with the same procedure. This diminished effect may have been due to the tDCS apparatus distracting participants.

Table 1. t-tests on CoRLEGO's results. The tests demonstrate that CoRLEGO was able to replicate the behavioural pattern found by Woodgate et al. (2015). The bottom lines show the results for initiation latency (IL) and the top lines the results for the maximum deviation (MD). The results in bold indicate a significant comparison at an α-level of 0.05. The abbreviations stand for the four conditions: switch (SW), streak (ST), anodal tDCS (AN) and cathodal tDCS (CA) conditions.

	SW	AN	CA
ST	t=3.07; p=0.015	t=0.08; p=0.94	t=0.83; p=0.22
	t=6.95; p<0.01	t=3.62; p=0.1	t=6.44; p<0.01
AN	t=3.42; p=0.01	–	t=−0.18; p=0.87
	t=9.01; p<0.01	–	t=−10.48; p<0.01
CA	t=3.40; p=0.01	–	–
	t=4.33; p<0.01	–	–

DISCUSSION

We presented an extension of a neurobiologically inspired robotics model (CoRLEGO). CoRLEGO models leakage of attentional selection into the motor system. The extension aimed to explain our findings that tDCS over the motor cortex modulates the colour priming effect (Woodgate et al., 2015). In particular, the extension aimed to explain why tDCS affected the maximum deviation of reaching movements and not the initiation latency. The new CoRLEGO suggests that colour priming based on the perceptual system is reflected in the initiation latency whereas the maximum deviation is affected by both perceptual- and motor-based priming. Furthermore, the detection of the odd colour leads to CoRLEGO initiating the movement (gating mechanism). In other words, the movement tends to begin once a task-relevant global property of the search display was determined. CoRLEGO's motor priming neurons support the spatial selection of the movement target via the feedback from motor cortex to the visual system. Hence in a more general sense, the purpose of the feedback connections is to enhance the detection of movement relevant information, i.e. the spatial location of the odd-colour target (as opposed to selecting information that is not directly relevant to the movement, i.e. the colour itself). Interestingly CoRLEGO's usage of feedback connections is consistent with Baldauf and Deubel's (2010) theory of visual preparation. Their theory suggests that the feedback connections guide visual attention to extract movement-relevant information from visual stimuli.

There are other predictions that can be derived from CoRLEGO. The moving blob activation in the velocity map should be found in animal studies or fMRI studies with high resolution. Interestingly such a prediction is feasible as the moving blob behaviour is similar to the travelling waves found in a variety of brain areas including the motor cortex (e.g. Takahashi, Saleh, Penn, & Hatsopoulos, 2011). In particular a recent animal study by Riehle, Wirtssohn, Grün, and Brochier (2013) found that activation travels from arm-related areas to hand-related areas during reach-to-grasp movements. In other words, the shift of activation mirrors the different phases of the movement. Hence CoRLEGO's moving blob behaviour suggests that these shifts also encode movement parameters. Finally the odd-colour detection threshold in the model's PRR also predicts results for an fMRI study. Here the activation flow, indexed with dynamic causal modelling or event-related fMRI, should be small prior to movement onset before increasing at the point of initiation (bar some signal latencies).

On a wider perspective CoRLEGO falls into the category of closed-loop control models of human movements. This type of model contrasts with open-loop control models. In these models the arm movement is planned, e.g. in terms of the sequence of muscles activations, prior to its execution. For a long time, this type of model dominated the explanations of human movements. However recent evidence and computational modelling favour the closed-loop approaches (see Todorov, 2004 for a review). But then again, this evidence is mainly based on experiments involving single targets. In multi-object situations such as the one discussed in this paper it is possible that the brain adopts a different strategy whereby the movements to different target positions are pre-planned and "averaged" trajectories are executed before the final target position is known to the motor system. In fact this interpretation has been put forward by Flanagan and colleagues (e.g. Stewart, Baugh, Gallivan, & Flanagan, 2013). In future work a computational model of this theory will have to be contrasted with CoRLEGO's closed-loop approach.

In the introduction of this paper we noted that choice reaching tasks have been used to tap into other cognitive processes such as language processing (Spivey et al., 2005) or representations of numbers (Song &

Nakayama, 2008b). For instance, Song and Nakayama (2008b) presented single digits on the centre of a screen and two boxes on either side of the digit. Participants were asked to reach for the box on the left side if the digit was smaller than five and if larger than five to the right side. Song and Nakayama (2008a) were able to show that the trajectories were more curved towards the centre of the screen the closer the digit was to five, thus pointing towards a spatial representation of numbers. Of course CoRLEGO would have to be extended and such spatial number representation introduced. However, we also expect that the central mechanism in CorLEGO, the moving blob mechanism, would be able to mediate the selection process in this spatial representation of numbers into the reaching movement. Similarly (Spivey et al., 2005) presented two pictures on either side of the screen and asked the participants to move a mouse pointer towards the picture with a dark frame. If the two pictures were phonologically similar (e.g. picture and pickle) the movement took a more central line whereas if the two pictures were phonologically dissimilar (e.g. picture and jacket) movement more biased towards the target image. Again a new version of CoRLEGO could include such phonological similarities, while the moving blob mechanism should be able to pass on these characteristics of the selection process. These discussions of future work highlight the potential of CoRLEGO's moving blob mechanism for modelling other leakage effects.

LESSONS FOR NEUROBIOLOGICALLY INSPIRED ROBOTICS

The work has several lessons for neurobiologically inspired robotics. Even though the set-up greatly simplified the visual environment and the complexity of movement control humans usually have to deal with, the set-up still contained enough noise to require the designer to rethink the architecture. As a result it was necessary to change the spatial organization of the velocity map and the influence of the perceptual system on the motor cortex. This experience demonstrates the importance of building a robot; in a possible simulation environment the designer would be more tempted to simply adjust the environment. It is also worth noting that the whole architecture including the dynamic neural fields must be robust to noise giving additional credence to the model.

One caveat of this approach is that running simulations can be time-consuming, as they are with CoRLEGO. Hence, extensive parameter optimization is not really possible in this framework. Similarly extensive model comparisons as suggested in the outlook section of the discussion are not feasible. In these applications it would be necessary to set up simulation of the robot's environment and any parameter optimization or model comparison can be done in the simulation environment. However, the final test for the success of the approach has to be made in a real-world scenario, as it is unlikely that the simulations can capture all characteristics of a real-world scenario. In other words, a combination of simulation and real-world scenario presents a way forward in neurobiologically inspired robotics.

Another perhaps more novel aspect of our robotics-based approach is the usage of a simple, cheap LEGO arm rather than an expensive industrial robot arm. Our arm is also very easy to control with its two degrees of freedom and is custom made to capture the crucial characteristics of experimental findings. In other words, LEGO allows the researcher to build a customized simple robot in order to answer a very specific scientific question. In contrast, an industrial robot arm would have a lot of baggage which the designer would need to deal with, but which are not relevant to the scientific question. Moreover the usage of LEGO will allow the complexity of the arm to increase (e.g. adding a gripper) while building on the present results. Nevertheless, the lessons learnt from the work with a LEGO robot should be transferable to an industrial robot.

CoRLEGO highlights yet again the power of maps in general and in particular the topological representation of movement parameters. It demonstrates that this framework can establish an elegant link between perceptual processing and movement control. However, it also indicates that the dynamic within the two systems may be very different, i.e. the perceptual system exhibits a decision style of dynamics while the motor system shows smooth transitions between states. Moreover the maps design an efficient and faster method to control movements as both systems can operate in parallel.

Finally, in the context of a technical solution it is not clear why it is necessary that the motor system needs to contribute to the detection of movement relevant information as suggested by CoRLEGO. It is possible that this solution is somehow owed to constraints imposed by the neurological substrate (e.g. speed signal transmission) and possibly not important for technical solutions. This is however a question for future research.

ACKNOWLEDGEMENTS

Philip J. W. Woodgate was supported by an EPSRC-DTC grant. Saber Sami was supported by a grant from the Wellcome Trust WT087554.

Appendix A. Mathematics of the extension of CoRLEGO

This section focuses on the mathematical description of the extension of CoRLEGO. Further details on the implementation of CoRLEGO can be found in Strauss and Heinke (2012). The maps involved in the extension of CoRLEGO are the target colour map (T_{col}), the target location map (T_{loc}), hand-target difference map (D), the velocity map (V), the green map (col_{green}) and the red map (col_{red}). To set the scene for the extensions, we also present the equations for CoRLEGO's dynamic neural fields.

DYNAMIC NEURAL FIELD

Our implementation follows Faubel and Schoener's (2008) approach which turned out to be a more robust than the original implementation by Amari (1977).

$$\tau \dot{u}(\mathbf{x}, t) = -u(\mathbf{x}, t) + h + s(\mathbf{x}, t) + exc_{loc}(\mathbf{x}, t) - inh_{loc}(\mathbf{x}, t) - inh_{glob}(\mathbf{x}, t) + q(\mathbf{x}, t). \quad \text{(A.1)}$$

Hereby, τ is a time parameter which defines how fast the DNF adapts towards changing inputs, u(x,t) stands for the field activation at time t and location x, h<0is the resting level of the field, s(x,t) is the external input of the field, w(x) is the activation kernel function and q(x,t) is normally distributed Gaussian noise. Note that x is two-dimensional for the most DNFs in my model. Thus, exc_{loc} and inh_{loc} define the local excitation and inhibition with the following equation:

$$exc_{loc}(\mathbf{x}, t) = \sum_{\mathbf{x}'} w_{exc}(\mathbf{x} - \mathbf{x}')f\left(u(\mathbf{x}', t)\right). \quad \text{(A.2)}$$

Next, the global inhibition inh_{glob} depends on the summarized field activation and the parameter g_{inh}:

$$inh_{glob}(\boldsymbol{x}, t) = g_{inh} \sum_{\boldsymbol{x}'} f\left(u(\boldsymbol{x}', t)\right). \tag{A.3}$$

Finally, the kernel w_k with its parameters σ_k (kernel width) and c_k (kernel strength) is defined with the following equation:

$$w_k(\boldsymbol{x}) = \frac{c_k}{\sigma_k \sqrt{2\pi}} \exp\left(-\frac{|\boldsymbol{x}|^2}{2\sigma_k^2}\right) \tag{A.4}$$

Kernel functions define how connections spread out through the DNF and how large the radius of influence of a neuron in its neighbourhood is. They are applied in all DNFs in the local excitation, inhibition and the noise term q.

Input to the Target Location Map

In CoRLEGO's original version the input to the target location map (T_{loc} map) is the topologically summed activation from the green colour map and the red colour map. The summation is weighted by the inverted output of the T_{col} map:

$$\begin{aligned} s_{T_{loc}}(\boldsymbol{x}, t) = {} & \left(1 - f_{T_{col}}\left(u_{T_{col}}(green, t)\right)\right) \cdot col_{green}(\boldsymbol{x}, t) \\ & + \left(1 - f_{T_{col}}\left(u_{T_{col}}(red, t)\right)\right) \cdot col_{red}(\boldsymbol{x}, t). \end{aligned} \tag{A.5}$$

To include the influence of motor priming (tDCS stimulation) the influence of the colour neurons is modulated with a factor (p):

$$\begin{aligned} s_{T_{loc}}(\boldsymbol{x}, t) = {} & \left(1 - p \cdot f_{T_{col}}\left(u_{T_{col}}(green, t)\right)\right) \cdot col_{green}(\boldsymbol{x}, t) \\ & + \left(1 - p \cdot f_{T_{col}}\left(u_{T_{col}}(red, t)\right)\right) \cdot col_{red}(\boldsymbol{x}, t). \end{aligned} \tag{A.6}$$

Gating Mechanism

The extension of CoRLEGO included a gating mechanism of the D map input (V_D) into the velocity map:

$$\begin{aligned} V_D(\boldsymbol{x}, t) = {} & \left(g\left(f_{T_{col}}\left(u_{T_{col}}(green, t)\right)\right) + g\left(f_{T_{col}}\left(u_{T_{col}}(red, t)\right)\right)\right) \\ & \cdot f_D\left(u_D(\boldsymbol{x}', t)\right) \end{aligned} \tag{A.7}$$

$g(\cdot)$ is a simple threshold function with the parameter Θ as threshold. If one of the two colour units in the colour target map (T_{col}) passes this threshold the activation output of the D map is passed onto the velocity map.

APPENDIX B. PARAMETERS

Image Processing Parameters

Map	hue	Δhue	sv	ero
colblue	230	50	20	1
colred	4	18	35	2
colgreen	110	50	20	2

DNF Parameters

Map	τ	β	h	ginh	cexc	σexc	cinh	σinh	cq	σq
B	25	12	−2	0.3	80	3	20	10	0.05	10
Hand map	2	12	−0.5	0.2	20	3	0	1	0.05	1
Tcol	1500	12	−1	7	5	0.1	10	0.1	0.05	1
Tloc	150	1.5	−0.5	0.3	35	4	0	8	0.05	1
D	15	12	−0.5	0.1	10	3	0	3	0.05	1
V	24	2	−1	0.4	20	5	0	1	0.05	5

Critical Parameters for Simulating Data

The threshold (θ) in the gating mechanism was set to 0.3. The factor p in the input to the location map was set to different depending on the condition CoRLEGO was set up to simulate:

Condition	**Cathodal tDCS**	**Anodal tDCS**	**Streak**	**Switch**
p	0.6	1	0.75	0.75

Note that the differing behaviour between the streak- and the switch-condition results from priming the same colour as the odd-colour in display (streak-condition) or priming the other colour as the odd-colour in display (switch-condition).

The priming (pre-activation) of the target colour map was implemented by setting the input of the to-be-primed colour units to 4 and then simulated the target colour map without any other map or input. This simulation continued until the map reached stable state. Only after this pre-activation phase was completed the whole of CoRLEGO resumed normal operations and the colour units of the target colour map received the total activation of the respective colour map.

REFERENCES

1. Amari, S. I. (1977). Dynamic of pattern formation in lateral-inhibition type neural fields. Biological Cybernetics, 27, 77–87.
2. Baldauf, D., & Deubel, H. (2010). Attentional landscapes in reaching and grasping. Vision Research, 50, 999–1013.
3. Desimone, R., & Duncan, J. (1995). Neural mechanisms of selective attention. Annual Review of Neuroscience, 18, 193–222.
4. Deubel, H., Schneider, W. X., & Paprotta, I. (1998). Selective dorsal and ventral processing: Evidence for a common attentional mechanism in reaching and perception. Visual Cognition, 5, 81–107.
5. Eckstein, M. P. (2011). Visual search: A retrospective. Journal of Vision, 11(5), 1–36.
6. Eisenberg, M., Shmuelof, L., Vaadia, E., & Zohary, E. (2011). The representation of visual and motor aspects of reaching movements in the human motor cortex. Journal of Neuroscience, 31, 12374–12377.
7. Erlhagen, W., & Schoener, G. (2002). Dynamic field theory of movement preparation. Psychology Review, 109(3), 545–572.
8. Faubel, C., & Schoener, G.(2008). Learning to recognize objects on the fly: A neutrally based dynamic field approach. Neural Networks, 21, 562–576.
9. Heinke, D. (2009). Computational modelling in behavioural neuroscience: Methodologies and approaches—minutes of discussions at the workshop in birmingham, UK in may 2007. In D. Heinke, & E. Mavritsaki (Eds.), Computational modelling in behavioural neuroscience: closing the gap between neurophysiology and behaviour. Psychology Press.

10. Heinke, D., & Backhaus, A. (2011). Modeling visual search with the selective attention for identifcation model (VS-SAIM)—a novel explanation for visual search asymmetries. Cognitive Computation, 3(1), 185–205.
11. Heinke, D., & Humphreys, G. W. (2003). Attention, spatial representation, and visual neglect: Simulating emergent attention and spatial memory in the selective attention for identification model (SAIM). Psychology Review, 110(1), 29–87.
12. Kristjansson, A., & Campana, C. (2010). Where perception meets memory: A review of repetition priming in visual search tasks. Attention, Perception, and Psychophysics, 72, 5–18.
13. Maljkovic, V., & Nakayama, K. (1994). Priming of pop-out: I. Role of features. Memory and Cognition, 22, 657–672.
14. Marr, D. (1982). Vision: A computational investigation into the human representation and processing of visual information. San Francisco, CA: W. H. Freeman.
15. Mavritsaki, E., Heinke, D., Allen, H., Deco, G., & Humphreys, G. W. (2011). Bridging the gap between physiology and behavior: Evidence from the ssots model of human visual attention. Psychological Review, 118(1), 3–41.
16. McClelland, J., Rumelhart, D., & Hinton, G. (1986). The appeal of parallel distributed processing. In Parrallel distributed processing: explorations in the microstructure of cognition. Volume I: foundations. Cambridge, MA: MIT Press/Bradford Books.
17. Müller, H. J., & Krummenacher, J. (2006). Visual search and attention. Special issues of visual cognition series. Psychology Press, URL: http://books.google.co.uk/books?id=Jw_rGwAACAAJ.
18. Nitsche, M. A., & Paulus, W. (2001). Sustained excitability elevations induced by transcranial dc motor cortex stimulation in humans. Neurology, 57, 1899–1901.
19. Notturno, F., Marzetti, L., Pizzella, V., Uncini, A., & Zappasodi, F. (2014). Local and remote effects of transcranial direct current stimulation on the electrical activity of the motor cortical network. Human Brain Mapping, 35, 2220–2232.
20. Ostry, D. J., Darainy, M., Mattar, A. A. G., Wong, J., & Gribble, P. L. (2010). Somatosensory plasticity and motor learning. Journal of

Neuroscience, 30, 5384–5393.

21. Pitt, M. A., Kim, W., Navarro, D. J., & Myung, J. I. (2006). Global model analysis by parameter space partitioning. Psychological Review, 113(1), 57–83.
22. Reis, J., Schambra, H. M., Cohen, L. G., Buch, E. R., Fritsch, B., Zarahn, E., Celnik, P. A., & Krakauer, W. J. (2009). Noninvasive cortical stimulation enhances motor skill acquisition over multiple days through an effect on consolidation.
23. Proceedings of the National Academy of Sciences of the United States of America, 106, 1590–1595.
24. Riehle, A., Wirtssohn, S., Grün, S., & Brochier, T. (2013). Mapping the spatiotemporal structure of motorcortical lfp and spiking activities during reach-tograsp movements. Frontiers in Neural Circuits, 7(48).
25. Rovamo, J., & Virsu, V. (1979). Estimation and application of the human cortical magnification factor. Experimental Brain Research, 37(3), 495–510.
26. Song, J.-H., & Nakayama, K. (2006). Role of focal attention on latencies and trajectories of visually guided manual pointing. Journal of Vision, 6, 982–995.
27. Song, J. H., & Nakayama, K. (2008a). Numeric comparison in a visually-guided manual reaching task. Cognition, 106, 994–1003.
28. Song, J. H., & Nakayama, K. (2008b). Target selection in visual search as revealed by movement trajectories. Vision Research, 48, 853–861.
29. Song, J. H., & Nakayama, K. (2009). Hidden cognitive states revealed in choice reaching tasks. Trends in Cognitive Sciences, 13(8), 360–366.
30. Spivey, M. J., Grosjean, M., & Knoblich, G. (2005). Continous attraction toward phonological competitors. Proceedings of the National Academy of Sciences of the United States of America, 102, 10393–10398.
31. Stewart, B. M., Baugh, L. A., Gallivan, J. P., & Flanagan, J. R. (2013). Simultaneous encoding of the direction and orientation

of potential targets during reach planning: evidence of multiple competing reach plans. Journal of Neurophysiology,.

32. Strauss, S., & Heinke, D. (2012). A robotics-based approach to modeling of choice reaching experiments on visual attention. Frontiers in Psychology, 3, 105.
33. Takahashi, K., Saleh, M., Penn, R. D., & Hatsopoulos, N. G. (2011). Propagating waves in human motor cortex. Frontiers in Neural Circuits, 5(40).
34. Todorov, E. (2004). Optimality principles in sensorimotor control. Nature Neuroscience, 7(9), 907–915.
35. Treisman, A. M., & Gelade, G. (1980). A feature-integration theory of attention. Cognitive Psychology, 12, 97–136.
36. Wills, A. J., & Pothos, E. M. (2012). On the adequacy of current empirical evaluations of formal models of categorization. Psychological Bulletin, 138(1), 102–125.
37. Woodgate, P. J. W., Strauss, S., Sami, S. A., & Heinke, D. (2015). Motor cortex guides selection of predictable movement targets. Behavioural Brain Research, 287, 238–246.
38. Yttri, E. A., Wang, C., Liu, Y., & Snyder, L. H. (2014). The parietal reach region is limb specific and not involved in eye-hand coordination. Journal of Neurophysiology, 111, 520–532.
39. Zach, N., Inbar, D., Grinvald, Y., Bergman, H., & Yaadia, E. (2008). Emergence of novel representations in primary motor cortex and premotor neurons during associative learning. Journal of Neuroscience, 28, 9545–9556.
40. Zeki, S. (1984). The construction of colours by the cerebral cortex. Proceedings of the Royal Instution of Great Britain, 56, 231–257.
41. Zhao, Y., Humphreys, G. W., & Heinke, D. (2012). A biased-competition approach to spatial cuing: Combining empirical studies and computational modelling. Visual Cognition, 20(2), 170–210.

CHAPTER 14

COMPUTATIONAL COGNITIVE MODELS OF SPATIAL MEMORY IN NAVIGATION SPACE: A REVIEW

Tamas Madl [a,c], Ke Chen[a], Daniela Montaldi [b], Robert Trappl [c]

[a] School of Computer Science, University of Manchester, Manchester M13 9PL, UK

[b] School of Psychological Sciences, University of Manchester, Manchester M13 9PL, UK

[c] Austrian Research Institute for Artificial Intelligence, Vienna A-1010, Austria

ABSTRACT

Spatial memory refers to the part of the memory system that encodes, stores, recognizes and recalls spatial information about the environment and the agent's orientation within it. Such information is required to be able to navigate to goal locations, and is vitally important for any embodied agent, or model thereof, for reaching goals in a spatially extended environment.

In this paper, a number of computationally implemented cognitive models of spatial memory are reviewed and compared. Three categories of models are considered: symbolic models, neural network models, and models that are part of a systems-level cognitive architecture. Representative models from each category are described and compared in a number of dimensions along which simulation models can differ (level of modeling, types of representation, structural accuracy, generality and abstraction, environment complexity), including their possible mapping to the underlying neural substrate.

Neural mappings are rarely explicated in the context of behaviorally validated models, but they could be useful to cognitive modeling research by providing a new approach for investigating a model's plausibility. Finally, suggested experimental neuroscience methods are described for verifying the biological plausibility of computational cognitive models of spatial memory, and open questions for the field of spatial memory modeling are outlined.

Keywords: Spatial memory models Computational cognitive modeling

INTRODUCTION

A wealth of neurophysiological results from human and animal experiments have, in recent years, helped shed light on the mechanisms and brain structures underlying spatial memory. Although it is possible to investigate spatial cognition purely from the point of view of one of the cognitive sciences, interdisciplinary analyses at the level of behavior as well as underlying neural mechanisms provide a more solid foundation and more evidence. Within the broader scope of cognitive sciences involved in investigating memory systems, such as psychology and neuroscience, computational models play a unique and important role in helping to integrate findings from different disciplines, as well as generating, defining, formalizing, and testing, and generating hypotheses, and thus helping to guide research in cognitive science.

There are multiple relevant reviews concerning the psychology of spatial cognition (Allen, 2003; Tommasi & Laeng, 2012) as well as its underlying neuroscience (Avraamides & Kelly, 2008; Burgess, 2008;

Moser, Kropff, & Moser, 2008; Tommasi, Chiandetti, Pecchia, Sovrano, & Vallortigara, 2012). Although some of these reviews also mention the occasional computational model, no systematic review of computational models of spatial memory has been published in the last decade (note that Trullier, Wiener, Berthoz, & Meyer, 1997 have reviewed biologically based artificial navigation systems, and Mark, Freksa, Hirtle, Lloyd, & Tversky, 1999 published a review of models of geographical space). The main contributions of the current paper lie in providing a review of computational cognitive models of spatial memory (taking into account implemented models of cognition across disciplines, including psychology, neuroscience, and AI); providing a comparison of these models; reporting possible underlying neural correlates corresponding to parts of these models to aid comparison and verification; and finally outlining open questions relevant to this field which have not been fully addressed yet.

Spatial Memory and Representations

Biological agents such as mammals, as well as embodied autonomous agents, exist within spatially extended environments. Given that these environments contain objects relevant to the agent's survival, such as nutrients or other agents, they need to take the positions of these objects into account. The purpose of spatial memory is to encode, store, recognize and recall spatial information about the environment, and the objects and agents within it.

Spatial representations can be categorized based on the reference frame used. Egocentric representations represent spatial information relative to the agent's body or body parts. In contrast, allocentric representations represent spatial information relative to environmental landmarks or boundaries, independent of their relation to the agent. We will return to these types of representations, and the way they are encoded in mammalian brains, in Section 2.

In addition to navigation space–the space of potential travel–other forms of spatial representation have also been considered in the literature (e.g. representations of the positions of body parts or external representations such as maps or diagrams—Tversky, 2005).

In this review, we will focus on representations of navigation space and the space around the body, because the largest number of computational cognitive models account for them, and also because they are the most ubiquitous and generalizable representations. Whereas information concerning the space of the body strongly depends on the specific form of embodiment (such as body size and shape), and the use of external spatial representations is exclusive to humans, the types of representations and strategies required for navigation space are similar for different kinds of bodies and agents.

Relevance of Computational Cognitive Models to Spatial Memory Research

Computational models attempt to formally describe a part (or parts) of cognition in a simplified fashion, allowing their simulation on computers (McClelland, 2009; Sun, 2008b), and providing more detail, precision, and possibly more clarity than qualitative descriptions. In addition, computational models might facilitate the understanding and clarification of the implications of a theory or idea, in ways that would be difficult for humans without simulation on computers (McClelland, 2009). Since spatial memory is an interdisciplinary research area (drawing on at least psychology, neuroscience, and artificial intelligence), involving multiple representations and processes, it is especially important to formulate theories precisely, using a common language. Computational models can provide such a common ground.

The development of computational cognitive models also requires making a large number of design decisions, possibly leading to novel hypotheses, which can then be evaluated. This process usually constitutes an ongoing cycle of development, testing, and revision. Critically, most of this is performed on a computer and thus can be quick and efficient.

This efficiency is especially important for modeling mechanisms with representations that are not easily explicated or measured directly, such as in the case of spatial cognition. Humans cannot easily report the structure of their spatial representations and the mechanisms operating on them. There are a large number of structures and mechanisms that could partially account for spatial skills (e.g. navigation), and a time-efficient way of defining them, and investigating their implications in an automated fashion is important to facilitate the evaluation of their plausibility.

Once a theory or hypothesis has been encoded computationally, generating predictions from it is a straightforward matter of providing model parameters and input data, and running the model on a computer. This is usually more efficient than obtaining experimentally verifiable predictions from a verbal/conceptual theory. The predictions can subsequently be tested or verified using data obtained from empirical experiments with humans or animals, and comparing this data with the model predictions (usually employing some statistical measure of model fit; Pitt, Myung, & Zhang, 2002).

Once in possession of the empirical data, both the prediction and the testing can be performed by running computer programs. Since this process is automated, it takes little human effort. This is a general advantage of computationally formulated models, but is especially useful for spatial memory models, since experiments investigating spatial cognition using the classical, iterative cycle of hypothesis formulation, prediction derivation and testing (Godfrey-Smith, 2003) usually require multiple, sometimes large environments (especially for navigation-scale spatial memory), and are thus impractical and time consuming to perform in the real world. In contrast, computational cognitive models of spatial memory can be run in a large number of different simulated environments, with different parameterizations, over a short period of time and with little effort.

Motivation for the Proposed Neural Mappings

Since cognitive modeling is concerned with describing and explaining cognitive phenomena, they should behave the same way as humans (or animals) do. Comparison of model predictions with behavioral evidence, 'goodness of fit', is the most widespread quantitative method of evaluating, judging and comparing cognitive models (Pitt et al., 2002). In addition to fit, model complexity and generalizability can (and should) also be analyzed qualitatively. Frequently employed qualitative criteria include explanatory adequacy, interpretability, and biological plausibility or realism (Cassimatis, Bello, & Langley, 2008; Myung, Pitt, & Kim, 2005).

Despite these criteria, the space of models possibly accounting for experimental data is under-constrained. There can be multiple models of comparable complexity achieving comparable goodness of fit, and there might not be enough empirical data available for full evaluation. Furthermore, it is often difficult to compare cognitive models along qualitative dimensions. For example, there is no consensus on which models are biologically plausible (there are large differences between different approaches, ranging from spiking neural network models with parameters derived directly from electrophysiological measurements to AI-based methods described as 'biologically inspired' based on vague functional similarity). Many authors of cognitive models describe their work without establishing how parts of their model relate to the functionally similar biological implementation, making it difficult to judge the degree of correspondence to the brain.

Since cognition is implemented by the brain, cognitive modelers would do well to take into account the known neuronal mechanisms underlying the cognitive phenomena they are trying to model, even if not aiming to be highly biologically accurate. We will propose tentative neural mappings of the models reviewed in this paper for the following reasons. First, such mappings might help assess the biological realism of models claiming to be biologically plausible, based on the degree of structural and functional correspondence between models and the neural areas implementing the cognitive mechanisms they account for. Since cognition is implemented by the brain, close similarity between cognitive models and their neural counterparts is desirable (whether structural, functional, paradigmatic, or otherwise). Clarifying neuronal correspondence might also help provide an additional quantitative evaluation criterion, by facilitating possible future verification using neuronal data—such as imaging data from humans or electrophysiological data from animals.

Interestingly, such neuronal data can help in substantiating a model even if there is very little similarity between the elementary units of a model and the brain (as is the case with symbolic models, which usually employ local and amodal symbols for representations, as opposed to the distributed and grounded representations of the brain). A good example is the ACT-R cognitive architecture, which is primarily symbolic but nevertheless has been shown to be capable of not only fitting brain imaging data, but roughly predicting activation levels of brain areas

(Anderson, Fincham, Qin, & Stocco, 2008; Qin, Bothell, & Anderson, 2007). This shows that it is possible even for high-level cognitive models which have little to do with biological neurons to contribute to and guide research in neuroscience; and that results in neuroscience can guide the development and parameter adjustment of such models despite their structural differences. Thus, the mapping between model components and brain areas might be interesting even for neuroscientists uninterested in pure cognitive modeling, or cognitive modelers uninterested in pure neuroscience.

Finally, relating models and their components to brain areas with known functions can facilitate their explanation, especially for readers with a background in cognitive neuroscience or psychology. Such mappings also help clarify and explicate structural differences and similarities between individual cognitive models.

NEURAL CORRELATES OF SPATIAL REPRESENTATIONS

Since this review is targeted mainly at researchers in cognitive modeling, who might not be deeply familiar with the details of the neurophysiology of spatial memory and spatial cognition, we briefly summarize the neuroscientific literature concerning how mammalian brains represent navigation space.[1]

This section is intended to provide a basis for the neural mappings of model parts (to provide further plausibility constraints, an additional basis for comparisons, and a functional guide for model parts). Our descriptions of the neural correlates of spatial representations are biased toward describing areas known to be important and with (more or less) known functions, and are not meant to be a complete review of all brain areas related to spatial cognition. See Burgess (2008) and Moser et al. (2008) for more comprehensive reviews of spatial cognition in the brain, and Kravitz, Saleem, Baker, and Mishkin (2011) for an overview of areas associated with visuospatial processing.

Allocentric Spatial Memory

Four types of cells play an important role in processing allocentric spatial representations in the mammalian brain, established mostly through single-cell electrophysiological recording studies from mammals (the following list is based on Madl, Franklin, Chen, Montaldi, & Trappl, 2014)—see also Fig. 1:

1. **Grid cells** in the medial entorhinal cortex (MEC) show increased firing at multiple locations, regularly positioned in a grid across the environment consisting of equilateral triangles (Hafting, Fyhn, Molden, Moser, & Moser, 2005). Grids from neighboring cells share the same orientation, but have different and randomly distributed offsets, meaning that a small number of them can cover an entire environment. It has been suggested that grid cells play a major role in path integration (PI),[2] since their activation is updated depending on the animal's movement speed and direction (Burgess, 2008; Hafting et al., 2005; McNaughton, Battaglia, Jensen, Moser, & Moser, 2006). There is evidence to suggest that grid cells exist not only in mammals, but also in the human entorhinal cortex (EC) (Doeller, Barry, & Burgess, 2011). In contrast to MEC, neurons in the lateral EC exhibit little spatial modulation, and are instead highly selective to sensory stimuli.
2. **Head-direction cells** (HD cells) fire whenever the animal's head is pointing in a certain direction. The primary circuit responsible for head direction signals projects from the dorsal tegmental nucleus to the lateral mammillary nucleus, anterior thalamus and postsubiculum, terminating in the entorhinal cortex (Taube, 2007). There is evidence that head direction cells exist in the human brain within the medial parietal cortex (Baumann & Mattingley, 2010).
3. **Border cells** and **boundary vector cells** (BVCs) are cells with boundary related firing properties. The former (Lever, Burton, Jeewajee, O'Keefe, & Burgess, 2009; Solstad, Boccara, Kropff, Moser, & Moser, 2008) seem to fire in proximity to environment boundaries, whereas the firing of the latter (Barry et al., 2006; Burgess, 2008) depends on boundary proximity as well as direction relative to the mammal's head. Cells with these properties have been found in the mammalian subiculum and entorhinal cortex (Lever et al., 2009; Solstad et al., 2008),

and there is also some behavioral evidence substantiating their existence in humans (Barry et al., 2006).

4. **Place cells** are pyramidal cells in the hippocampus which exhibit strongly increased firing when the animal is in specific spatial locations, largely independent from orientation in open environments (Burgess, 2008; O'Keefe & Dostrovsky, 1971), thus providing a representation of an animal's (or human's Ekstrom et al., 2003) location in the environment. A possible explanation for the formation of place cell firing fields is that they emerge from a combination of grid cell inputs on different scales (Moser et al., 2008; Solstad, Moser, & Einevoll, 2006). It has also been proposed that place fields might be mainly driven by environmental geometry, arising from a sum of boundary vector cell inputs (Barry et al., 2006; Hartley, Burgess, Lever, Cacucci, & O'Keefe, 2000); or by a combination of grid cell and boundary vector cell inputs (Madl et al., 2014). Apart from information about the current spatial location, hippocampal place cells also participate in place–object associations (Kim, Delcasso, & Lee, 2011; Manns & Eichenbaum, 2009), associating place cell representations of specific locations with the representations of specific objects in recognition memory (the perirhinal cortex, among others, is heavily involved in recognition memory for objects—Brown & Aggleton, 2001; Yonelinas, Otten, Shaw, & Rugg, 2005). In addition, in the primate hippocampus, view-dependent instead of place-dependent cells have also been identified (dubbed spatial-view cells Rolls & Xiang, 2006). Finally, an interesting cell type with spatially localized firing activity has been found in the medial prefrontal cortex (mPFC), representing **goal** or **reward** locations (Hok, Save, Lenck-Santini, & Poucet, 2005).

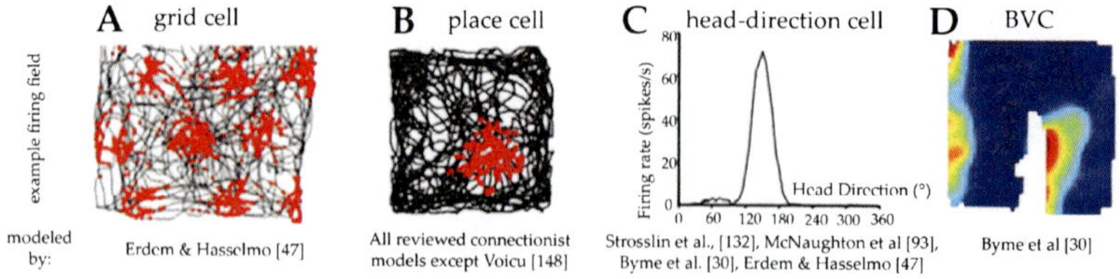

Figure. 1. Grid cells, place cells, boundary-related cells, head-direction cells, and the neuronal basis of self-motion information. A.–D.: Four cell type firing fields associated with allocentric spatial representation; as well as reviewed models accounting for them. A. Regular grid cell firing pattern from rat intracranial recording (black lines: rat trajectory, red dots: places where grid cell showed increased firing). B. Hippocampal place cell firing pattern (A and B from Burgess, 2008). C. Firing pattern of a head-direction cell tuned to about 150 allocentric direction (relative to distal landmarks or boundaries). D. Firing fields of 'boundary vector cells' identified in the rat entorhinal cortex. In specific areas of the environment (highlighted with hot colors) these cells exhibit increased firing rates (from Solstad et al., 2008). (For interpretation of the references to color in this figure legend, the reader is referred to the web version of this article.)

Hippocampal place cells seem to encode long-term allocentric spatial representations of environments (this is suggested by the spatially localized firing of place cells, the observation that this firing did not depend on heading direction and remains stable in an environment for several weeks, and finally the associations between place cells and specific objects). It has been argued that multiple such representations are learned for different environments, with different frames of reference and on different scales. Evidence for this includes the observation that place cells 're-map' when rats enter a new environment (the firing fields of the same cells reflect a completely different map in different environments), and the observation that their firing field sizes can significantly differ (Derdikman & Moser, 2010).

Allocentric representations allow not only the storage and subsequent recall of remembered routes, they also allow the calculating of novel routes, shortcuts or detours (important especially after changes in the environment, e.g. when a known route is blocked). Furthermore, it is possible to keep track of more allocentrically encoded object positions than egocentric positions—since the latter are encoded relative to the agent and thus require updates as the agent moves through the environment, making accurate egocentric representations of large numbers of objects intractable.

Such allocentric representations of physical locations in the environment have been called 'cognitive maps'–a term coined by Tolman (1948)–and there is substantial evidence that the hippocampal–entorhinal complex is the main neural correlate involved in their storage and recall (Moser et al., 2008).

Another proposed form of allocentric representation is a topological map. Topological maps lack metric information (such as distances or directions), but provide adjacency and containment information and thus allow route planning as well (although planning optimal routes can be difficult) (Booij, Terwijn, Zivkovic, & Krose, 2007). There is no well-established neural correlate of possible topological representations in the brain; although computational models with topological assumptions have successfully accounted for some hippocampal experimental data (Chen, Kloosterman, Brown, & Wilson, 2012; Dabaghian, Cohn, & Frank, 2011) (and there is some neural evidence for the involvement of posterior parietal cortex (PPC) Calton & Taube, 2009 and retrosplenial cortex (RSC) Epstein, 2008).

Egocentric Spatial Memory

For humans and primates, vision is the primary perceptual modality, having the largest cortical area associated with its processing. There are multiple pathways originating from the visual cortices. Apart from a pathway supporting object vision along ventral areas (the 'what' pathway), two others have been proposed which are relevant for spatial memory.

The primary visual cortex (V1) located in the occipital lobe projects visual information through higher visual cortices to the Posterior Parietal Cortex (PPC). The parieto-medial temporal pathway connects this occipito-parietal circuit with areas in the medial temporal lobe including hippocampal, entorhinal and subicular areas involved in processing long-term allocentric spatial representations supporting spatial navigation (see above) (Kravitz et al., 2011).

On the other hand, many brain areas involved in the representation of egocentric space reside in the posterior parietal cortex. Posterior parietal areas can be said to extract object positions relative to the agent from sensory information. Patients with parietal lesions might have intact primary sensory and motor representations, but often suffer from spatial neglect—they are unable to perceive one side of space (Husain, 2008).

Evidence suggests that the **precuneus** is the main brain area concerned with multiple types of egocentric representations, as well as transformations between them (Kravitz et al., 2011; Vogeley et al., 2004; Zaehle et al., 2007). The precuneus seems to coordinate spatial processing in the reference frames of the eyes and the head with controlling body and limb-centered actions (in addition to the intraparietal and postcentral sulci and the parieto-occipital region; Plank, 2009; Vogeley et al., 2004)—for example, area 5d within this parietal area seems to represent reach vectors (hand position relative to reach target).

Neuropsychological studies have also implicated the lateral intraparietal area (LIP) in representing visual stimuli in the reference frame of the body (Snyder, Grieve, Brotchie, & Andersen, 1998), the ventral intraparietal area (VIP) containing receptive fields with head-centered reference frames Duhamel, Colby, and Goldberg (1998), the medial intraparietal area (MIP) in the encoding of object locations in eye-centric coordinates (Pesaran, Nelson, & Andersen, 2006), and area 6a (Marzocchi, Breveglieri, Galletti, & Fattori, 2008). The latter two areas have also been called the 'parietal reach region' and seem to encode the location of reach targets in an eye-centered reference frame (Bhattacharyya, Musallam, & Andersen, 2009). (See Kravitz et al., 2011 for a detailed review of visuospatial processing in the brain.)

Finally, the retrosplenial cortex (RSC) and the parahippocampal place area (PPA) in the parahippocampal cortex both seem to be involved in the visual representations of places, since they respond strongly to scenes such as landscapes or cityscapes but weakly to non-scene objects (such as animals or small objects).

Apart from visuospatial representations, the basal ganglia also play an important role in egocentric navigation, and are thought to associate a cue with a reward (Packard & McGaugh, 1996), triggering guidance behavior along a known route. The basal ganglia can thus encode the body turns/directions to take when landmarks are recognized, depending on the spatial relationship between the landmark and the body (e.g. turn left at the big tree). This encoding allows navigation based on simple associations between actions and egocentric spatial relations (also called 'taxon navigation', as opposed to 'locale navigation' which requires allocentric spatial representations). This taxon strategy seems to be in use mainly when a route is well-known (Hartley, Maguire, Spiers, & Burgess, 2003). In contrast, novel route planning requires additional allocentric representations (see previous section).

Structures Involved in Transformation

Since sensory information is perceived from the reference frame of the observing agent, allocentric spatial representations must be built via transformation of the sensory input. Furthermore, allocentric information has to be transferred back into an egocentric reference frame in order to allow spatial actions.

Because of its interconnections with brain areas associated with both egocentric and allocentric spatial representations, it has been suggested that the **RSC** is involved with translations of frames of reference. The RSC receives direct inputs from visual areas V2 and V4, and egocentric sensory information from parietal areas 7a and LIP, among others; as well as inputs from the hippocampal formation and the anterior thalamus usually associated with allocentric position and heading information (Vann, Aggleton, & Maguire, 2009).

Area 7a in the posterior parietal cortex is another area strongly connected to both the medial temporal areas associated with allocentric representations, and the parietal areas associated with egocentric representations. Thus, area 7a could also play a role in transforming between reference frames. For example, neurons in area 7a can transform viewer- to object-centered spatial information (Byrne, Becker, & Burgess, 2007; Crowe, Averbeck, & Chafee, 2008).

Structures Involved In Associative and Reward-Based Learning

Hebb's rule is a prevalent and frequently modeled associative learning rule, which is based on the idea of activity-dependent synaptic modification, and proposing that a change in the strength of a connection is a function of the neural activities of the connected neurons. Hebbian learning is often summarized as 'neurons that fire together, wire together'. There is strong empirical evidence for such a learning mechanism ubiquitously occurring in brains (Song, Miller, & Abbott, 2000). This learning rule is critical for associative learning in spatial memory paradigms—for example, for learning associations between the representation of a rat's current location, and sensory stimuli at that location. In a variant of Hebbian learning, called competitive learning, neurons of one population compete with each other to respond to the pattern appearing in another population from which they receive input (the more strongly a neuron responds to the input, the more it inhibits other neurons, and the more its connection strengths to highly active input neurons increase) (Grossberg, 1987; Kaski & Kohonen, 1994; Rumelhart & Zipser, 1985).

As opposed to the unsupervised, associative Hebbian learning rule, reward-based learning is also frequently observed in spatial memory experiments. As mentioned above, the mPFC seems to be involved in representing goal or reward locations (Hok et al., 2005) (and has also been suggested to be involved in responding to rewards). Animals including humans have a propensity to seek out rewards, and are able to learn the spatial locations of such rewards. The primary neural correlates of reward-learning include the orbitofrontal cortex (OFC, which seems to encode stimulus reward value), the amygdala, and the ventral striatum; all three show increased activity during the expectation of a reward. The dopamine system also plays an important role, being involved in the signaling of error in the prediction of reward (presumably aiding learning and facilitating the improvement of reward predictions). To select an action based on an expected reward, stimulus–response or response–reward associations have to be learned; empirical evidence implicates the dorsal striatum in this process (which exhibits increased activity when a contingency is established between responses and reward) (Maia, 2009; O'Doherty, 2004).

Reinforcement learning theory has been used in attempts to mathematically formalize the process of reward-based learning through interacting with an environment. Reinforcement learning (RL) agents represent the world as a set of states S, a set of actions A possible in each state and leading to a new state, and possible rewards r. They learn from the consequences of their actions, and try to select actions based on past experiences (exploitation) as well as novel choices (exploration). The name comes from the reinforcement signal–a numerical reward–used in such models; RL agents aim to choose actions that maximize the reward they obtain over time (Woergoetter & Porr, 2008). It has been argued that mathematically derived solutions to RL can plausibly be implemented in brains, based on the reward-relevant brain areas listed in the previous section (explaining RL and its correspondence to brains would exceed the scope of this paper; see Maia, 2009 for an explanation and review of evidence). Reinforcement learning can be used to learn which action to take in each location, e.g. to learn how to navigate to a food source (see also Fig. 4 and some of the models reviewed below).

COMPUTATIONAL COGNITIVE MODELS OF SPATIAL MEMORY

Introduction

Computational models attempt to formally describe an aspect (or aspects) of cognition in a simplified fashion, allowing their simulation on computers (McClelland, 2009; Sun, 2008b). Computational cognitive modeling is concerned with achieving a better understanding of various cognitive functionalities through computational models of representations, mechanisms, and processes.

Cognitive models should be functional—they should perform well at the task they were designed for (which can be difficult, especially for challenging tasks such as trying to robustly map real-world environments).

Psychological or cognitive plausibility are also important—these models aim to model cognitive phenomena (spatial memory and associated processes), and should correspond to them as closely as possible in terms of the mechanisms, processes, and representations employed, and behavioral measures produced. They should account for empirical data as well as possible (high 'goodness of fit'), and should do so in the simplest possible way (low complexity), making as few unsubstantiated assumptions as possible. They should also have the ability to generalize to new data, not only account for the data provided to the model during development and training (Myung et al., 2005).

Cognitive models should also be as biologically plausible as possible within their paradigm. Although many cognitive models are not concerned with the physiological details of neural functioning (with the exception of biological/spiking neural networks—see Section 3.4), the underlying neuroscience of the modeled cognitive phenomena is nevertheless arguably relevant. Functions of the mind are implemented in brains; thus, neuroscience can provide valuable input regarding the structure and function of plausible models, even for those not intending to model the neuron level. Further advantages of taking neural implementation into consideration include constraining the model space (reducing the large number of algorithms possibly accounting for given behavior data), providing additional evaluation criteria, and facilitating model comparison by establishing analogies between representations in models and in brains (see also Section 1.3).

Clarification of the elemental units used by models, and how they relate to neural substrate, is critical in evaluating biological plausibility. The correspondence does not need to be on the neuron level—symbolic cognitive models can also structurally correspond to brains on a higher level (e.g. on the level of brain areas and their connectivity). Explicating the correspondence between model components and brain areas, as done by the researchers of ACT-R (Anderson et al., 2008) (who have also performed brain imaging experiments for validation), helps to verify structural similarity between the model and the corresponding neural substrate, and thus also to evaluate claims of biological plausibility. Describing such neural mappings is one of the aims of this section, as well as establishing tentative mappings based on functional correspondence in cases where the authors did not explicitly describe them in their work, as is the case for the majority of models outside of computational neuroscience.

Clarification of the following properties is also important in characterizing computational models of spatial memory (partially based on O'Reilly, 1998 and Webb, 2001[3]).

- The level of modeling (characterizing the elemental units),
- The types of representation accounted for (e.g. egocentric, allocentric, metric, topological)
- The learning mechanism, if any (e.g. Hebbian learning, reinforcement learning)
- The generality and abstraction of the models (the range of phenomena accounted for, and complexity relative to the modeled phenomena)
- Structural similarity (how well models represent the underlying neural mechanisms)
- Performance match or 'goodness of fit' to behavior data (to what extent the model can match target behavior; useful for comparing different models of the same phenomena).

It is important to note that this review is limited to computational models of cognition concerned with navigation space, that were published in the last two decades,[4] and as such excludes models of diagrammatic spatial reasoning, models of low-level sensory representations, robotic models unconcerned with biological cognition, and other models which might include spatial information on a different scale or for a different purpose. Furthermore, we exclude reactive navigation models without representations, which might allow agents to solve problems in space, but cannot be said to model spatial memory.

Finally, we do not claim to review every single model involving spatial memory (such an endeavor could fill a book); the aim of this review is to summarize representative models for major modeling directions (of any set of models which are highly similar in terms of paradigm, structure and functionality, only the most recent one is reviewed; similarly, if the same first author publishes multiple times on a model, only the most recent version of the model is included).

Overview

The spatial memory models reviewed in this section are divided into three categories, inspired by major modeling paradigms in the field of computational psychology (Sun, 2008a). The section 'symbolic spatial memory models' describes models emphasizing explicit rules and localist representations based on symbolic logic (Bringsjord, 2008). In contrast, 'neural network-based spatial memory models' are based on a number of simple processing units affecting each other via weighted connections, operate in parallel, usually employ distributed representations, and commonly learn rules from training data instead of encoding explicit rules (Thomas & McClelland, 2008). Finally, we also review a number of spatial memory models that are a part of cognitive architectures (which are concerned with modeling a wide range of cognitive phenomena in addition to spatial memory, and are often employing a combination of the mentioned paradigms).

We have confined our survey to these relevant categories and model types to keep it within the limited space available.

Each of these types of models have different strengths and roles in modeling and understanding spatial memory. Symbolic models operate on a high-level of abstraction (they are usually not concerned with neuron level phenomena), and are often functionally more powerful than neural network models (they can often perform more complex tasks). They usually have less structural similarity to brains, and are thus less constrained (even if validated against behavior data, it is difficult to evaluate multiple symbolic models performing a similar task with comparable goodness-of-fit). In contrast, neural network models are often more similar to the neurophysiological implementation (both in terms of representation and mechanism) and are thus easier to constrain by established neuroscientific knowledge and by additional types of data (such as neural recordings or brain imaging). However, this paradigm often makes it difficult to implement complex cognitive processes, especially those requiring serial processing steps (for example, none of the neural network models are able to perform spatial reasoning or loop closure, in contrast to some symbolic approaches). Finally, cognitive architectures can follow either or both of these paradigms, and have the additional advantage of incorporating multiple cognitive mechanisms—thus, they can perform, and be evaluated against, different tasks and datasets.

Apart from categorizing the models based on their underlying modeling paradigm, we will also group them into models evaluated in simplified, simulated environments, and into models which are capable of dealing with–and being evaluated in–real world environments (such as robotic implementations). In general, robotics emphasizes high-performance solutions to low-level 'sensor problems' (e.g. dealing with sensory uncertainty/noise or processing or recognizing complex sensory data), and aims for high performance (accuracy, efficiency, etc.) instead of cognitive plausibility (Jefferies & Yeap, 2008). However, as Gallistel (2008) points out, the nature of the computational problems of navigation and map making based on limited information does not depend on whether one is studying biological or artificial systems. Thus, the latter could help in understanding the former.

Robots and animals must perform similar computations when trying to make sense of space. Computational models of cognition operating in similar environments to the modeled biological agent, and dealing with similar difficulties posed by the real world (such as complexity, limited knowledge, uncertainty, or noise), can be regarded as being more plausible than models not accounting for such difficulties (Webb, 2000). This is the main motivation for dedicating subsections to cognitive models evaluated in the real-world (but excluding systems concerned with practical robot performance rather than investigating cognition).

The following list presents an overview of the models reviewed below. Models embodied on robots capable of running in the real world are printed in bold, and, for clarity, the first mention of a model in each subsection below is underlined. A comparative table of all reviewed models, with additional properties for comparison, can be found at the end of this section (Table 1).

- Symbolic models (Section 3.3)
  - Allocentric models
    - (Yeap, Wong, & Schmidt, 2008)
    - (Jefferies, Baker, & Weng, 2008)
    - perceptual wayfinding model (Raubal, 2001)
  - Egocentric models
    - NAVIGATOR (Gopal & Smith, 1990)
  - Allocentric + egocentric
    - **HSSH** (Beeson, Modayil, & Kuipers, 2010)
    - (Franz, Stürzl, Hübner, & Mallot, 2008)
    - DP-model (Brom, Vyhnánek, Lukavský, Waller, & Kadlec, 2012)

- Neural network-based models (Section 3.4)
  - Allocentric models
    * (Burgess, Jackson, Hartley, & O'Keefe, 2000)
      - (later extended in simulation as the BVC model by Barry et al., 2006)
    * (Strösslin, Sheynikhovich, Chavarriaga, & Gerstner, 2005)
    * (Barrera, Cáceres, Weitzenfeld, & Ramirez-Amaya, 2011)
    * (Schölkopf & Mallot, 1995)
    * (Voicu, 2003)
    * (McNaughton et al., 1996)
    * (Erdem & Hasselmo, 2012)
  - Allocentric + egocentric
    * (Byrne et al., 2007)
- Cognitive Architectures (Section 3.5)
  - Allocentric models
    * LIDA (Madl, Franklin, Chen, & Trappl, 2013)
  - Egocentric models
    * ACT-R/S (Harrison et al., 2003)
    * CLARION (Sun & Zhang, 2004)
  - Allocentric + egocentric
    * Casimir (Schultheis & Barkowsky, 2011)

Table 1. Characteristics of the reviewed models. The table consists of seven columns, showing model name and citation, the elemental spatial position representation, reference frames accounted for (Ref.), learning mechanisms, if any (Learn.), structural similarity to corresponding brain areas (Sim.), complexity of the environment the model is able to operate in (C.), and tasks in which the model has been evaluated. The following further abbreviations are used: • Superscripts in the model name denote whether they are symbolic (s), neural network-based (n), a combination of the two (hybrid—h) (necessary in the case of cognitive architectures). • In the reference frames accounted for: ego...egocentric, allo...allocentric, vs...visuo-spatial, +...ego + allo, *...all three (superscripts denote whether the maps are t...topological, m...metric, n...metric but non-Euclidean, o...containing orientation information) • In the learning mechanism: prob...probabilistic, det...deterministic, SLAM...probabilistic Simultaneous Localization and Mapping, Hebb…Hebbian, RL…reinforcement learning • In the structural similarity: numbers range from 5—strong similarity to 1—no clear similarity (5…biological ANN or SNN evaluated against neural data, 4…biological ANN or SNN, 3…non-biological ANN, 2…symbolic mechanism with clear correspondences to modeled biological mechanism, 1…no clear structural similarity) • In the complexity of the environment: numbers range from 5—highly complex to 1—simple (5…large-scale real-world env., 4…small real-world env. mostly within agent's sensory horizon, 3…simulation with multiple objects or obstacles, 2…simulation with no objects or obstacles, 1…finite number of discrete states) • In the tasks in which the model has been evaluated: brief task names are used; they are further described in the text (the superscript denotes the type of data compared against: h…human behavior, a…animal behavior, n…animal neural data, q…no quantitative, only qualitative comparison against human behavior data, and -…no biological or behavior data, only functional evaluation).

		Name/citation	Position repr.	Ref.	Learn.	Sim.	C.	Evaluation tasks
Symbolic	*Real-world*	Yeap et al. (2008) [based on ASRs by Yeap, 1988]	Boundary element triplets	allom	det	1	5	Mapping$^-$ Homing$^-$
		Jefferies et al. (2008) [based on ASRs by Yeap, 1988]	Boundary element triplets	allom+t	det	1	5	Globalmapping$^-$ Localmapping$^-$
		HSSH Beeson et al. (2010)	Occupancy grids, topo-logical places & paths	+m+t	SLAM	2	5	Global&localmaps$^-$ Pathplanning$^-$
		Franz et al. (2008)	Disparity signatures	+m+t	det	1	4	Mapping$^-$,homing$^-$ Shortcuts$^-$
	Simulated	NAVIGATOR Gopal and Smith (1990)	Semantic networks (spatial + non-spatial info)	egon	det	1	1	Wayfinding errorsq
		perceptual wayfinding model Raubal (2001)	Topological graph (from fully observable env.)	allot	none	1	1	Wayfinding$^-$
		DP-model Brom et al. (2012)	Weighted ego + allo vectors	+m	det	1	3	Pointing accuracyh

Neural network-based	*Real-world*	Burgess et al. (2000)	Place & goal cells	allom	Hebbian	4	4	Navigation−[h]a
		[later extended in simulation as the BVC model Barry et al., 2006]			Competitive	[5a]		Place fieldsq[hn]a
		Strösslin et al. (2005)	'Step', place & HD cells	allom	Hebbian	4	4	Navigation−
		[based on Arleo & Gerstner, 2000]			RL			Map learning−
		Barrera et al. (2011)	Place cellsm	allom+t	Hebbian	4	5	Navigationa
		[based on Barrera & Weitzen-feld, 2008]	Actor unitst		RL			Map learning−
	Simulated	Schölkopf and Mallot (1995)	Place specific SOM units	allot	Self-organized	3	1	Navigation−
								Mapping(discrete maze)−
		Voicu (2003)	Hierarchical SOM units	allom	Self-organized	3	3	Map learningh
		[extending Voicu & Schmajuk, 2000]						Distance judgmentsh
								Navigation & detour planning−
		McNaughton et al. (1996)	Place & HD cells	allom	Hebbian	5	2	Path integration−
		[evaluated by Samsonovich & McNaughton, 1997]						PF learning & stretchingn
								Slow rotationn
		Byrne et al. (2007)	BVCs, place & HD cells	+m	Hebbian	4	3	Mapping−,PI−,HD lesionsq
		[based on the BVC mod-el Barry et al., 2006]						Hemispheric neglectq
		Erdem and Hasselmo (2012)	Recency, topology, reward, Place, HD & grid cells	allom+t	Hebbian	5	3	Mapping−,PI−,navigation−
		[based on Hasselmo, 2008]						T-Maze,'look-ahead'n

Cog. archi-tectures		ACT-R/Ss	Egocentric vectors in 'configural chunks'	egom	det (but prob. retrieval)	1	3	Map learning–,PI–,navigation–
		Harrison et al. (2003)						
		Casimirh	'Spatio-analogical frag-ments'	*m+t	det	1	3	Sp. reasoningh,reinterpretationh
		Schultheis and Barkowsky (2011)	(semantic network-like rep.)					Mental scanningh&recall effectsh
		CLARIONh	Chunks (explicit, sym-bolic)	egom	RL	3	3	Complex
		Sun and Zhang (2004)	ANN (implicit)		Backprop			minefield navigationh
		LIDAh	'place nodes' in PAM	allom	det	3	3	Map learning–,navigation (TSP)h
		Madl et al. (2013)						Map errorsh,WM capacityh

aThe later extension was substantiated against neural as well as behavior data (Barry et al., 2006), however, it was not implemented in a real-world robot.

Symbolic Spatial Memory Models

Symbolic models of spatial memory are concerned with explicitly representing spatial knowledge in a declarative form as facts and rules. They are based on the assumption that cognition consists of discrete mental states (representations), which can be modeled as localist symbols (in contrast, in neural network-based models the representations are not discrete, but constitute distributed and potentially overlapping patterns of activation—see next section). A number of processes operate on these representations, creating, modifying, or deleting them (Smolensky, 1987). One of the earliest definitions of such symbolic models has been put forth by Newell and Simon (1976), coining the term of a 'physical symbol system', a class of systems having symbols, being capable of manipulating them, and being realizable within our physical universe.

Symbolic models are often based on cognitive science theories (most frequently information processing models), and thus are able to claim a degree of cognitive plausibility. There is usually very little similarity between the elementary representations of symbolic models and biological neurons (mainly because of the choice of localist and amodal representations, in contrast to the distributed and more modal representations of the brain; Barsalou, 2008; Martin & Chao, 2001). However, they can still correspond to the brain on a higher level (e.g. functional correspondence to brain areas, as established for ACT-R). Despite the structural and paradigmatic difference, and for reasons mentioned in the Introduction, brain areas corresponding to model parts will be pointed out based on such functional correspondences where applicable.

Models Evaluated in Real-World Environments

A few cognitive models of spatial memory have been implemented in robotic systems capable of navigating in the real world. Jefferies and Yeap (2008) provides a survey of such cognitive mapping approaches that have been designed to work on robots. Usually, robotic implementations following the symbolic[5] approach build metric representations of the local environment using an approach called SLAM (Simultaneous Localization and Mapping). Recent SLAM approaches are capable of recognizing a place the robot has seen before (this is called 'loop closing'), and correcting errors in the map representation by exploiting and correcting for the difference between expected and observed location on the map. SLAM is usually implemented by a probabilistic state estimation method, integrating self-motion information and landmark observations in a statistically optimal fashion (Jefferies & Yeap, 2008; Thrun & Leonard, 2008).

The core ideas of SLAM–using probabilistic inference to deal with uncertainty and noise and to infer near-optimal estimates of the locations of the agent, and objects in its environment–do not contradict the cognitive sciences. They fit in well with the recent 'Bayesian brain' hypothesis (Knill & Pouget, 2004); the idea that the brain integrates information in a statistically optimal fashion. There is evidence that spatial cues might be integrated statistically optimally in humans (Nardini, Jones, Bedford, & Braddick, 2008) and animals (Cheng, Shettleworth, Huttenlocher, & Rieser, 2007) on the behavioral level. Computational models resembling SLAM–using probabilistic state estimation–have been proposed to explain spatial orientation and cognitive mapping (Cheung, Ball, Milford, Wyeth, & Wiles, 2012; Fox & Prescott, 2010). It has also been suggested that hippocampal place cells might be able to perform approximate Bayesian inference on the neuronal level, based on electrophysiological recording evidence (Madl et al., 2014).

However, the representation implementation is highly important for judging the plausibility of such probabilistic models (e.g. in terms of their structural accuracy, and levels of abstraction and modeling). In SLAM approaches in robotics, maps are stored in different ways, most commonly as covariance matrices, or as occupancy grids (two-dimensional matrices with entries storing the probability of occupancy), or tree-based representations (Thrun & Leonard, 2008). In the absence of psychological or neuroscientific data substantiating the existence of explicit covariance representations in human or animal cognition, and of biologically realistic implementations, it would be difficult to argue for the plausibility of covariance matrices as cognitive models of spatial memory. Here we only include models where authors explicitly address the relationship of their models to cognitive science or neurobiology (unfortunately, although citing empirical evidence, few of these authors evaluate their models against empirical data from humans or animals). For reviews of robotic SLAM, see e.g. Bailey and Durrant-Whyte (2006), Durrant-Whyte and Bailey (2006) and Thrun and Leonard (2008).

Building on work by Yeap (1988)–one of the first symbolic computational models of cognitive maps–a number of robotic systems have been built (many of which have departed from the original claim of being computational theories of cognitive maps and will therefore be omitted). Yeap suggests the computation of abstract allocentric maps of a region (from the shape and disposition of surfaces/boundaries relative to the agent) which the author calls 'absolute space representation' or ASR (see Fig. 2(A)). Multiple ASRs can be interconnected as a traversable network to form a cognitive map of the entire environment, and afford the notion of 'places'; a network of ASRs can model a network of places, with exits leading from one to the other, such as rooms in a building. The elemental representation in ASRs is a list of triplets, each representing a boundary element (BE), and containing its size, angle to the next adjacent BE, and whether it is empty space, not empty, or occluded.

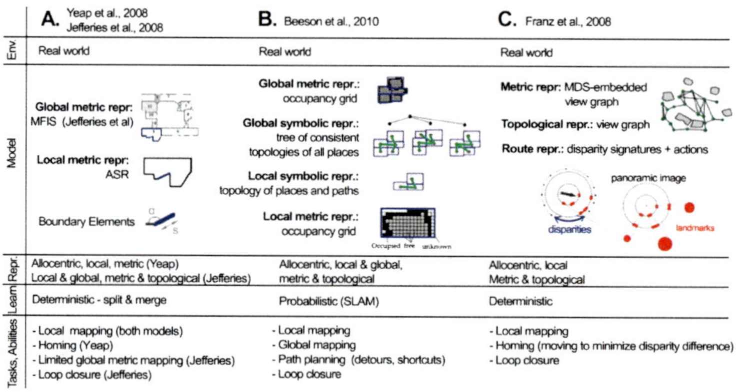

Figure. 2. Overview of symbolic models evaluated in real-world environments. A: (Jefferies et al., 2008; Yeap et al., 2008); both models create local metric maps (absolute space representations–ASRs–consisting of boundary elements); the latter model also builds a global metric map (Memory for Immediate Surroundings—MFIS) with which it can perform loop closing. B: HSSH (Beeson et al., 2010). C: (Franz et al., 2008). *Deterministic learning* is a collective term for all mechanisms that learn by adding new symbolic representations to memory upon perceiving a new object (as opposed to probabilistic or neural network learning mechanisms). *Local maps*represent spaces appearing to enclose the agent (such as a room). *Global maps* can represent and align multiple local maps in the same reference frame.

Based on this model, Yeap et al. (2008) developed a robotic system capable of building allocentric maps. The robot uses a simple exploration strategy (move forward in a straight line, stop when encountering an obstacle, turn away from the obstacle but maintain forward direction), after which it has to find its way 'home' (back to its starting location).

It used 8 simple sonar sensors to measure distances to obstacles and boundaries, and built a metric map based on both the robot path, and linear surfaces around it approximated from sonar data. This map was subsequently split, or merged, into distinct regions (e.g. corridors and rooms) using features such as average width (e.g. corridors are long and narrow), and employing the split and merge algorithm (Pavlidis & Horowitz, 1974) to find continuous regions. Each continuous motion segment of the robot (without stopping or turning) was represented as an ASR (consisting of multiple boundary elements from sonar data). The robot was able to use the final network of ASRs it has built using the split and merge algorithm to find its way back 'home' by backtracking the distances traveled.

The robot could localize itself using 'confidence maps' computed from the similarity between the currently perceived region or ASR (current sonar readings), and all stored ASRs. The authors reported that the localization was accurate with respect to the occupied region (i.e. the error was smaller than the size the regions). The robot could also robustly estimate a homing vector and return to its starting position even when the ASRs computed during the outward and inward journey were inconsistent—not requiring correct and consistent metric representations for homing is the main strength of the model. However, it could not match re-observed boundaries with those in its memory, and thus it is unable to 'close the loop'.

The model does not make any claims of structural accuracy with regard to its neurobiological equivalent. Based on functional similarity, ASR regions contain some of the information represented by place cells ('confidence maps' on ASR regions, similarly to place cells, carry information regarding the currently occupied space), goal cells (ASRs regions, like goal cells, can constitute goal representations) and by boundary vector cells (boundary elements carry boundary size and angle information).

Also drawing on the ideas of Jefferies et al. (2008) and Yeap (1988) proposed that a cognitive map might consist of a topological global map containing metric local space representations, aiming to benefit from the advantages of both—simple localization and metric consistency of the local maps, and easier 'loop closing' with the help of global maps (as well as the confinement of location errors to the local maps). The idea of separate local and global representations is consistent with most empirical cognitive science research (Hirtle & Jonides, 1985; Poucet, 1993; although there is some debate regarding whether and which mechanisms/areas are metric or topological). In contrast to the previously described model, the robot by Jefferies et al. (2008) used laser rangefinders as sensors (which provide more accuracy and resolution than simple sonar sensors).

Their approach turns the raw laser data (distance measurements) into lines representing boundaries, finds the exits (gaps in the boundary), and then computes ASRs based on this information. Different ASRs representing local regions can subsequently be connected topologically via the identified exits to form a global map.

Finally, with the help of this topological map, they also build a global map of limited extent containing the last few local spaces visited (called 'Memory for the Immediate Surroundings', MFIS), providing easier recognition that the robot has re-entered a previously observed part of the environment (loop closing). This model is one of only two reviewed models capable of building a global map and of loop closing (the other being Beeson et al., 2010—see below).

The authors argue for the psychological plausibility of their model using the empirical evidence for local and global spatial maps (Poucet, 1993) and multiple reference frames for different parts of an environment (Derdikman & Moser, 2010; McNaughton et al., 2006).

The model does not aim for structural resemblance to the brain. As it is also based on Yeap (1988), tentative arguments of functional similarity can be made between ASR regions and place cells, and boundary elements and boundary vector cells. No equivalent of a consistent, metric, global map has been found in the brain (the same place cells participate in representing very different locations in different environments; there is no one-to-one mapping as in the MFIS).

- Beeson et al. (2010) also propose a spatial memory model combining the strengths of topological and metrical approaches, calling it *HSSH* (Hybrid Spatial Semantic Hierarchy), an extension of the SSH model proposed by Kuipers (2000). The HSSH has four major levels of representation: a local metrical level (in which the agent builds a metric Local Perceptual Map—LPM), a local topological level (in which the agent identifies discrete places in a large-scale environment and describes paths in it), a global topological level (for resolving structural ambiguities and determining how the environment is best described as a graph of places, paths and regions), and a global metrical level (describing the environment in a single metric global map using the same reference frame).

On the first level, the LPM is built using probabilistic SLAM (Thrun & Leonard, 2008) based on laser rangefinders, and represented as an occupancy grid (a discretized grid in which each cell contains the probability of being occupied by an obstacle). On the local topological level, a discrete set of 'places' and 'path segments' connecting them are identified (using an approach based on Voronoi graphs and recognized gateways/doors). The global topological map is built by creating a tree of all possible topological maps (map hypotheses) consistent with current experience. After each travel action, every map hypothesis is extended; if it leads to a predicted transition to a known state, the hypothesis can be updated or refuted based on the subsequent observation. This allows 'closing the loop' and pruning the tree of topological maps when places are revisited.

Finally, on the global metrical level, a metric map of the entire environment in a global reference frame can be assembled on the structural skeleton provided by the global topological map (and based on the known robot trajectory and the displacements between places to appropriately translate local frames of reference). HSSH is the only model except for Jefferies et al. (2008) capable of building a global map and closing the loop.

Although not aiming for structural similarity to the brain, the HSSH and its predecessors claim to be 'theories of robot and human commonsense knowledge of large-scale space: the cognitive map' (Kuipers, 2008).

Unfortunately, no comparisons of the model's performance with human data have been performed. In terms of functional similarity, the occupancy grid employed as the low-level metric representation bears some resemblance to hippocampal place cells, as both can be used to infer the most likely location of the agent, as well as the most likely locations of boundaries (Barry et al., 2006). However, there are also significant differences, including the resolution (1 cm in some SLAM approaches, as opposed to the sizes of place cell firing fields,[6] which range from 20 cm or less to multiple meters, Kjelstrup et al., 2008; O'Keefe & Burgess, 1996), constancy (place fields can be destroyed or changed by adding barriers or making other changes in the environment), shape properties (occupancy grid cells are square, place cells can have multiple firing fields of different round shapes), representation (occupancy grid cells contain probabilities, place cell firing rates almost certainly do not, since they strongly depend on factors such as running speed), among others (Moser et al., 2008). Independently of the differences in the representation employed, probabilistic inference (the mechanism which SLAM is based on) has been argued to be plausible based on empirical data (see above).

- Franz et al. (2008) have developed a robotic system on a Khepera miniature robot that accounts for egocentric route navigation, as well as allocentric topological navigation and global metric navigation (with the first two working on the robot and the latter implemented in a simulation), building on their earlier work (Franz & Mallot, 2000).

Route navigation (or taxon navigation) works by storing simple associations of actions to egocentric spatial relations. Several such associations can be concatenated to routes that might lead from the current location to a goal location (see Section 2). Franz et al. (2008) use a panoramic stereo camera to calculate the disparities of N=72 image sectors, after identifying each sector in both images (disparities are defined as how much an image sector appears shifted in the second image relative to the first; from these disparities, distances can be computed using elementary trigonometry). They represent a place using a 'disparity signature', a list of disparities and their corresponding reliabilities. Storing such place representations allows a simple homing by using a strategy of calculating the disparity signatures for several possible movements, and then choosing the movement that minimizes the difference between the current and the goal disparity signature. Sequences of distinguishable disparity signatures can constitute a route and allow taxon navigation.

Topological navigation integrates routes leading through the same place to a representation that can be used for navigating to multiple goals. In this model, topological navigation is afforded by a 'view graph', which is built by measuring similarities between views (using maximal pixel-wise cross-correlation), and connecting two routes whenever two views are sufficiently similar and whenever the robot succeeds in homing to this similar view. This system could successfully explore an environment, and perform homing and shortcut planning in the real world. However, it requires views to be unique (since it connect routes when views match)—thus, it cannot close the loop in environments with non-unique views.

The model was also extended by an approach to survey navigation in a simulation. This requires a representation in a common frame of reference. The model attempts to construct a global metric map by metrically embedding the view graph using an approach based on multi-dimensional scaling (MDS).

Franz et al. (2008) argue that their navigation strategies are 'biomimetic'; citing behavioral evidence from studies with insects, which lend strong support to the claim that insects seem to use mainly view-based homing for navigation (Graham & Collett, 2002)—a strategy resembling the 'disparity minimization' approach of the authors. However, they do not claim any structural similarity to brains, whether mammalian or insectile. The taxon navigation strategy can be implemented in principle by the basal ganglia storing routes as stimulus–reaction mappings, in combination with neurons encoding views, such as spatial-view cells or PPA neurons (see Section 2). However, even on the functional level, this similarity is highly tentative, since mammals can robustly navigate to goals even in dynamic environments, or after changes in the environment (which would interfere with the simple correlation-based similarity measure of this model), and also because it is highly unlikely that mammals recognize views based purely on disparities (for example, scene recognition works almost as well on computer screens as in real scenes, suggesting that stereo vision does not play a major role).

Models Evaluated in Simulations

Computational experiments in simulations have to deal with fewer issues such as complexity, sensory inaccuracy, or noise. Thus, their developers often have the resources to endow them with a larger range of abilities and to account for more tasks and paradigms (at the expense of less similarity to the actual environment of the modeled biological cognition).

- One of the first symbolic models of spatial memory in urban environments, growing out of the symbolic AI paradigm of the last century, was the *NAVIGATOR* model by Gopal, Klatzky, and Smith (1989), implemented in LISP. It runs in a simple environment consisting of horizontal and vertical streets, as well as 'plots'—locations and associated sets of objects (such as houses)—and associated decision points—points at which navigational decisions can be made. The environment is represented in a predicate calculus-based language. The agent (called NS, navigating system) can perceive information from the plot associated with its location, as well as other plots visible in each feasible direction of view; and can either turn in the four modeled directions (to perceive in that direction) or move in one of those directions.

Upon receiving an input, the NS selects the most salient objects, and stores them in a working memory (WM). WM has the function of processing perceived information, transferring it to long-term memory (LTM), monitoring instructions, and planning paths to a goal through pattern matching. LTM in turn permanently stores conceptual, spatial, and goal knowledge; in the form of semantic network representations (e.g. decision-point 2 *associated-with* house, house *color-of* red) which can decay ('forgetting'). These representations are connected by 'links' which represent spatial relations, and can be learned either from perceptual input (when two locations are present in WM at the same time), or from explicit instructions connecting two locations (e.g. 'go from A to B').

Based on its semantic network representation, NAVIGATOR is able to find goal locations and plan novel paths. The agent runs in a very simple (simulated, discrete and static) environment. The model is claimed to qualitatively replicate several aspects of human spatial behavior, such as way-finding errors (three types of errors made by NAVIGATOR appear similar to humans—errors made at locations with more information, at locations requiring complex navigational actions, and errors due to misidentification of the goal) (Gopal & Smith, 1990). No quantitative comparison against human data is performed.

The NAVIGATOR model is based on information processing theories of cognitive psychology and thus can claim a degree of cognitive plausibility. No structural similarity to neural architecture is claimed. The spatial parts of the semantic networks constituting the representations in NAVIGATOR bear some resemblance to hippocampal representations (plots and decision points to place cells), but are too simplistic for an actual functional correspondence (e.g. not every point of the environment is represented, and the distance metric is city-block, not Euclidean).

- Raubal (2001) describe the *perceptual wayfinding model*, a cognitively based computational model for wayfinding which, unlike NAVIGATOR, considers the information needs of navigators at each decision point. The model is based on the 'Sense-Plan-Act' framework, as well as affordance theory (affordances are possibilities for action)—the idea that *animals perceive the environment in terms of what they can do with it and in it* (Gibson, 1986). It is a goal-based agent—given a current state description, goal information, and the results of possible actions, it chooses actions to achieve a goal.

Its main components are its observation schema (containing spatial and temporal location, goal, and measuring limitations, in fixed frame-like structures), a wayfinding strategy (decision rules for wayfinding), and 'commonsense knowledge' (including procedural knowledge—how to move in a direction, what to do upon reading specific symbols such as arrows on signs). The implemented agent runs in a very simple simulated environment which is static and discrete, having a limited number of possible percepts and actions at each point. The agent can observe the entire environment at any given time (unlike NAVIGATOR, which also used static and discrete environments but accounted for partial observability). The environment is represented as a graph of decision points, where each node has a position and a state, and each edge represents a transition between positions and states. Since the evaluation scenario is set in an airport, each position has information regarding how to reach goals (signs containing arrows to gates). The agent first perceives the environment (senses), then decides which action leads it toward its goal (a trivial decision given the signs at each node), and then carries out the action (acts).

The perceptual wayfinding model is evaluated in an airport wayfinding task (successfully finding gates), but not compared against any human or animal data. Because of the amount of information pre-programmed into the implemented agent, and because of the fully observable static environment, the agent needs and has no learning mechanism. There is little functional similarity of the components of this model to the brain.

• Due to more recent improvements in computer graphics, it has become possible to simulate virtual agents in more complex, three-dimensional environments. In a recent model, Brom et al. (2012) have proposed a computational model of both egocentric and allocentric spatial memory for intelligent virtual agents (IVAs), calling their spatial model the *DP-model* since it was evaluated in a disorientation paradigm (see below). IVAs can be considered to be embodied, although in a much simpler and more predictable environment than the real world.

The information flow in the DP-model is as follows: sensory systems assemble information in the 'perception field', based on which egocentric representations (spatial vectors to objects in the agent's own reference frame) are built in the 'egocentric subsystem', which has both an STM and LTM component. Egocentric representations can consolidate into the LTM component of the egocentric subsystem, as well as allocentric representations in the long-term 'allocentric subsystem'. Both egocentric and allocentric representations are weighted, and weights serve as a representation of accuracy—how well a representation was learned (they are required to model errors, since the vectors are represented precisely, without modeling sensory inaccuracies or noise). The agent's perception field contains all objects in the agent's visual field (which is 120 wide). Eye movements, foveation, attention, and visual recognition are not modeled; objects are represented as state-less and static symbols. The egocentric component contains the agent's current heading (with respect to the south–north axis), a set of weighted egocentric vectors from the agent to objects, and the egocentric updating configuration (containing the rates of increasing or decreasing the weights of egocentric vectors). The allocentric component contains a set of weighted allocentric vectors between all objects, and an allocentric updating configuration (specifying the speed of increasing weights of the allocentric vectors). Egocentric vector weights are increased at every time step if the associated object is still part of the perceptual field, and decreased if it is not. The vectors themselves are updated whenever the agent moves to point correctly from the agent's position to the associated object. Allocentric vectors are learned from egocentric vectors.

The agent is also endowed with an action selection mechanism, enabling it to follow a specified trajectory to learn a representation of space during the learning phase, as well as to perform pointing tasks. In these tasks, the agent first observes and learns a number of object locations, and subsequently has to point to these locations after the objects have been removed. The pointing error in this task is a function of the vector weights (themselves depending on how often and how long the associated object has been seen during the learning phase). An advantage of this model compared to the previous two models is that it runs in a more complex (continuous, dynamic, three-dimensional) simulated environment.

Brom et al. (2012) successfully replicate human data from two pointing paradigm experiments previously performed using their model, experiment 7 of Holmes and Sholl (2005), in which subjects learned the locations of objects in a room and then had to point to the remembered locations of the objects with their eyes closed after a 45 rotation left or right (both in an oriented and in a disoriented condition induced by slow rotation on a swiveling chair), and experiment 1 of Waller and Hodgson (2006), a similar pointing paradigm.

This model builds on theories from cognitive psychology and produces error patterns consistent with humans in pointing paradigms, but does not claim structural similarity to brains. Some tentative functional correspondence between egocentric vectors and representations in the parietal reach region (and other correlates of egocentric spatial memory) might be identified, since they encode the positions of targets in an egocentric reference frame.

Neural Network-Based Spatial Memory Models

Unlike symbolic systems, neural network models usually employ non-local and distributed representations (also called subsymbolic representations), within interconnected networks of simple units. NNs are simplified models of the brain composed of a number of units (analogs of neurons) with weighted connections between them. Mental states are represented as numeric activation values of the units (or subsets of the units), and learning is usually implemented by modifying connection strengths between the units (Thomas & McClelland, 2008).

There is a variety of flavors and implementations of neural networks, ranging from the simplest perceptrons (which sum up a number of inputs multiplied by incoming weights and threshold the result to yield a binary output) over the commonly used feed-forward artificial neural networks, networks of perceptrons without cycles such as feed-forward ANNs and self-organizing maps,[7] and recurrent neural networks allowing feed-back connections and cycles (such as attractor networks[8]), over neural networks aiming to make only biologically plausible assumptions (BNNs, 'biological NNs'), to spiking neural networks (SNNs, which are the most biologically realistic, and are the most computationally expensive to run; Jain, Mao, & Mohiuddin, 1996).

Of these, only the latter two (BNNs and SNNs) explicitly aim to be biologically realistic, with this claim being extensively verified only for SNNs (they are able to account for electrophysiological recording data from biological brains). In addition to modeling neuronal and synaptic state, they also model temporal dynamics, and use short and sudden increases in voltage ('spikes') to transmit information (Ghosh-Dastidar & Adeli, 2009). BNNs, although not directly modeling electrophysiology, also aim to be biologically realistic in terms of brain connectivity and their learning mechanisms. We shall collectively refer to all other types of neural networks (the ones not aiming to closely model biological neurons) as artificial neural networks (ANNs). ANNs, unlike BNNs and SNNs, are usually driven by mathematical reasoning instead of biological accuracy.

Because of the biological inspiration and the clear analogy between units of neural networks and neurons in brains, neural networks have been claimed to be more biologically plausible than symbolic models. This is verifiably true for many SNNs (spike trains, firing rates, membrane potentials etc. can be compared with biological neurons). For ANNs, the claim of biological realism can be cast in doubt, since they make undefended design decisions (e.g. elements not having clear biological counterparts such as fixed biases, nonmonotonic activation functions, or the commonly used back-propagation learning algorithms) (Dawson & Shamanski, 1994).

Still, even if their degree of realism is debatable, ANNs are structurally more similar to brains than symbolic cognitive models—the representations employed by both are mostly distributed, grounded and modal Barsalou (2008). Furthermore, on a higher level, neural network-based models incorporate properties characteristic of biological cognition, such as content-addressable memory, context-sensitive processing, and graceful degradation under damage or noise Thomas and McClelland (2008). Finally, such models can accommodate the anatomical connections and functional distinctions known from neuroscience in a more straightforward fashion than symbolic models. Fig. 1 depicts anatomical connections between the spatially relevant regions described in Section 2, and shows some example recorded firing fields of cells with spatially localized firing. Most neural network models reviewed below attempt to be consistent with at least a subset of these results. For example, all of them model place cells, except for the SOM model by Voicu (2003). The model by Byrne et al. (2007)accounts for all of these cell types (with a simplified anatomy).

Models Evaluated in Real-World Environments

• A large number of biological ANN-based models have been proposed based on the hippocampus and other neuroanatomical bases of spatial memory. Burgess et al. (2000) proposed one such model that was implemented on a Khepera robot, based on the influential idea that place cell firing is driven by inputs with Gaussian responses tuned to the presence of walls at particular distances (O'Keefe & Burgess, 1996) (later expanded and called Boundary Vector Cell model Barry et al., 2006, which successfully accounted for rat neural and human behavior data, but was not implemented in a real-world robot). The model is mainly designed to account for the place specificity of hippocampal cells and their contribution to behavior.

It consists of a population of 'sensory cells', projecting to 'entorhinal cells', which map to 'place cells' via competitive learning, which in turn map to 'goal cells' by one-shot Hebbian learning. 'Goal cells' also receive inputs from a reward signal and from four 'head-direction cells' (north, south, east, west). Sensory cells are a rectangular array of cells, each representing a different possible distance and allocentric direction to a wall, just like BVCs (Barry et al., 2006) (however, unlike BVCs, only the four orthogonal compass directions are represented). Each entorhinal cell receives hard-wired connections from two sensory cells related to two orthogonal walls. Entorhinal cells are connected to place cells, with the connection weights being adjusted by competitive learning in order to increase the spatial specificity of place cells. Finally, connections between place cells and goal cells are learned by one-shot Hebbian learning—when the agent encounters a location with a reward, a goal cell is excited, and the connection between it and the corresponding place cells increased. When the rat moves away from the reward location, the activity of these place cells will decrease; thus, the activation of goal cells will encode the proximity to the reward, allowing a gradient ascent based navigation strategy.

The robot running the model is able to navigate to local goals. It is running in a single small environment without objects, and cannot plan novel paths. However, the modeled place cell firing fields resemble empirically observed firing fields (including changes in their amplitude and shape when the environment is changed in size or shape—these firing field changes are reported to be consistent with experimental data).

The model is largely based on the neural basis of allocentric spatial memory. Although the goal learning model is speculative, both the anatomical connections and arising firing fields of the 'place cells' in the model are plausible, and qualitatively resemble empirically recorded firing fields. Later extensions of the model–which however have not been implemented on a real-world system–include comparison to empirical data, electrophysiological data recorded from rats as well as human behavior data (Barry et al., 2006) (the model could successfully account for the effect of changed environment size on both the firing fields of rat place cells and on object locations remembered by humans).

• Another biological ANN model that is also capable of controlling a real-world Khepera robot was proposed by Strösslin et al. (2005), building on earlier modeling work (Arleo & Gerstner, 2000). Unlike (Burgess et al., 2000), this model includes full visual processing, not just distance measurements to boundaries. The model consists of multiple interconnected populations of neurons (subnetworks).

The 'local view' (LV) processes and stores visual stimuli, and contains rotation cells and step cells. The 'head direction system' (HD), corresponding to the postsubiculum, contains head-direction cells (driven by rotation cells in LV). The 'allothetic place cells' (APC) represent the agent's position in the environment (driven by step cells in LV). The 'position integrator' (PI) is a path integration system (driven by step cells in LV). Both the APC and PI project to the 'combined place code' (CPC), corresponding to the hippocampus and subiculum. Finally, 'action cells' in nucleus accumbens perform navigation learning based on place cells in CPC. The model uses V1-inspired ensembles of units with Gabor wavelet-like receptive fields (filters) to represent visual input in LV. Rotation cells (RCs) in LV discriminate headings regardless of position, based on average relative distance between stored and current filter activity; whereas step cells (RCs) discriminate positions–regardless of headings–based on perceived angular differences between landmarks (firing rates of SCs depend on the most similar column difference of the associated filters, similarly to the 'disparities' in the model by Franz et al. (2008) described above). The HD system updates head directions based on both idiothetic cues (dead reckoning) and allothetic cues (from the rotation cells). APC place cells are driven by multiple step cells (connections are set by one-shot Hebbian learning), and thus their firing is based on the current view. APC place cells help calibrate PI cells using allothetic information to correct accumulating errors Etienne et al. (1996). Finally, information from APC (allothetic) and PI (idiothetic) converge in the CPC place cells.

Connections between APC and CPC are modified using Hebbian learning. Goal-driven actions are learned in AC using Q-learning, a variant of reinforcement learning (the ACs would correspond to neurons in the nucleus accumbens). Each action cell encodes a motor command, determining the allocentric direction of the next movement.

The model is capable of learning a map in the form of a consistent place cell code, and is able to solve navigation tasks and learning tasks such as the Morris water-maze task.[9] It cannot plan novel routes.

Although not using spiking dynamics, the model incorporates insights from the neuroscience of spatial cognition known at the time of its development, and, unlike many ANNs, does not include neuroscientifically questionable design decisions. Furthermore, it is consistent with the neuroanatomy of the hippocampal–entorhinal complex. Thus, it can claim a high amount of neural plausibility. In addition to the neurally plausible models reviewed in the next section, it also functions in the real world, with realistic input. However, it is not evaluated against neural or behavior data.

• Barrera et al. (2011) proposed another biological ANN model based on brain neurophysiology, which they evaluated against rat behavior data, unlike the previously reviewed models (extending their earlier work Barrera & Weitzenfeld, 2008). Similarly to the model by Strösslin et al. (2005) above, they use modeled 'place cells' to represent spatial locations, and use reinforcement learning to learn appropriate reward-oriented actions at spatial locations. Their model receives four kinds of sensory inputs: incentives (providing the motivation/reinforcement signal), kinesthetic self-motion information, visual landmark information (driving the place cell representation), and affordances information (providing possible actions to the action selection module).

These kinds of input are processed by four corresponding modules, a 'motivation' module (calculating a reward signal from the incentives), a 'path integration' module (updating position based on self-motion), a 'landmarks processing' module (representing the current view of the animal, based on all perceived landmarks; suggested to correspond to the EC), and an 'affordance processing' module (encoding possible turns the rat can perform at a given location and orientation). The reward signal from 'motivation' drives the 'learning' module (learning by reinforcement; corresponding to the VTA, NA and striatum in brains), and the outputs of the path integration and landmarks processing modules drive the 'place representation module', which in turn project to the 'action selection' module. The 'place representation' module includes 'place cells' (PCs, the activity of which arises from a weighted linear combination of the path integration and landmark inputs; corresponding to the hippocampus) as well as a 'world graph layer' (WGL, suggested to correspond to the prelimbic cortex). The WGL learns a topological map by learning associations between overlapping place fields, as well as learning actor units representing actions with high expected rewards associated with place cells (actor unit weights are learned by reinforcement learning). The WGL also performs place recognition, by classifying the currently active PCs.

Finally, the action selection module computes a motor output (the next moving displacement and direction), given the current possible affordances, current location (place cells), and the expectations of maximum reward from the actor units in WGL. The model is able to learn metric (PCs) as well as topological maps (WGL) in the place representation layer, and is able to navigate to reward locations.

The authors evaluated their model against rat behavior data in a simple maze navigation paradigm, in which water-deprived rats were looking for a water dispenser, learning its location during a number of training sessions. They used AIBO robots in the same paradigm, in similar mazes. The robots could learn near-accurate metric and topological maps of the mazes, and exhibited learning curves (during learning the reward location) and numbers of incorrect trials and optimal trials (during test trials) similar to those of the rats.

The model is based on rat neurophysiology, and thus is neuronally plausible. It is also able to function in the real world, and has also been evaluated against rat behavior data (the learning curves in a simple maze were comparable), lending credence to the authors' claim that their model can be used by experimentalists to predict rodents' spatial behavior, and test neuroscientific hypotheses. Additionally, although not replicating neural data, the authors present results verifying the engagement of the proposed neural correlates of their models (reporting gene expression data) in the rats they used in their experiments (Barrera et al., 2011).

A further neural network based model of mapping very successful in robotics which was also inspired by rat neurophysiology is RatSLAM (Milford & Wyeth, 2010). It will not be reviewed here, since RatSLAM is not a cognitive model, and is not compared to or intended to model either behavior or biology (the authors aim for practical robot performance instead of plausibility).

Models Evaluated in Simulations

• Schölkopf and Mallot (1995) proposed a neural network model of cognitive map learning in a maze, a model aiming for cognitive rather than biological plausibility (but nevertheless pointing out similarities to neural substrate). Their agent employs a central perception–action cycle (Fuster, 2002) (similarly to the sense–plan–act cycle of the symbolic perceptual wayfinding model; Raubal, 2001). The model assumes it is dealing with a maze environment consisting of at least two places, with corridors connecting the places; and also assumes a direct correspondence between these corridors and 'views' (a view is thought of as being attached to the wall opposite to the entry of the respective corridor); and that views are uniquely distinguishable.

The model is based on the idea of a 'place graph' (an allocentric graph of places, connected by corridors) and a 'view graph' (a graph of local views connected by edges with labels representing egocentric movements; and connected only if they can be experienced in immediate temporal sequence). The view graph is learned using a SOM-type (self-organizing sequence map) neural network (Kohonen, 1990), which has three layers: an input layer (with units representing views), a movement layer (representing the movements left, right or back; with only one of these three units active at each time), and a 'map layer'. The map layer receives sequences as inputs, from both the movement layer (a sequence of movements), and the view layer (a sequence of views represented by the activity of the view layer units). A map of the current maze is learned by 'random exploration', i.e. a large number of random movements and views are passed to the network, which uses learning by self-organization (Kohonen, 1990) to assign map units in a way that they closely resemble the view graph (i.e. near views are represented by near units, and distant views by distant units). After learning, path planning to arbitrary views can be performed by a gradient ascent strategy (spreading activation from the goal, and then at each map unit, progressing to the adjacent map unit with the highest activation), a planning strategy that the authors implemented algorithmically (not in a neural network).

Unlike the previously reviewed neural network models, this model is able to plan novel routes algorithmically. It is also one of only three neural network models implementing topological maps (the other two being Barrera et al., 2011 and Erdem & Hasselmo, 2012).

Since there is little direct correspondence between this model and neuroanatomy, and since planning is implemented algorithmically, this model cannot be called biologically plausible. However, it is argued by the authors to functionally resemble some aspects of biological spatial memory (such as free/passive exploration and expectations of future views).

• A model also based on self-organized learning was proposed by Voicu (2003), extending their earlier work (Voicu & Schmajuk, 2000). Unlike the model above, it is capable of running in a full two-dimensional metric simulation instead of a restricted maze-like environment. A further difference is that it learns hierarchical instead of flat spatial representations—which is frequently argued to be the structure of cognitive maps (see Derdikman & Moser, 2010, Hirtle & Jonides, 1985 and McNamara, 1986, for behavioral and Derdikman & Moser, 2010, for neural evidence).

The model architecture consists of a hierarchical allocentric cognitive map and four additional modules (a localization system providing landmark representations, a working memory for planning paths, a motor system translating them into actions and a control system supervising information flow between these modules). The cognitive map itself uses types of SOM (recurrently connected hetero-associative networks; Kohonen, 1990) to build associations. There are three different networks representing associations between all landmarks, associations between landmarks having the largest number of associations at the first level, and associations between landmarks having the largest number of associations at the second level, respectively. The map is learned in two stages: an exploration stage for building the first level at the highest resolution (moving randomly at the beginning, avoiding previous places, and then, over ten acquired landmarks, moving toward those having the fewest associations), and a second stage, building the hierarchical cognitive map (selecting the landmarks with the largest number of associations and associating them). Weights are adjusted depending on distance (lower distances yielding lower weights), so that activation gradients can serve to plan a path toward a goal.

The model can learn hierarchical metric maps, and can plan novel paths. It succeeded in reproducing the empirically observed hierarchical cognitive maps by Hirtle and Jonides (1985), and also produced similar distance judgment errors as humans (distances spanning multiple clusters or submaps are overestimated both by humans and by the model).

This model uses SOMs, types of ANNs, and does not aim to be neurobiologically plausible. The spatial specificity of its SOM units is also a property of hippocampal place cells, but its units correspond to much larger areas than the observed PFs of place cells. However, it is able to reproduce human behavior data, and thus can make empirically validated claims for cognitive (if not biological) plausibility.

- The *map-based path integrator (MPI)* model by McNaughton et al. (1996) was an influential and still highly relevant model of spatial representation and path integration in brains, implemented as a SNN. It was tested and evaluated by Samsonovich and McNaughton (1997) and later reviewed and argued to be plausible based on neural evidence by McNaughton et al. (2006). The model is based on 'attractor maps', continuous attractor networks in which the mobility threshold for transitions between neighboring attractors is negligibly small, as opposed to the large thresholds for jumps between distant points (and with global feedback inhibition limiting total activity in the network)—this leads to activity focused on one maximum unit and declining with distance from that unit (i.e. an activity packet), tending to move toward the maximal input into the network or staying stationary in the absence of input.

The two most important modules are H, a one-dimensional cyclic attractor map (CANN) encoding the head direction of an agent (containing 'HD cells' arranged on a circle in the order of their head-direction preference), and a P, a two-dimensional attractor map used to encode the agent's current position, as well as for path integration (containing 'place cells' arranged in a plane, with weights that decrease with distance). The head direction estimate and position estimate correspond to the maxima of the activity packets on the circular CANN and two-dimensional CANN, respectively. To implement path integration for the HD cells, two additional layers are required, one with units representing angular velocity (H′), and a conjunctive layer representing both current head direction and velocity (R—receiving connections from H and H′), and projecting back to the appropriate HD units. The R layer drives the HD activity packet in the right direction whenever the agent is turning, since R units project to the right of the currently most active HD cell for positive angular velocity, and to the left for negative velocity (and with below-threshold activity if the velocity is zero).

Similarly, path integration for 'place cells' in P works by employing a number of intermediate 2D CANN layers in the I module, each layer corresponding to a different possible head direction (and receiving activation from that HD cell in H), with connections that project to units in the P layer, but displaced in the respective head direction (e.g. if the 'north' HD cell activates the corresponding I layer, units of this layer would project to a place cell that is associated with more northern locations in the P layer, instead of equivalent units in P corresponding to the same locations as units in I). Thus, the projection to P from the currently active I layer (depending on the most active HD cell) can move the 'place cell' activity packet in the correct direction.

Finally, HD cells and place cell firing is not only driven by path integration, but also by associated sensory representations, encoded in an additional module called V. Associating spatially localized place cells with sensory representations can correct accumulating path integration errors, as well as represent stimuli encountered in a specific location. Such sensory associations can be learned by Hebbian learning, whereas the weights driving the path integration mechanism (such as from H to I) which are preconfigured and fixed.

The model is implemented as an integrate and fire SNN, and is able to numerically reproduce several single-cell experimental findings, such as place field stretching upon changing environment size, dependence of place field location on the entry site, slow rotation of place fields in disoriented rats, and learning in novel environments; and also makes a novel prediction which was verified experimentally after publication of the model (activity jumps in P upon significant unexpected changes in sensory input). However, navigation or path planning or the representation of objects on the map is not explicitly modeled by the authors. The model's main strength lies in proposing the first plausible neural network model of path integration.

The model and its elements are neuroanatomically plausible; MEC might perform path integration (passing activation to hippocampal place cells), and the analogy between modeled and biological HD cells is clear (see McNaughton et al., 2006 for evidence). Despite the anatomical plausibility of the elements, and the common use of attractor networks to model head direction, it should be noted that no empirically validated mechanism has been proposed yet that could result in the very specific connectivity required by continuous attractors in brains.[10] The model is implemented as a SNN, and is thus biologically more realistic than the reviewed ANN models. Finally, it also succeeds in reproducing and even predicting empirical data, further substantiating its plausibility.

• Another influential model was the *'BBB' model* proposed by Byrne et al. (2007) and based on the BVC model (Barry et al., 2006) (the predecessor of which was implemented on a robot, and reviewed in the previous subsection; Burgess et al., 2000). The model is based on the brain areas involved in allocentric spatial representations in the medial temporal lobe, as well as the egocentric areas in the parietal lobe (see Section 2), and thus accounts for both kinds of reference frames.

In the model, egocentric maps are represented by a set of neurons in a grid, each tuned to respond most strongly to an object at a particular distance and direction from the agent's head. Allocentric maps are represented similarly, using neurons with specific preferred distances and directions, with the difference that the neurons reference direction is fixed to features of the environment, instead of the agent's current head direction (these are equivalent to BVCs). The model consists of an 'egocentric frame' module (representing egocentric maps, corresponding to the precuneus), a 'HD cells' module representing head direction, a 'transformation' module (translating between egocentric and allocentric maps, corresponding to RSC), an 'allocentric frame' module (representing allocentric maps, suggested to correspond to BVCs), a 'place cell modules' (representing current location and associating sensory representations with locations), and an 'object identity module' (for sensory representations, with each unit representing an object or landmark; corresponding to the perirhinal cortex).

The network has a 'top down' (temporal to parietal) and a 'bottom up' (parietal to temporal) phase, during which the allocentric map updates the egocentric one and vice versa (the information flow in the opposite direction is blocked in each phase). Similarly to the BVC model and its predecessors (Burgess et al., 2000), place cell firing is driven by BVCs (the firing of which in turn depends on the distances and directions of boundaries). The 'transformation' module contains N identical subpopulations, each tuned to a specific head-direction, and connected to the egocentric map so as to rotate it by the angle of that head direction (to translate it to a north-oriented allocentric reference frame). At each time step, only the subpopulation corresponding to the currently active HD cell is active. Just like in the previous model, HD cell activities are updated using CANN dynamics and angular velocity input; however, unlike the MPI, linear path integration is not performed by the allocentric representation. Instead, the 'transformation module' performs this function as well, by having an alternative set of pre-trained weights that result not only in the rotation but also in the translation of a map by a constant amount (the model only accounts for constant velocities).

.The model is able to learn allocentric as well as egocentric representations of the local space surrounding the agent in a simulation, and is the only reviewed neural network-based model with the ability to translate between the two. It is also able to mentally explore representations, and to plan routes, by mentally generating velocity signals ('mock motor efference') which are decoupled from the motors. However, it cannot plan novel routes (e.g. shortcuts/detours).

Because of the clear correspondence of model parts and brain areas, the authors are able to simulate 'lesions' (by selectively deactivating model parts or connections) and to account for lesion studies (failure to identify landmarks in half of the egocentric space hemispheric neglect patients; and place cell firing with HD cell lesions). They could also model mapping, path integration, and a paradigm in which visual and path-integrative inputs were conflicting.

The model was implemented as a biological neural network (with rate-coded instead of spiking neurons). Its modules and connections are based on neuroscientific, and psychological evidence, and are highly plausible. The model was further strengthened by evaluating it in lesion study paradigms and qualitatively comparing the results with human and rat data.

Most reviewed neural network models accounting for navigation make use of either place cell-like units associated with units representing motor actions, or a gradient ascent strategy, propagating activation from a goal location in a heavily interconnected place cell-like network, and always selecting directions that increase the current activation, until eventually reaching the goal. There is no direct evidence for either of these strategies actually being used by brains (no action representations monosynaptically connected to place cells have been found; and except for area CA3, place cells do not seem to be heavily interconnected—and in any case, such activity diffusion is inherently limited in range due to signal decay in biologically realistic networks).

• In contrast to these navigation strategies, Erdem and Hasselmo (2012) have proposed a SNN model of navigation based on probing linear look-ahead trajectories in several candidate directions to find a trajectory leading to the goal location.[11] This model is also based on the neural correlates of allocentric spatial memory in the medial temporal lobe, and incorporates hierarchical spatial representations. It incorporates four modeled medial temporal cell types, and an additional three cell types in a 'PFC' module.

Suggested to correspond to the entorhinal cortex, it models 'head-direction cells', 'persistent spiking cells', and 'grid cells', and corresponding to the hippocampus, it models 'place cells'. The prefrontal module in turn contains 'recency cells', 'topology cells', and 'reward cells' (presumably corresponding to mPFC). HD cells are modeled to have a receptive field at a specific preferred angle from an anchor cue (they are only driven by sensory input, not by self-motion, unlike CANN models of HD cells). Modeled grid cell firing is based on the persistent spiking cell model (briefly, grid fields arise from an interference oscillation in persistent spiking cells) (Hasselmo, 2008). Place cells are driven by grid cells in the model, as suggested before by theoretical models (Moser et al., 2008; Solstad et al., 2006); place fields arise from a thresholded product of the grid fields (the multiplication is implemented using coincidence detection in the model).

In contrast to the metric place cell map, a topological map is created in the PFC module. Each place cell is associated with a corresponding recency, topology, and reward cell; and topology cells are laterally interconnected. The activity of recency cells decays exponentially in time; their firing depends on the time elapsed since the last visit of the associated place cell. Each time the agent visits a place cell, the topology layer's lateral connections are reinforced by Hebbian learning, depending on thresholded current activities of recency cells, with the threshold controlling what time window is considered 'recently visited' and which topological weights should be reinforced. Finally, reward cells are also associated with place cells, and fire persistently if their corresponding place cell marks the location of a goal or reward.

During goal-directed navigation, the agent decides on what direction to choose by probing several linear look-ahead trajectory probes with different directions starting from its current location. Each probe engages the HD cell–persistent spiking cell–grid cell–place cell circuit as if the agent was physically moving along the probe trajectory. If the probe leads to the activation of a reward cell at the goal location, associated with a place cell, the rat proceeds to move in the direction of the probe. In order to avoid the probes missing the goal location, and to allow reaching intermediate goals, the reward signal is diffused in the PFC module.

Thus, secondary goals associated with place cells close to the reward cell (and thus receiving diffused activation from it) can be navigated to first, until the agent gets close enough to find the actual, highest-activated reward cell with a probe. Finally, since only directions not obstructed by an obstacle can be probed, the agent can navigate around obstacles (but also find a novel shortcut once an obstacle is removed and a novel probe direction to the reward becomes possible). The model was able to produce grid cell ensemble activity resembling recorded rat medial entorhinal neurons demonstrating 'look-ahead' activity in a T-maze navigation task (Gupta, Erdem, & Hasselmo, 2013).

The model is able to learn both metric and topological maps, and can perform path planning on the learned maps, including planning novel routes such as shortcuts or detours.

This model is implemented as a SNN, using biologically realistic modules and connectivity; furthermore, its look-ahead mechanism results in activity patterns resembling data from biological neurons.

Spatial Memory Models in Cognitive Architectures

In contrast to computational cognitive models focused on accounting for one or few specific processes, systems-level cognitive architectures aim to comprehensively model a wide range of cognitive phenomena, attempting to account for behavior and structural properties of minds (Sun, 2007). Cognitive models of specific processes can be implemented within the framework of a systems-level cognitive architecture. Such models also play an important role in cognitive science, providing detailed, formal explanations, providing hypotheses, and guiding research (see Introduction). However, the goal of being integrated with a broadly scoped, domain generic model–and desirability of being able to function using the same mechanisms and internal parameters as an agent running the same cognitive architecture in a completely different task–sets the task of modeling with cognitive architectures apart from the task of developing cognitive models.

A large number of cognitive architectures have been proposed (many of which deal with modeling spatial representations in some way), too many to review here; we will aim to outline a representative sample contributing to spatial memory modeling instead of exhaustiveness, and only include architectures explicitly claiming to model human or animal cognition (we omit the large number of robotic or AI architectures uninterested in biological cognition). More comprehensive reviews can be found in Duch, Oentaryo, and Pasquier (2008), Goertzel, Lian, Arel, de Garis, and Chen (2010)and Samsonovich (2010).

There is some intersection here with the previous two categories, since there are cognitive architectures that are exclusively symbolic, exclusively neural network-based, or hybrid (combinations of symbolic and neural network parts); we shall point out the corresponding paradigm in the text, as well as in the comparison in Table 1. To the authors knowledge, there exists no cognitive architecture explicitly aiming to be cognitively plausible (i.e. model humans or animals) which would account for navigation-space spatial memory as well as being implemented on a real-world robot in current literature. Thus we omit the 'real-world' category from this section—all reviewed models run in simulations.

The popular ACT-R (Adaptive Control of Thought Rational) cognitive architecture by Anderson, Matessa, and Lebiere (1997) follows a production-rule based approach (productions consist of sensory preconditions or 'IF' statements, and associated actions or 'THEN' statements executed when the precondition matches the state of the world). It utilizes two types of memory: declarative memory, encoding factual knowledge about the world (as symbolic entities called 'chunks'), and procedural memory, containing procedural knowledge in the form of productions (IF-THEN rules). The general usefulness of these chunks and production rules is stored in a neural network reflecting previous usage (which has led some researchers to categorize ACT-R as a hybrid cognitive architecture, despite it being primarily symbolic Duch et al., 2008).

Apart from memory, the central components of ACT-R are perceptual-motor modules interfacing with the environment, buffers, and a central pattern matcher for productions (matching, selecting and executing production rules). This central module is hypothesized to correspond to the basal ganglia in the brain. ACT-R has been used to replicate a large number of psychological experiments (Anderson et al., 2004). Although the original version did not explicitly account for spatial cognition, it has later been extended to include spatial memory models.

• One such extension, called *ACT-R/S* was proposed by Harrison et al. (2003), adding two additional systems to ACT-R: a 'manipulative system' (representing spatial characteristics of objects facilitating manipulation), and a 'configural system' (representing the relative, approximate configuration of objects in space). The latter consists of a 'path integrator' and a buffer containing a number of spatial chunks called 'configurals', each storing an egocentric vector to an object along with its identity (ACT-R/S only includes egocentric representations). Objects attended to enter this configural buffer, which holds the two or three most recent objects—when this capacity is exceeded, the least recent chunk will be discarded from this buffer (but will still exist in 'declarative memory' for later retrieval).

The 'path integrator'–instead of updating an allocentric location representation–updates all egocentric representations in the configural buffer after each movement by the motor system (this is feasible due to the small number of configurals actively maintained in the buffer). Apart from object identity, configurals store multiple vectors, to all edges of an object—in the implemented model, which was two-dimensional, objects were approximated by their bounding box, and four vectors were stored to the edges of that bounding box (to the left, top, right, and bottom sides of an object). Multiple configurals referring to the same object from different points of view can be present in the model, which would have different edge vectors but the same identity tag.

The authors implemented a food search model, which can randomly explore an environment, try to recall a food location, or visually search for food. The search is performed by requesting unattended objects from the configural system, identifying it using the visual system, and continuing the search if it is not food, or setting it as a goal if it is. In the latter case, the agent orients itself toward the food location, and begins another search (this time for obstacles—any object that intersects its path to the food location). If obstacles are found, the agent adds a subgoal to move to the left or right of it, depending on which brings it closer to the goal. If no obstacles are left, it moves to the goal location. During navigation, the agent repeatedly checks if it has arrived at its destination, and also repeatedly corrects path integration errors using its visual system (that is, if the egocentric representations updated by path integration do not match their perceived correct location, they are corrected).

Furthermore, it encodes 'episodic traces' (current contents of the configural buffer) at each step. If visual search fails to find food, these episodic traces can be recalled to find previously identified food locations as well as nearby objects (after which it can perform another visual search for those nearby objects and navigate to them to get closer to the food location). The authors functionally evaluate this model of path integration and navigation, and point out functional similarities between configural chunks and primate spatial-view cells.

The psychological plausibility of the ACT-R model and its parameters (buffer capacities, timings etc.) have been extensively strengthened in a large number of different paradigms. There is also functional similarity between the egocentric representations in this model, and egocentric representations in the brain (e.g. spatial-view cells). However, there is no clear structural similarity between this model and neurobiology.

- *Casimir* by Schultheis and Barkowsky (2011) is a cognitive architecture explicitly devised as a framework for computationally modeling human spatial knowledge processing. Its main parts are a long-term memory (LTM), working memory (WM), and a diagram interaction component (externalizing WM representations on diagrams, or visually inspecting diagrams to build WM representations).

The LTM stores hierarchical, semantic network-like representations (nodes and connections between them; categories and objects as well as spatial relations are represented as nodes, whereas connections signify associations; e.g. three nodes and two connections could represent the relation 'Paris'-'south of'-'London'). The WM can be split into three parts, one concerned with retrieving representations from LTM based on a 'problem representation', one performing memory updates of WM and LTM, and a 'visuo-spatial WM' part storing and manipulating short-term representations relevant to the current problem. The problem representation also takes the form of a semantic network, and allows the specification of a query (such as the cardinal direction to a location, or a distance between locations).

Retrieval from LTM works by spreading activation over the nodes in LTM from the problem representation; the subnet ('fragment') with the highest sum of activation is retrieved to the visuo-spatial WM (retrieved subnets also have to be directly or indirectly interlinked). This LTM structure and retrieval process can account for some human memory phenomena. Knowledge from different sources can enter visuo-spatial WM, including knowledge retrieved from LTM, built by visual inspection, or constructed from previous representations; and is represented not symbolically but in a spatio-analogical form (i.e. there is a structural correspondence between the representations and what they represent in the world).

Casimir assumes that there is no strict division between spatial and visual representations, but, rather, a continuum between the extremes of simple nonmodal spatial mental models (spatial) and mental images (visual). Representations are deemed more visual with increasing numbers of relations, involved knowledge types (such as distance, direction, topological knowledge), specificity, and exemplarity (concrete exemplars or prototypes). A 'conversion' process in working memory can construct and extend representations, adding retrieved fragments if necessary, or converting fragments to spatial mental models. An 'exploration' process in turn can extract spatial information from existing representations, or infer knowledge using spatial reasoning.

Because of its emphasis on structural modeling (spatio-analogical instead of symbolic representations), Casimir is argued to exceed the modeling capabilities of other cognitive architectures in the spatial domain (Schultheis & Barkowsky, 2011). The architecture was tested on paradigms involving eye movements in a spatial reasoning task (Sima, Lindner, Schultheis, & Barkowsky, 2010), mental scanning (the effect of the time to scan between entities in a mental image increasing linearly with the distance between them), mental reinterpretation of spatial relations (Sima, 2011), and recall effects (Schultheis, Lile, & Barkowsky, 2007). The model has a simple visual perception implementation facilitating the replication of such experiments. However, navigation has not been implemented.

The model is heavily based on prevalent cognitive science theories of mental representations (e.g. analogical representations Barsalou, 2008, mental models Mani & Johnson-Laird, 1982, mental images Shepard & Metzler, 1971), and replicates human behavior data in a number of paradigms. However, it does not aim to be biologically plausible, and its parts do not clearly correspond to brain areas or neurons.

- *CLARION* by Sun and Zhang (2004) is a hybrid cognitive architecture accounting for spatial representations. It incorporates explicit (symbolic) as well as implicit (subsymbolic) knowledge through its four memory modules: the action-centered subsystem (regulating procedural knowledge and actions), non-action-centered subsystem (maintaining general declarative knowledge), motivational subsystem (providing motivation for action), and metacognitive subsystem (monitoring and directing the operations of the other subsystems).

Each module has a localist-distributed representation (explicit knowledge) and a distributed section stored in a neural network (implicit knowledge). Spatial representations can be acquired by associating explicit knowledge in the form of 'chunks' (similarly to ACT-R chunks—e.g. a chunk representing a reward) with the corresponding implicit representation of sensory input.

CLARION's ability to represent and navigate in space is shown in the complex minefield navigation (MN) task implemented by Sun, Merrill, and Peterson (2001). In this task, an agent has to navigate through a two-dimensional minefield to reach a target. The agent only has access to limited sensory information (short-range sonar readings to mines, range and bearing gauges showing distance and direction to the target, and the remaining time), and has to reach the target in a limited amount of time. Only egocentric spatial relations were used (distances and directions to nearby mines). The agent used a type of reinforcement learning called Q-learning (with a gradient reward depending on target distance, and a second reward at the end depending on the agents success—depending on how close it got to the target) to learn an optimal action policy.

The model was evaluated against human behavior data, and produced trajectories and learning curves similar to humans in this paradigm. It does not learn an allocentric map; rather, it uses reinforcement learning to learn the optimal actions to reach its goal given the obstacles in the environment. Information about the current obstacles is represented as implicit knowledge in the 'state' layer of CLARION's neural network (see Fig. 9).

Since the model uses very general modules (there is no specialized spatial memory module), and since it consists of both symbolic and neural network parts, it is difficult to identify structural correspondences to neurobiology. CLARION has succeeded in modeling human behavior data from a large number of paradigms–including the above mentioned minefield navigation task–and thus can be called cognitively plausible (Sun & Zhang, 2004).

• Another hybrid cognitive architecture is *LIDA*[12] by Franklin, Madl, D'Mello, and Snaider (2014), with recently developed spatial capabilities Madl et al. (2013). Although not modeling neurons, LIDA is biologically inspired, with each major part of the model functionally mapped to brain areas (Franklin et al., 2014; Goertzel et al., 2010), and is largely based on the Global Workspace Theory of functional consciousness (Baars & Franklin, 2009; Baars, Franklin, & Ramsoy, 2013), as well as a number of psychological and neuropsychological theories including grounded cognition (Barsalou, 2008), working memory (Baddeley, 1992), and Slomans H-CogAff cognitive framework (Sloman, 1998) among others. It is a recent architecture and only partially implemented, but has replicated a number of psychological experiments (Franklin et al., 2014).

LIDA's cognitive cycles, corresponding to the action–perception cycles in neuroscience Fuster (2002), consist of three phases. The 'understanding' phase includes sensing the environment, detecting features, recognizing objects and categories, and building internal representations. The 'attending' phase is responsible for deciding what portion of this representation should be attended to and broadcast to the rest of the system, making it the current contents of consciousness. This portion allows the agent to choose an appropriate action to execute in the 'action' phase. During the understanding phase, percepts are recognized based on LIDA's perceptual knowledge base, the Perceptual Associative Memory (PAM), which is a connectionist structure containing nodes with activation connected by links. Recognized objects, categories, etc. are stored in LIDA's preconscious 'Working Memory', and are represented by structures of PAM nodes and links between them.

These PAM node structures–parts of the PAM network–are hierarchical, modal representations similar to Barsalou's perceptual symbols Barsalou (2008). Since they are hierarchical and associative, they are well-suited to represent 'hierarchical cognitive maps', by associating PAM nodes representing objects or landmarks with 'place nodes'. Place nodes are special kinds of PAM nodes representing a spatial location; they are arranged in layers of two-dimensional rectangular grids with different resolutions (distances between the place nodes). The layers are interconnected, multiple high-resolution place nodes project to a single low-resolution place node (with overlap); which implements spatial clustering. This can account for systematic position errors in humans due to hierarchical representation (Madl et al., 2013).

LIDA agents use a gradient ascent based navigation strategy (passing activation from a goal location through the place node network), similarly to some of the neural network models above. However, a significant difference is that hierarchical map representation is used during navigation (first a rough route is planned using the lowest resolution layer, and then successively refined on the higher resolution layers). It can be shown that in multi-goal navigation tasks, gradient ascent on a single map leads to a sub-optimal nearest-neighbor strategy (as does the 'look-ahead' approach (Erdem & Hasselmo, 2012) and RL with simple goal-distances as rewards (Barrera et al., 2011; Strösslin et al., 2005), although RL with different reward functions can improve this). Humans significantly outperform the nearest-neighbor strategy in multi-goal paradigms such as the traveling salesperson problem[13] (TSP), planning near-optimal routes. The gradient ascent strategy on a hierarchical cognitive map in LIDA significantly improves route optimality, without sacrificing the biological plausibility of a connectionist map for a symbolic planning mechanism.

LIDA-based agents have been shown to be able to perform mapping and navigation, and model human behavior in different tasks, including modeling map recall errors, capacity limits of spatial working memory, and errors in the TSP paradigm (Madl et al., 2013) (work is underway to embody LIDA on a robot (Franklin et al., 2014) and to extend it with both egocentric and allocentric real-world spatial memory). Although not a biological neural network, spatial memory in LIDA is connectionist; and there is similarity between 'place nodes' and hippocampal place cells (also accounting for hierarchies in an empirically substantiated fashion, unlike most other models).

Comparative Table

Table 1 shows a comparison of the reviewed models, characterizing them according to the criteria outlined in Section 3. It compares the level of modeling by stating the elemental position representation for each model, as well as the reference frames or types of representations accounted for, the learning mechanism, the structural similarity between models and underlying neural mechanisms, and the complexity of the environments and types of tasks in which the models have been evaluated (to help assess their generality and complexity). Quantitative 'goodness of fit' was not included because most models did not perform quantitative statistical evaluations against data; and the exceptions that did used different tasks.

DISCUSSION

Direct comparison of the reviewed models is made difficult by their very different goals and paradigms. Although computational cognitive models should be evaluated quantitatively as well as qualitatively, the majority of the reviewed models were not quantitatively evaluated against actual behavior data. Exceptions include:

- (Symbolic—Brom et al., 2012): replication of human accuracies in pointing tasks (subjects/agent had to remember locations of several objects in a room, and subsequently asked to point to the locations after the objects have been removed)
- (Neural network-based—Barry et al., 2006; Burgess et al., 2000): This model is the only reviewed model which was compared to both electrophysiological data from rat place cells, and behavioral data human subjects. It could successfully account for the effects of changed environment size on both place fields and on remembered locations of objects.
- (Neural network-based—Barrera et al., 2011): The model's learning curve when learning to reach a goal in a maze was comparable to that of rats in an experiment
- (Neural network-based—Voicu, 2003): The model imposed hierarchies comparable to human hierarchical cognitive maps, and resulted in comparable distance estimation biases
- (Cognitive architecture-based—Schultheis & Barkowsky, 2011): Replication of eye movements in spatial reasoning, mental scanning, mental reinterpretation of spatial relations, and recall effects
- (Cognitive architecture-based—Sun & Zhang, 2004): Replication of human data in a minefield navigation task

- (Cognitive architecture-based—Madl et al., 2013): Replication of human performance in the traveling salesman problem and of map representation errors

Apart from psychological plausibility in terms of comparable behavior, the functional advantages of the models are also important aspects. Although all models represent spatial information in some form, there is a large difference in terms of the complexity of the environments they can handle, the accuracy of these representations, and the range of tasks they can be used for.

It should be noted that although all of these models can be said to create maps (of different kinds and different accuracies), only a few of them can be said to be modeling 'cognitive maps' in the sense of Tolman (1948), who has pointed out that cognitive maps can be used to plan novel routes such as shortcuts or detours (for known routes, no allocentric map would be necessary). In this sense, only 7 models are accounting for cognitive maps—those that can perform path planning (see also the 'Abilities' row in Figs. 2–9): Beeson et al. (2010), Byrne et al. (2007), Erdem and Hasselmo (2012), Gopal and Smith (1990), Madl et al. (2013), Schölkopf and Mallot (1995) and Voicu (2003).

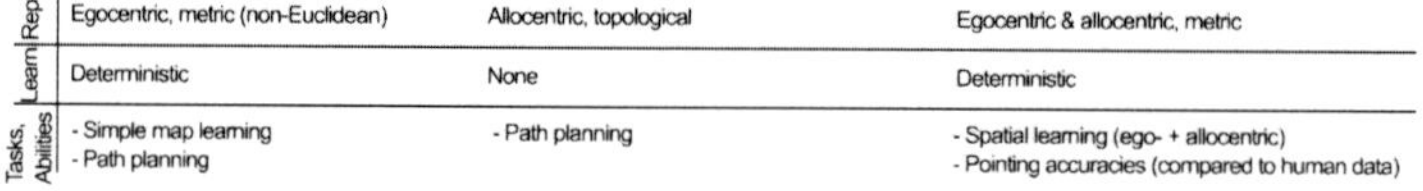

Figure. 3. Overview of symbolic models evaluated in simulated environments.

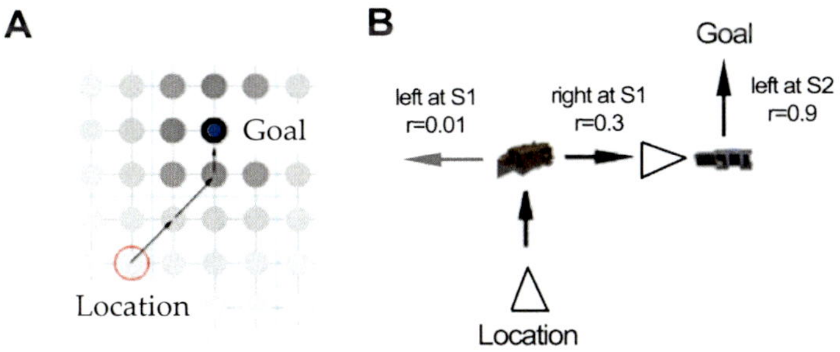

Figure. 4. Two navigation strategies. A: Allocentric navigation using a gradient ascent strategy on a heavily interconnected network of place representations, as used by the biological ANN model by Burgess et al. (2000), the ANN model by Schölkopf and Mallot (1995), as well as the LIDA hybrid cognitive architecture (on a hierarchical network). B: Egocentric navigation by always executing the action associated with the highest reward r at each state S, learned by reinforcement learning (used in the neural network models by Barrera et al., 2011, and Strösslin et al., 2005, as well as in the CLARION cognitive architecture).

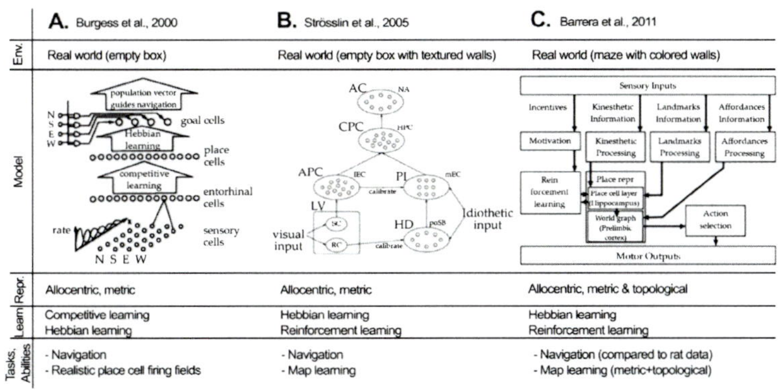

Figure. 5. Overview of neural network models evaluated in real-world environments.

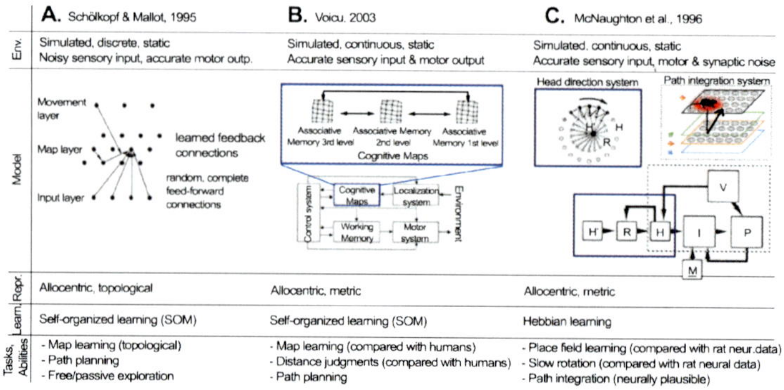

Figure. 6. Overview of neural network models evaluated in simulated environments.

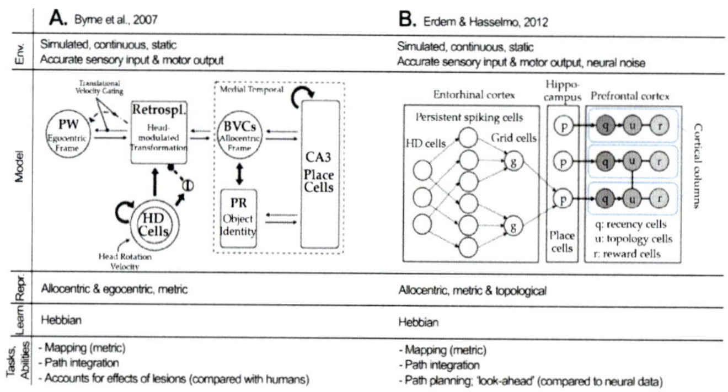

Figure. 7. Overview of neural network models evaluated in simulated environments.

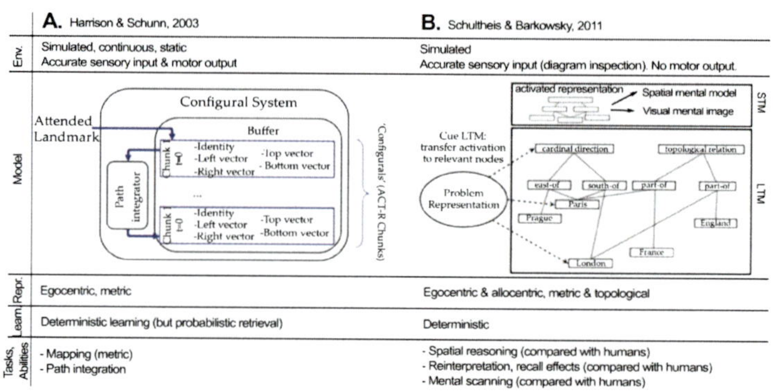

Figure. 8. Overview of cognitive architectures evaluated in simulations.

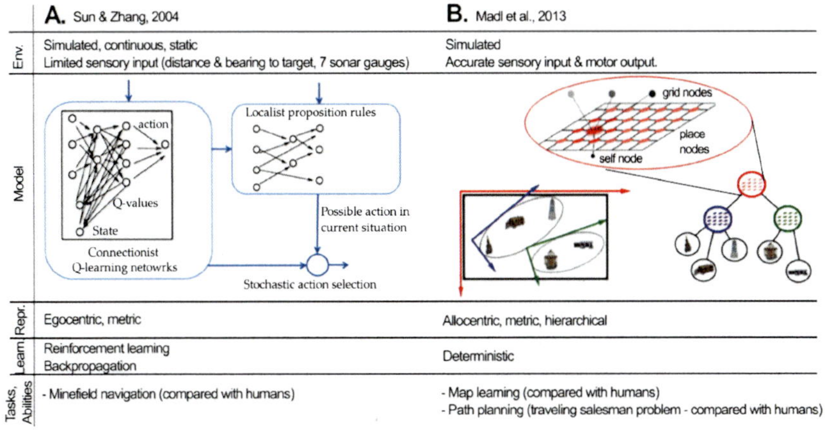

Figure. 9. Overview of cognitive architectures evaluated in simulations.

In general, models capable of handling a higher environmental complexity in Table 1should be regarded as functionally more powerful. Models capable of running in the real world face greater challenges and are more difficult to implement than simulated models, since they need to cope with noise and errors both in their sensory input and motor output, as well as with the usually greater complexity and unpredictability of real environments.

Global mapping, i.e. correctly aligning multiple maps of local surroundings in the same reference frame, and loop closing, i.e. the problem of recognizing a place the agent has seen before (and correcting representation errors), are particularly difficult tasks in the real world. The main reason for this is that different places can look very similar (perceptual aliasing), and the same place can also look different at various times in dynamic environments. Only two of the reviewed models are able to perform both global mapping and loop closing in the real world (Beeson et al., 2010; Jefferies et al., 2008).

Looking at the structural similarities (which roughly translate to biological plausibility) and the environmental complexities in the table, it can be seen that in most cases there is a tradeoff between the two. Models with high biological realism (SNNs, e.g. McNaughton et al., 1996; or Erdem & Hasselmo, 2012) usually have trouble handling highly complex real-world environments (due mainly to their high computational demands, but also to the observation that it is easier to model high-level cognitive tasks such as planning with simpler–such as symbolic–models). In contrast, models built to work well on real-world robots (such as HSSH) usually cannot be called biologically realistic, and also have difficulties fitting human behavior data (due mainly to the abstractions and methodological shortcuts employed to quickly develop efficient algorithms that can tackle complex input, and also due to computational restrictions of robots).

It is very difficult to implement and run a model that incorporates both high psychological and biological plausibility and the ability to handle real-world environments. The model by Barrera et al. (2011) is notable because although it cannot close the loop and cannot perform global mapping, it can learn a real-world maze with a learning curve similar to rats, using a model that is highly structurally similar to rat brains.

The line of research attempting to implement real-world capable cognitive models can be expected to yield important insights in the cognitive sciences. First, because of the desirability of realistic input and output for accurate models of biological cognition (sticking to overly simplistic environments causes similar difficulties for a mechanistic understanding of cognition as studying spherical wooden balls or the solar system model would for nuclear physicists). Second, robotics and machine learning research has already provided significant insights and facilitated breakthroughs in cognitive neuroscience, and there is reason to believe it will continue to do so. Examples are the development of statistical methods to deal with sensory uncertainty (which later proved to help explain behavioral and neural data, starting the 'Bayesian brain' movement; Knill & Pouget, 2004), machine learning approaches for learning optimal action policies in unpredictable environments (reinforcement learning, which has contributed to understanding the neuroscientific study of conditioning; Maia, 2009), or dynamical systems and control theory (which have inspired dynamical systems approaches to cognition; Beer, 2000).

Open Questions

It is interesting to note that the vast majority of the reviewed models incorporate allocentric representations (every reviewed real-world capable model does), and that a majority of the models capable of handling large-scale real-world environments represent both metric and topological spatial maps. The first point–the importance of allocentric spatial representations–has been known to cognitive science for many decades (Tolman, 1948). However, surprisingly little psychology and neuroscience research effort has been invested in identifying the mechanisms involved in topological mapping (for example, there is still no well-established neural correlate of topological maps in the brain—see Section 2; furthermore, the computational mechanism of how humans might partition space into topological maps is not well understood).

Models incorporating topological spatial representations such as the ones reviewed above might provide inspiration and insight for such research (unfortunately, none of them empirically validate their model with regard to topological mapping). Using empirically verified computational cognitive models to try out hypotheses regarding topological representation or the topology building mechanism in humans or animals would be an interesting and mostly unexplored line of research.

Along similar lines, it has long been suspected that the 'cognitive map' might be hierarchical (Derdikman & Moser, 2010; Hirtle & Jonides, 1985; McNamara, 1986), and multiple models incorporate hierarchies in their maps (such as HSSH, the model by Voicu, 2003, LIDA, and Casimir). Plausible neural correlates of hierarchical maps have also been identified in hippocampal and entorhinal cortical neurons with significantly varying firing field sizes (Derdikman & Moser, 2010). However, the mechanism which humans or animals use to cluster spatial representations into maps and sub-maps and organize them into a hierarchy is not yet understood (it is likely that the simple distance-based clustering mechanisms employed by most existing hierarchical models are insufficient to explain the error patterns caused by hierarchical maps; for example, perceptual or functional similarity almost certainly play a role in the mechanism organizing landmarks hierarchically in brains).

A further not fully understood part of spatial memory is the transformation process converting between egocentric and allocentric representations. Some of the reviewed models include both types of representations (Beeson et al., 2010; Brom et al., 2012; Byrne et al., 2007; Franz et al., 2008; Schultheis & Barkowsky, 2011). However, none of these models have evaluated their transformation mechanism against empirical data, with the exception of the neural network model by Byrne et al. (2007) (which seems to predict heavily coordinated and correlated activity in the neural correlates of transformation, i.e. the RSC; but such activity has been not observed).

A question that has yielded significant progress–but still no mature models explaining empirical data–regards the identification of 'landmarks' (how does the perceptual system identify landmarks, using them for orientation, as opposed to navigationally irrelevant stimuli?). Factors such as distance, stability, uniqueness, perceptual salience, and functional relevance seem to play a role. However, most existing spatial memory models either focus on localizing and navigating based on geometry, or are tested in sparse environments where a strategy of using every encountered object as a landmark is viable.

Finally, progress in the field of modeling spatial memory could be made by integrating the insights of individual models accounting for various phenomena (egocentric/allocentric, metric/topological, local/global, associative/reinforcement learning, geometric/landmark based, etc.) and tasks within the same model. Both the task of integrating these disparate processes, and evaluating them in a large number of tasks and settings, could yield new insights. Cognitive architectures would be in a uniquely suitable position to incorporate such an integration due to their generality and pre-existing non-spatial cognitive mechanisms.

Methods for Verifying the Biological Plausibility of Cognitive Spatial Memory Models

The overview of Section 3 has outlined a number of qualitative and quantitative ways to evaluate computational models. In this section, we shall focus on describing recent methods for judging the biological plausibility of a model. Apart from qualitative evaluations of structural similarity to the underlying neurobiology (such as the similarities in Table 1), it is also possible to empirically validate biological plausibility by comparing model predictions with neuroscientific data.

For biologically realistic neural network models, the most straightforward way of empirical verification is comparison with in-vivo electrophysiological single-unit recordings (in which microelectrodes are used to measure the action potentials of individual neurons in the brain of a live animal performing a task, preferably the same task in a similar environment as that of the model). For ANNs, a mapping function can be designed converting their numeric activation value to a spike rate; in the case of SNNs, the comparison is straightforward (spike trains or even voltage traces can be compared). The BVC model (Barry et al., 2006) is an example computational model successfully predicting the firing activity of spatially relevant neurons in single-unit recordings.

However, for most models, this is not viable; most often because they do not contain representations analogous to single biological neurons. In this case, higher-level brain-imaging data can be used for evaluation, which shows the time-dependent activity of brain areas involved in performing a task. Most frequently employed examples are fMRI (functional magnetic resonance imaging, a technique with high spatial but low temporal resolution) and EEG (electroencephalography, with low spatial but high temporal resolution). For models whose modules have been mapped to brain areas, it is possible to convert the activity of model parts into predicted brain area activations, and thus compare the model with neuroscientific data. Since the mapping function is arbitrary and does not place structural requirements on the underlying model, this procedure is possible even for models with little or no biological realism.

The ACT-R cognitive architecture is an example model that has used this approach successfully. ACT-R's major modules have been mapped to brain areas (such as the imaginal module to the posterior parietal region, or the central pattern matcher to the basal ganglia), and a suitable mapping function has been devised that converts activity in these modules into activation patterns resembling fMRI data (Qin et al., 2007), and more recently, EEG data (Motomura, Ojima, & Zhong, 2009), successfully predicting brain activity in novel circumstances (Anderson et al., 2008).

CONCLUSION

Having briefly summarized the basis of spatial memory in brains, we then reviewed a number of computational cognitive models of spatial memory, and presented a comparative table to help overview the major modeling directions taken within this large and highly fragmented topic. Although focusing on models concerned with human or spatial cognition, we have attempted to bring the fields of cognitive science, robotics, and neuroscience closer together by highlighting sources of overlap and interaction, and the modeling approaches most closely matched to each. We have pointed out what robotics and neuroscience can contribute to the field of cognitive modeling, and proposed some novel potential mappings between parts of existing models and relevant brain areas, in the hope of facilitating understanding, comparison, and evaluation. We have also outlined some open questions in the field, and how current (and future) models could address these questions. Computational cognitive modeling has much to offer spatial memory research (and cognitive science research in general), verifying existing hypotheses, yielding new ones, and guiding research.

ACKNOWLEDGMENTS

We are grateful to Prof. Stan Franklin for his helpful comments. This work has been supported by EPSRC (Engineering and Physical Sciences Research Council) grant EP/I028099/1, and FWF (Austrian Science Fund) grant P25380-N23.

REFERENCES

1. Allen, G. L. (2003). Human spatial memory: remembering where. Psychology Press.
2. Anderson, J. R., Bothell, D., Byrne, M. D., Douglass, S., Lebiere, C., & Qin, Y. (2004). An integrated theory of the mind. Psychological Review, 111, 1036.
3. Anderson, J. R., Fincham, J. M., Qin, Y., & Stocco, A. (2008). A central circuit of the mind. Trends in Cognitive Sciences, 12, 136–143.
4. Anderson, J. R., Matessa, M., & Lebiere, C. (1997). ACT-R: a theory of higher level cognition and its relation to visual attention. Human–Computer Interactions, 12, 439–462.
5. Arleo, A., & Gerstner, W. (2000). Spatial cognition and neuro-mimetic navigation: a model of hippocampal place cell activity. Biological Cybernetics, 83, 287–299.
6. Avraamides, M. N., & Kelly, J. W. (2008). Multiple systems of spatial memory and action. Cognitive Processing, 9, 93–106.

7. Baars, B. J., & Franklin, S. (2009). Consciousness is computational: the LIDA model of global workspace theory. International Journal of Machine Consciousness, 1, 23–32.
8. Baars, B. J., Franklin, S., & Ramsoy, T. Z. (2013). Global workspace dynamics: cortical 'binding and propagation' enables conscious contents. Frontiers in Psychology, 4.
9. Baddeley, A. (1992). Working memory. Science, 255, 556–559.
10. Bailey, T., & Durrant-Whyte, H. (2006). Simultaneous localization and mapping (SLAM): part II. IEEE Robotics & Automation Magazine, 13, 108–117.
11. Barrera, A., Cáceres, A., Weitzenfeld, A., & Ramirez-Amaya, V. (2011). Comparative experimental studies on spatial memory and learning in rats and robots. Journal of Intelligent & Robotic Systems, 63, 361–397.
12. Barrera, A., & Weitzenfeld, A. (2008). Biologically-inspired robot spatial cognition based on rat neurophysiological studies. Autonomous Robots, 25, 147–169.
13. Barry, C., Lever, C., Hayman, R., Hartley, T., Burton, S., O'Keefe, J., et al. (2006). The boundary vector cell model of place cell firing and spatial memory. Reviews in the Neurosciences, 17, 71–97.
14. Barsalou, L. W. (2008). Grounded cognition. Annual Review of Psychology, 59, 617–645.
15. Baumann, O., & Mattingley, J. B. (2010). Medial parietal cortex encodes perceived heading direction in humans. Journal of Neuroscience, 30, 12897–12901.
16. Beer, R. D. (2000). Dynamical approaches to cognitive science. Trends in Cognitive Sciences, 4, 91–99.
17. Beeson, P., Modayil, J., & Kuipers, B. (2010). Factoring the mapping problem: mobile robot map-building in the hybrid spatial semantic hierarchy. The International Journal of Robotics Research, 29, 428–459.
18. Bhattacharyya, R., Musallam, S., & Andersen, R. A. (2009). Parietal reach region encodes reach depth using retinal disparity and vergence angle signals. Journal of Neurophysiology, 102, 805–816.
19. Booij, O., Terwijn, B., Zivkovic, Z., & Krose, B.(2007). Navigation

using an appearance based topological map. In 2007 IEEE international conference on robotics and automation (pp. 3927–3932). IEEE.

20. Bringsjord, S. (2008). Declarative/logic-based computational cognitive modeling. In The handbook of computational cognitive modeling. Cambridge: Cambridge University Press.
21. Brom, C., Vyhnánek, J., Lukavský, J., Waller, D., & Kadlec, R. (2012). A computational model of the allocentric and egocentric spatial memory by means of virtual agents, or how simple virtual agents can help to build complex computational models. Cognitive Systems Research, 17–18, 1–24.
22. Brown, M. W., & Aggleton, J. P. (2001). Recognition memory: what are the roles of the perirhinal cortex and hippocampus? Nature Reviews Neuroscience, 2, 51–61.
23. Brown, E. N., Frank, L. M., Tang, D., Quirk, M. C., & Wilson, M. A. (1998). A statistical paradigm for neural spike train decoding applied to position prediction from ensemble firing patterns of rat hippocampal place cells. The Journal of Neuroscience, 18, 7411–7425.
24. Burak, Y., & Fiete, I. R. (2009). Accurate path integration in continuous attractor network models of grid cells. PLoS Computational Biology, 5, e1000291.
25. Burgess, N. (2008). Spatial cognition and the brain. Annals of the New York Academy of Sciences, 1124, 77–97.
26. Burgess, N., Jackson, A., Hartley, T., & O'Keefe, J. (2000). Predictions derived from modelling the hippocampal role in navigation. Biological Cybernetics, 83, 301–312.
27. Byrne, P., Becker, S., & Burgess, N. (2007). Remembering the past and imagining the future: a neural model of spatial memory and imagery. Psychological Review, 114, 340.
28. Calton, J. L., & Taube, J. S. (2009). Where am i and how will i get there from here? A role for posterior parietal cortex in the integration of spatial information and route planning. Neurobiology of Learning and Memory, 91, 186–196.
29. Cassimatis, N. L., Bello, P., & Langley, P. (2008). Ability, breadth, and parsimony in computational models of higher-order cognition.

Cognitive Science, 32, 1304–1322.

30. Chen, Z., Kloosterman, F., Brown, E. N., & Wilson, M. A. (2012). Uncovering spatial topology represented by rat hippocampal population neuronal codes. Journal of Computational Neuroscience, 33, 227–255.
31. Cheng, K., Shettleworth, S. J., Huttenlocher, J., & Rieser, J. J. (2007). Bayesian integration of spatial information. Psychological Bulletin, 133, 625–637.
32. Cheung, A., Ball, D., Milford, M.,Wyeth, G., &Wiles, J.(2012). Maintaining a cognitive map in darkness: the need to fuse boundary knowledge with path integration. PLoS Computational Biology, 8, e1002651.
33. Crowe, D. A., Averbeck, B. B., & Chafee, M. V. (2008). Neural ensemble decoding reveals a correlate of viewer-to object-centered spatial transformation in monkey parietal cortex. The Journal of Neuroscience, 28, 5218–5228.
34. Dabaghian, Y., Cohn, A. G., & Frank, L. (2011). Topological coding in hippocampus. In Computational modeling and simulation of intellect: current state and future prospectives (pp. 293–320).
35. Dawson, M. R., & Shamanski, K. S. (1994). Connectionism, confusion and cognitive science. Journal of Intelligent Systems, 4, 215–262.
36. Derdikman, D., & Moser, E. I. (2010). A manifold of spatial maps in the brain. Trends in Cognitive Sciences, 14, 561–569.
37. Doeller, C. F., Barry, C., & Burgess, N. (2011). From cells to systems: grids and boundaries in spatial memory. The Neuroscientist,.
38. Duch, W., Oentaryo, R.J., & Pasquier, M. (2008). Cognitive architectures: where do we go from here? In AGI, Volume 171 (pp. 122–136).
39. Duhamel, J.-R., Colby, C. L., & Goldberg, M. E. (1998). Ventral intraparietal area of the macaque: congruent visual and somatic response properties. Journal of Neurophysiology, 79, 126–136.
40. Durrant-Whyte, H., & Bailey, T. (2006). Simultaneous localization and mapping: part I. IEEE Robotics & Automation Magazine, 13, 99–110.

41. Ekstrom, A. D., Kahana, M. J., Caplan, J. B., Fields, T. A., Isham, E. A., Newman, E. L., et al. (2003). Cellular networks underlying human spatial navigation. Nature, 424, 184–187.
42. Eliasmith, C. (2007). Attractor network. Scholarpedia, 2(10), 1380.
43. Epstein, R. A. (2008). Parahippocampal and retrosplenial contributions to human spatial navigation. Trends in Cognitive Sciences, 12, 388–396.
44. Erdem, U. M., & Hasselmo, M. (2012). A goal-directed spatial navigation model using forward trajectory planning based on grid cells. European Journal of Neuroscience, 35, 916–931.
45. Etienne, A. S., Maurer, R., & Séguinot, V. (1996). Path integration in mammals and its interaction with visual landmarks. Journal of Fish Biology, 199, 201–209.
46. Fox, C. W., & Prescott, T. J. (2010). Hippocampus as unitary coherent particle filter. In IJCNN (pp. 1–8). IEEE Press.
47. Franklin, S. (2003). IDA, a conscious artifact? Journal of Consciousness Studies, 10, 4–5.
48. Franklin, S., Madl, T., D'Mello, S., & Snaider, J. (2014). LIDA: a systemslevel architecture for cognition, emotion, and learning. IEEE Transactions on Autonomous Mental Development, 6, 19–41.
49. Franz, M. O., & Mallot, H. A. (2000). Biomimetic robot navigation. Robotics and Autonomous Systems, 30, 133–153.
50. Franz, M. O., Stürzl, W., Hübner, W., & Mallot, H. A. (2008). A robot system for biomimetic navigation—from snapshots to metric embeddings of view graphs. In Robotics and cognitive approaches to spatial mapping (pp. 297–314). Springer.
51. Fuster, J. M. (2002). Physiology of executive functions: the perception–action cycle. In D. T. Stuss, & R. T. Knight (Eds.), Principles of frontal lobe function (pp. 96–108). New York: Oxford University Press.
52. Gallistel, C. R. (2008). Dead reckoning, cognitive maps, animal navigation and the representation of space: an introduction. In Robotics and cognitive approaches to spatial mapping (pp. 137–143). Springer.
53. Ghosh-Dastidar, S., & Adeli, H.(2009). Spiking neural networks.

International Journal of Neural Systems, 19, 295–308.

54. Gibson, J. J. (1986). The ecological approach to visual perception. Psychology Press.
55. Godfrey-Smith, P. (2003). Theory and reality. Science Education, 88, 236.
56. Goertzel, B., Lian, R., Arel, I., de Garis, H., & Chen, S. (2010). A world survey of artificial brain projects, part II: biologically inspired cognitive architectures. Neurocomputing, 74, 30–49.
57. Gopal, S., Klatzky, R. L., & Smith, T. R. (1989). Navigator: a psychologically based model of environmental learning through navigation. Journal of Environmental Psychology, 9, 309–331.
58. Gopal, S., & Smith, T. (1990). Human way-finding in an urban environment: a performance analysis of a computational process model. Environment and Planning A, 22, 169–191.
59. Gorchetchnikov, A., & Hasselmo, M. (2005). A biophysical implementation of a bidirectional graph search algorithm to solve multiple goal navigation tasks. Connection Science, 17, 145–164.
60. Graham, P., & Collett, T. S. (2002). View-based navigation in insects: how wood ants (Formica rufa L.) look at and are guided by extended landmarks. Journal of Experimental Biology, 205, 2499–2509.
61. Grossberg, S. (1987). Competitive learning: from interactive activation to adaptive resonance. Cognitive Science, 11, 23–63.
62. Gupta, K., Erdem, U., & Hasselmo, M. (2013). Modeling of grid cell activity demonstrates in vivo entorhinal 'look-ahead' properties. Neuroscience, 247, 395–411.
63. Hafting, T., Fyhn, M., Molden, S., Moser, M., & Moser, E. (2005). Microstructure of a spatial map in the entorhinal cortex. Nature, 436, 801–806.
64. Harrison, A.M., & Schunn, C.D. et al. (2003). ACT-R/S: look ma, no 'cognitive-map'. In International conference on cognitive modeling (pp. 129–134).
65. Hartley, T., Burgess, N., Lever, C., Cacucci, F., & O'Keefe, J. (2000). Modeling place fields in terms of the cortical inputs to the hippocampus. Hippocampus, 10, 369–379.

66. Hartley, T., Maguire, E. A., Spiers, H. J., & Burgess, N.(2003). The well-worn route and the path less traveled: distinct neural bases of route following and wayfinding in humans. Neuron, 37, 877–888.
67. Hasselmo, M. E. (2008). Grid cell mechanisms and function: contributions of entorhinal persistent spiking and phase resetting. Hippocampus, 18, 1213–1229.
68. Hirtle, S., & Jonides, J. (1985). Evidence of hierarchies in cognitive maps. Memory & Cognition, 13, 208–217.
69. Hok, V., Save, E., Lenck-Santini, P., & Poucet, B. (2005). Coding for spatial goals in the prelimbic/infralimbic area of the rat frontal cortex. Proceedings of the National Academy of Sciences of the United States of America, 102, 4602–4607.
70. Holmes, M. C., & Sholl, M. J. (2005). Allocentric coding of object-to-object relations in overlearned and novel environments. Journal of Experimental Psychology: Learning, Memory, and Cognition, 31, 1069.
71. Husain, M. (2008). Hemineglect. Scholarpedia, 3(2), 3681.
72. Jain, A. K., Mao, J., & Mohiuddin, K. M. (1996). Artificial neural networks: a tutorial. IEEE Computer, 29, 31–44.
73. Jefferies, M., Baker, J., & Weng, W. (2008). Robot cognitive mapping—a role for a global metric map in a cognitive mapping process. In Robotics and cognitive approaches to spatial mapping (pp. 265–279).
74. Jefferies, M., & Yeap, W. (2008). Robotics and cognitive approaches to spatial mapping. Vol. 38. Springer Verlag.
75. Jensen, O., & Lisman, J. E. (2000). Position reconstruction from an ensemble of hippocampal place cells: contribution of theta phase coding. Journal of Neurophysiology, 83, 2602–2609.
76. Kaski, S., & Kohonen, T. (1994). Winner-take-all networks for physiological models of competitive learning. Neural Networks, 7, 973–984.
77. Kim, J., Delcasso, S., & Lee, I. (2011). Neural correlates of object-in-place learning in hippocampus and prefrontal cortex. The Journal of Neuroscience, 31, 16991–17006.
78. Kjelstrup, K. B., Solstad, T., Brun, V. H., Hafting, T., Leutgeb, S.,

Witter, M. P., et al. (2008). Finite scale of spatial representation in the hippocampus. Science, 321, 140–143.

79. Knill, D. C., & Pouget, A. (2004). The Bayesian brain: the role of uncertainty in neural coding and computation. Trends in Neurosciences, 27, 712–719.
80. Kohonen, T.(1990). The self-organizing map. Proceedings of the IEEE, 78, 1464–1480.
81. Kravitz, D. J., Saleem, K. S., Baker, C. I., & Mishkin, M.(2011). A new neural framework for visuospatial processing. Nature Reviews Neuroscience, 12, 217–230.
82. Kuipers, B. (2000). The spatial semantic hierarchy. Artificial Intelligence, 119, 191–233.
83. Kuipers, B. (2008). An intellectual history of the spatial semantic hierarchy. In Robotics and cognitive approaches to spatial mapping (pp. 243–264). Springer.
84. Lever, C., Burton, S., Jeewajee, A., O'Keefe, J., & Burgess, N. (2009). Boundary vector cells in the subiculum of the hippocampal formation. Journal of Neuroscience, 29, 9771–9777.
85. Madl, T., Franklin, S., Chen, K., Montaldi, D., & Trappl, R. (2014). Bayesian integration of information in hippocampal place cells. PLoS One, e89762.
86. Madl, T., Franklin, S., Chen, K., & Trappl, R. (2013). Spatial working memory in the LIDA cognitive architecture. In Proc. international conference on cognitive modelling (pp. 384–390).
87. Maia, T. V. (2009). Reinforcement learning, conditioning, and the brain: successes
88. and challenges. Cognitive, Affective, & Behavioral Neuroscience, 9, 343–364.
89. Mani, K., & Johnson-Laird, P. N. (1982). The mental representation of spatial descriptions. Memory & Cognition, 10, 181–187.
90. Manns, J. R., & Eichenbaum, H. (2009). A cognitive map for object memory in the hippocampus. Learning & Memory, 16, 616–624.
91. Mark, D. M., Freksa, C., Hirtle, S. C., Lloyd, R., & Tversky, B. (1999). Cognitive models of geographical space. International Journal of Geographical Information Science, 13, 747–774.

92. Martin, A., & Chao, L. L. (2001). Semantic memory and the brain: structure and processes. Current Opinion in Neurobiology, 11, 194–201.
93. Marzocchi, N., Breveglieri, R., Galletti, C., & Fattori, P. (2008). Reaching activity in parietal area V6A of macaque: eye influence on arm activity or retinocentric coding of reaching movements? European Journal of Neuroscience, 27, 775–789.
94. McClelland, J. L. (2009). The place of modeling in cognitive science. Topics in Cognitive Science, 1, 11–38.
95. McNamara, T. P. (1986). Mental representations of spatial relations. Cognitive Psychology, 18, 87–121.
96. McNaughton, B., Barnes, C., Gerrard, J., Gothard, K., Jung, M., Knierim, J., et al. (1996).
97. Deciphering the hippocampal polyglot: the hippocampus as a path integration system. Journal of Fish Biology, 199, 173–185.
98. McNaughton, B. L., Battaglia, F. P., Jensen, O., Moser, E. I., & Moser, M.-B. (2006).
99. Path integration and the neural basis of the 'cognitive map'. Nature Reviews Neuroscience, 7, 663–678.
100. Milford, M., & Wyeth, G. (2010). Persistent navigation and mapping using a biologically inspired SLAM system. The International Journal of Robotics Research, 29, 1131–1153.
101. Moser, E. I., Kropff, E., & Moser, M.-B. (2008). Place cells, grid cells, and the brain's spatial representation system. Annual Review of Neuroscience, 31, 69–89.
102. Motomura, S., Ojima, Y., & Zhong, N. (2009). EEG/ERP meets ACT-R: a case study for investigating human computation mechanism. In Brain informatics (pp. 63–73). Springer.
103. Myung, I. J., Pitt, M. A., & Kim, W. (2005). Model evaluation, testing and selection. In Handbook of cognition (pp. 422–436).
104. Nardini, M., Jones, P., Bedford, R., & Braddick, O. (2008). Development of cue integration in human navigation. Current Biology, 18, 689–693.
105. Newell, A., & Simon, H. A. (1976). Computer science as empirical inquiry: symbols and search. Communications of the ACM, 19,

113–126.

106. O'Doherty, J. P. (2004). Reward representations and reward-related learning in the human brain: insights from neuroimaging. Current Opinion in Neurobiology, 14, 769–776.
107. O'Keefe, J., & Burgess, N. (1996). Geometric determinants of the place fields of hippocampal neurons. Nature, 381, 425–428.
108. O'Keefe, J., & Dostrovsky, J. (1971). The hippocampus as a spatial map. Preliminary evidence from unit activity in the freely-moving rat. Brain Research, 34, 171–175.
109. O'Reilly, R. C. (1998). Six principles for biologically based computational models of cortical cognition. Trends in Cognitive Sciences, 2, 455–462.
110. Packard, M. G., & McGaugh, J. L. (1996). Inactivation of hippocampus or caudate nucleus with lidocaine differentially affects expression of place and response learning. Neurobiology of Learning and Memory, 65, 65–72.
111. Pavlidis, T., & Horowitz, S. L. (1974). Segmentation of plane curves. IEEE Transactions on Computers, 23, 860–870.
112. Pesaran, B., Nelson, M. J., & Andersen, R. A. (2006). Dorsal premotor neurons encode the relative position of the hand, eye, and goal during reach planning. Neuron, 51, 125–134.
113. Pitt, M. A., Myung, I. J., & Zhang, S. (2002). Toward a method of selecting among computational models of cognition. Psychological Review, 109, 472.
114. Plank, M. (2009). Behavioral, electrocortical and neuroanatomical correlates of egocentric and allocentric reference frames during visual path integration. (Ph.D. thesis), Ludwig-Maximilans-Universität München.
115. Poucet, B. (1993). Spatial cognitive maps in animals: new hypotheses on their structure and neural mechanisms. Psychological Review, 100, 163.
116. Qin, Y., Bothell, D., & Anderson, J. R. (2007). ACT-R meets fMRI. In Web intelligence meets brain informatics (pp. 205–222). Springer.
117. Raubal, M. (2001). Human wayfinding in unfamiliar buildings: a simulation with a cognizing agent. Cognitive Processing, 2, 363–

388.

118. Rolls, E. T., & Xiang, J.-Z. (2006). Spatial view cells in the primate hippocampus and memory recall. Reviews in the Neurosciences, 17, 175–200.
119. Rumelhart, D. E., & Zipser, D. (1985). Feature discovery by competitive learning. Cognitive Science, 9, 75–112.
120. Samsonovich, A. V. (2010). Toward a unified catalog of implemented cognitive architectures. In BICA, 221 (pp. 195–244).
121. Samsonovich, A., & McNaughton, B. L. (1997). Path integration and cognitive mapping in a continuous attractor neural network model. The Journal of Neuroscience, 17, 5900–5920.
122. Schölkopf, B., & Mallot, H. A. (1995). View-based cognitive mapping and path planning. Adaptive Behavior, 3, 311–348.
123. Schultheis, H., & Barkowsky, T. (2011). Casimir: an architecture for mental spatial knowledge processing. Topics in Cognitive Science, 3, 778–795.
124. Schultheis, H., Lile, S., & Barkowsky, T. (2007). Extending ACT-R's memory capabilities. In Proc. of EuroCogSci, Vol. 7 (pp. 758–763).
125. Shepard, R., & Metzler, J. (1971). Mental rotation of three-dimensional objects. Science (New York, NY), 171, 701.
126. Sima, J. F. (2011). The nature of mental images—an integrative computational theory. In Proceedings of the 33rd annual conference of the cognitive science society (pp. 2878–2883). Citeseer.
127. Sima, J. F., Lindner, M., Schultheis, H., & Barkowsky, T. (2010). Eye movements reflect reasoning with mental images but not with mental models in orientation knowledge tasks. In Spatial cognition VII (pp. 248–261). Springer.
128. Sloman, A. (1998). What sort of architecture is required for a human-like agent? In Charles Ling, & Ron Sun (Eds.), Cognitive modeling workshop, at AAAI 1998 (pp. 1–8). AAAI.
129. Smolensky, P. (1987). Connectionist AI, symbolic AI, and the brain. Artificial Intelligence Review, 1, 95–109.
130. Snyder, L. H., Grieve, K. L., Brotchie, P., & Andersen, R. A. (1998). Separate body-and world-referenced representations of visual space

in parietal cortex. Nature, 394, 887–891.

131. Solstad, T., Boccara, C. N., Kropff, E., Moser, M.-B., & Moser, E. I. (2008).
132. Representation of geometric borders in the entorhinal cortex. Science, 322, 1865–1868.
133. Solstad, T., Moser, E. I., & Einevoll, G. T. (2006). From grid cells to place cells: a mathematical model. Hippocampus, 1031, 1026–1031.
134. Song, S., Miller, K. D., & Abbott, L. F. (2000). Competitive hebbian learning through spike-timing-dependent synaptic plasticity. Nature Neuroscience, 3, 919–926.
135. Strösslin, T., Sheynikhovich, D., Chavarriaga, R., & Gerstner, W. (2005). Robust selflocalisation and navigation based on hippocampal place cells. Neural Networks, 18, 1125–1140.
136. Sun, R. (2007). The importance of cognitive architectures: an analysis based on clarion. Journal of Experimental & Theoretical Artificial Intelligence, 19, 159–193.
137. Sun, R. (2008a). The Cambridge handbook of computational psychology. Cambridge: Cambridge University Press.
138. Sun, R. (2008b). Introduction to computational cognitive modeling. In Cambridge handbook of computational psychology (pp. 3–19).
139. Sun, R., Merrill, E., & Peterson, T. (2001). From implicit skills to explicit knowledge: a bottom–up model of skill learning. Cognitive Science, 25, 203–244.
140. Sun, R., & Zhang, X. (2004). Top–down versus bottom–up learning in cognitive skill acquisition. Cognitive Systems Research, 5, 63–89.
141. Taube, J. S.(2007). The head direction signal: origins and sensory-motor integration. Annual Review of Neuroscience, 30, 181–207.
142. Thomas, M. S., & McClelland, J. L. (2008). Connectionist models of cognition. In The Cambridge handbook of computational psychology (pp. 23–58).
143. Thrun, S., & Leonard, J. J. (2008). Simultaneous localization and mapping. In Springer handbook of robotics (pp. 871–889).
144. Tolman, E. C. (1948). Cognitive maps in rats and men. Psychological Review, 55, 189.

145. Tommasi, L., Chiandetti, C., Pecchia, T., Sovrano, V. A., & Vallortigara, G. (2012). From natural geometry to spatial cognition. Neuroscience & Biobehavioral Reviews, 36, 799–824.
146. Tommasi, L., & Laeng, B. (2012). Psychology of spatial cognition. Wiley Interdisciplinary Reviews: Cognitive Science, 3, 565–580.
147. Trullier, O., Wiener, S. I., Berthoz, A., & Meyer, J.-A. (1997). Biologically based artificial navigation systems: review and prospects. Progress in Neurobiology, 51, 483–544.
148. Tversky, B. (2005). Functional significance of visuospatial representations. In Handbook of higher-level visuospatial thinking (pp. 1–34).
149. Vann, S. D., Aggleton, J. P., & Maguire, E. A. (2009). What does the retrosplenial cortex do? Nature Reviews Neuroscience, 10, 792–802.
150. Vogeley, K., May, M., Ritzl, A., Falkai, P., Zilles, K., & Fink, G. R. (2004). Neural correlates of first-person perspective as one constituent of human selfconsciousness. Journal of Cognitive Neuroscience, 16, 817–827.
151. Voicu, H. (2003). Hierarchical cognitive maps. Neural Networks, 16, 569–576.
152. Voicu, H., & Schmajuk, N. (2000). Exploration, navigation and cognitive mapping. Adaptive Behavior, 8, 207–223.
153. Waller, D., & Hodgson, E. (2006). Transient and enduring spatial representations under disorientation and self-rotation. Journal of Experimental Psychology: Learning, Memory, and Cognition, 32, 867.
154. Webb, B. (2000). What does robotics offer animal behaviour? Animal Behaviour, 60, 545–558.
155. Webb, B. (2001). Can robots make good models of biological behaviour? Behavioral and Brain Sciences, 24, 1033–1050.
156. Willshaw, D. J., & Von Der Malsburg, C. (1976). How patterned neural connections can be set up by self-organization. Proceedings of the Royal Society of London. Series B. Biological Sciences, 194, 431–445.

157. Woergoetter, F., & Porr, B. (2008). Reinforcement learning. Scholarpedia, 3, 1448.
158. Yeap, W. K. (1988). Towards a computational theory of cognitive maps. Artificial Intelligence, 34, 297–360.
159. Yeap, W. K., Wong, C. K., & Schmidt, J. (2008). Using a mobile robot to test a theory of cognitive mapping. In Robotics and cognitive approaches to spatial mapping (pp. 281–295). Springer.
160. Yonelinas, A. P., Otten, L. J., Shaw, K. N., & Rugg, M. D. (2005). Separating the brain regions involved in recollection and familiarity in recognition memory. The Journal of Neuroscience, 25, 3002–3008.
161. Yoon, K., Buice, M. A., Barry, C., Hayman, R., Burgess, N., & Fiete, I. R. (2013).
162. Specific evidence of low-dimensional continuous attractor dynamics in grid cells. Nature Neuroscience, 16, 1077–1084.
163. Zaehle, T., Jordan, K., Wüstenberg, T., Baudewig, J., Dechent, P., & Mast, F. W. (2007).
164. The neural basis of the egocentric and allocentric spatial frame of reference. Brain Research, 1137, 92–103.

REFERENCES

1. E. Barnadó, A. Mekaouche, J.M. Garrell, A Study of a Genetic Classifier System based on the Pittsburgh Approach on a Medical Domain, IEA/AIE-99, 1999, Springer Verlag, 175-184
2. J. Baxter, "A Bayesian / information theoretic model of learning via multiple task sampling," Machine Learning, vol. 28, pp. 7–40, 1997.
3. Y. Bengio and O. Delalleau, "Justifying and generalizing contrastive divergence," Neural Computation, vol. 21, no. 6, pp. 1601–1621, 2009.
4. Y. Bengio, O. Delalleau, and N. Le Roux, "The Curse of highly variable functions for local kernel machines," in Advances in Neural Information Pro cessing Systems 18 (NIPS'05), (Y. Weis s, B. Sch¨ olkopf, and J. Platt, eds.), pp. 107– 114, Cambridge, MA: MIT Press, 2006.

5. Y. Bengio, O. Delalleau, and C. Simard, “Decision trees do not generalize to new variations,” Computational Intelligence, Computational Intelligence, Volume 26, Number 4, 2010.
6. Y. Bengio, P. Lamblin, D. Popovic, and H. Larochelle, “Greedy layer-wise training of deep networks,” in Advances in Neural Information Processing Systems 19 (NIPS’06), (B. Sch¨ olkopf, J. Platt, and T. Hoffman, eds.), pp. 153– 160, MIT Press, 2007.
7. Y. Bengio, N. Le Roux, P. Vincent, O. Delalleau, and P. Marcotte, “Convex neural networks,” in Advances in Neural Information Pro cessing Systems 18 (NIPS’05), (Y. Weiss, B. Sch¨ olkopf, and J. Platt, eds.), pp. 123–130, Cambridge, MA: MIT Press, 2006.
8. Y. Bengio and Y. LeCun, “Scaling learning algorithms towards AI,” in Large Scale Kernel Machines, (L. Bottou, O. Chap elle, D. DeCoste, and J. Weston, eds.), MIT Press, 2007.
9. Y. Bengio, P. Simard, and P. Frasconi, “Learning long-term dep endencies with gradient descent is difficult,” IEEE Transactions on Neural Networks, vol. 5, no. 2, pp. 157–166, 1994.
10. J. Bergstra, Y. Bengio, “Algorithms for Hyp er-Parameter Optimization”, in Journal of Machine Learning Research, pp 201-210 (2012)
11. J. Bergstra, Y. Bengio, “Random Search for Hyper-Parameter Optimization”, in Journal of Machine Learning Research, pp 281-305 (2011)
12. J. Bergstra and Y. Bengio, “Slow, decorrelated features for pretraining complex cell-like networks,” in Advances in Neural Information Pro cessing Systems 22 (NIPS’09), (D. Schuurmans, Y. Bengio, C. Williams, J. Lafferty, and A. Culotta, eds.), 2010.
13. H. Bourlard and Y. Kamp, “Auto-asso ciation by multilayer perceptrons and singular value decomp osition,” Biological Cyb ernetics, vol. 59, pp. 291–294, 1988.
14. L. Breiman, “Random forests,” Machine Learning, vol. 45, no. 1, pp. 5–32, 2001.
15. L. Breiman, J. H. Friedman, R. A. Olshen, and C. J. Stone, “Classification and Regression Trees”. Belmont, CA: Wadsworth International Group, 1984.

16. G. Calzada, J.M. Garrell, R. Barti, ModelGP: A GP-based platform for predictive model construction, Proceedings of SOCO'99, 1999
17. J.M. Campanario, "Using neural networks to study networks of scientific journals". Scientometrics. Vol. 33, n°1, p. 23-40, 1995.
18. H. Chen, "Machine learning for information retrieval: neural networks, symbolic learning, and genetic algorithms". Journal of the American Soclety for Information Science. Vol. 46, n° 3, p. 194-216, 1995.
19. R. Collobert and J. Weston, "A unified architecture for natural language pro cessing: Deep neural networks with multitask learning," in Proceedings of the 25th International Conference on Machine Learning (ICML'08), (W. W. Cohen, A. McCallum, and S. T. Roweis, eds.), pp. 160–167, ACM, 2008.
20. J. L. Elman, "Learning and development in neural networks : The importance of starting small," Cognition, vol. 48, pp. 781–799, 1993.
21. D. Erhan, P.-A. Manzagol, Y . Bengio, S. Bengio, and P. Vincent, "The difficulty of training deep architectures and the effect of unsup ervised pretraining," in Proceedings of the 20th International Conference on Artificial Intelligence and Statistics (AISTATS'09), pp. 153–160, 2009.
22. M. A. F. Figueiredo and A. K. Jain, "Unsup ervised learning of finite mixture models," IEEE Trans. Pattern Anal. Machine Intell., vol. 24, no. 3, pp. 381–396, 2002.
23. K. Fukushima, "Neocognitron: A self-organizing neural network model for a mechanism of pattern recognition unaffected by shift in position," Biological Cybernetics, vol. 36, pp. 193–202, 1980.
24. J.M. Garrell, E. Golobardes, E. BeANNdó and X. Llorà, Automatic Classification of Mammary Biopsy Images with Machine Learning Techniques, Proceedings EIS'98, Volume 3: Artificial Intelligence, ICSS Academic Press, 411-418, 1998
25. J.M. Garrell, E. Golobardes, E. BeANNdó and X. Llorà, Automatic diagnosis with genetic algorithms and case-based reasoning, Artificial Intelligence in Engineering, Elsevier Science Ltd. Vol. 13, No. 4, 367-372, 1999.

26. Simon Haykin, Neural Networks: a comprehensive foundation, (2nd edition), USA, Prentice Hall, ISBN: 0-13-908385-5, 2000
27. K. A. Heller and Z. Ghahramani, "A nonparametric bayesian approach to modeling overlapping clusters," in Proceedings of the Eleventh International Conference on Artificial Intelligence and Statistics (AISTATS'07), pp. 187–194, San Juan, Porto Rico: Omnipress, 2007.
28. G. E. Hinton, "Learning distributed representations of concepts," in Proceedings of the Eighth Annual Conference of the Cognitive Science Society, pp. 1–12, Amherst: Lawrence Erlbaum, Hillsdale, 1986.
29. G. E. Hinton, S. Osindero, and Y. Teh, "A fast learning algorithm for deep belief nets," Neural Computation, vol. 18, pp. 1527–1554, 2006.
30. G. E. Hinton and R. Salakhutdinov, "Reducing the dimensionality of data with neural networks," Science, vol. 313, pp. 504–507, 2006.
31. G. E. Hinton and T. J. Sejnowski, "Learning and relearning in Boltzmann machines," in Parallel Distributed Processing: Explorations in the Microstructure of Cognition. Volume 1: Foundations, (D. E. Rumelhart and J. L. McClelland, eds.), pp. 282–317, Cambridge, MA: MIT Press, 1986.
32. T. Kohonen, (1982). „Self-organized formation of topologically correct feature maps". *Biological Cybernetics.* N° 43, p. 59-69.
33. T. Kohonen; et *al* . „*SOM_PAK: the self-organizing map program package*". Helsinki University of Technology, Laboratory of Computer and Information Science. (Technical Report, A31), 1996.
34. T. Kohonen et *al.* „*LVQ_PAK: the learning vector quantization program package*". Helsinki University of Technology, Laboratory of Computer and Information Science. (Technical Report, A30), 1996.
35. H. Larochelle, Y. Bengio, J. Louradour, and P. Lamblin, "Exploring strategies for training deep neural networks," Journal of Machine Learning Research, vol. 10, pp. 1–40, 2009.

36. H. Lee, R. Grosse, R. Ranganath, and A. Y. Ng, "Convolutional deep belief networks for scalable unsup ervised learning of hierarchical representations," in Pro ceedings of the 26th International Conference on Machine Learning (ICML'09), (L. Bottou and M. Littman, eds.), Montreal (Qc), Canada: ACM, 2009.

37. S. Lawrence, C. Lee Giles, and Ah Chung, "What Size Neural Network Gives Optimal Generalization", Technical Rep ort, Institute for Advanced Computer Studies, University of Maryland, 1996

38. Y. Le Cunn, J. Denker and S. Solla, "Optimal Brain Damage", AT&T Bell Lab oratories, 1990

39. X. Llorà, J.M. Garrell, GENIFER: A Nearest Neighbour based Classifier System using GA, Proceedings of GECCO'99, 1999, CA: Morgan Kaufmann, pp. 797

40. Ludovic, P. Helene and S. Michele, "Unsupervised Layer-Wise Model Selection in Deep Neural Networks", H. Coelho et al., IOS Press, 2010

41. K. S. Narendra and K. Parthasarathy, "Gradient methods for optimization of dynamical systems containing neural networks", IEEE Transactions o d Neural Networks, pp. 252-262, 1991

42. Rahimi and B. Recht, "U ns up ervised Regression with applications to Nonlinear System Identification", Intel Research, Seattle, 2010

43. D. Stathakis, "How many hidden layers and nodes?", in International Journal of Remote Sensing, Vol. 30, No. 8, 2133-2147, 2009

44. Witten, IH, E. Frank. 2005. Data Mining: Practical Machine Learning Tools and Techniques. 2nd ed. Kaufmann M.

45. Wilkinson, R.; Hingston, P. „Incorporating the vector space model in neural network used for document retrieval". *Library Hi Tech.* Vol. 10, n° 1-2, p. 69-75, 1992.

46. Zimmer, U.; Fischer, C; Von Puttkamer, E. „Navigation on topologic feature-maps", *Proceedings of the 3"1. InteANNtional Conference on Fuzzy Logic, Neural Nets and Soft Computing.* lizuka (Japan), August 1 - 7. P. 131-132. 1994.

CITATIONS

CHAPTER 8

Ken Takiyama, Yutaka Sakai, A balanced motor primitive framework can simultaneously explain motor learning in unimanual and bimanual movements, https://doi.org/10.1016/j.neunet.2016.10.013

CHAPTER 9

Naoya Oosugi, Keiichi Kitajo, Naomi Hasegawa, Yasuo Nagasaka, Kazuo Okanoya, Naotaka Fujii, A new method for quantifying the performance of EEG blind source separation algorithms by referencing a simultaneously recorded ECoG signal, Neural Networks, http://dx.doi.org/10.1016/j.neunet.2017.01.005

CHAPTER 10

Diego G. Loyola R, Mattia Pedergnana, Sebastián Gimeno García, Smart sampling and incremental function learning for very large high dimensional data, Neural Networks, Volume 78, 2016, Pages 75-87, ISSN 0893-6080, http://dx.doi.org/10.1016/j.neunet.2015.09.001.

CHAPTER 11

Victoria J. Hodge, Simon O'Keefe, Jim Austin, Hadoop neural network for parallel and distributed feature selection, Neural Networks, Volume 78, 2016, Pages 24-35, ISSN 0893-6080, http://dx.doi.org/10.1016/j.neunet.2015.08.011.

CHAPTER 12

Alicia Costalago Meruelo, David M. Simpson, Sandor M. Veres, Philip L. Newland, Improved system identification using artificial neural networks and analysis of individual differences in responses of an identified neuron, Neural Networks, Volume 75, 2016, Pages 56-65, ISSN 0893-6080, http://dx.doi.org/10.1016/j.neunet.2015.12.002.

CHAPTER 13

Soeren Strauss, Philip J.W. Woodgate, Saber A. Sami, Dietmar Heinke, Choice reaching with a LEGO arm robot (CoRLEGO): The motor system guides visual attention to movement-relevant information, Neural Networks, Volume 72, 2015, Pages 3-12, ISSN 0893-6080, http://dx.doi.org/10.1016/j.neunet.2015.10.005.

CHAPTER 14

Tamas Madl, Ke Chen, Daniela Montaldi, Robert Trappl, Computational cognitive models of spatial memory in navigation space: A review, Neural Networks, Volume 65, 2015, Pages 18-43, ISSN 0893-6080, http://dx.doi.org/10.1016/j.neunet.2015.01.002.

INDEX

D

E

F

G

H

I

J

T

U

V